Lecture Notes in Computer Science 9645

Commenced Publication in 1973
Founding and Former Series Editors:
Gerhard Goos, Juris Hartmanis, and Jan van Leeuwen

More information about this series at http://www.springer.com/series/7409

Hong Gao · Jinho Kim
Yasushi Sakurai (Eds.)

Database Systems for Advanced Applications

DASFAA 2016 International Workshops:
BDMS, BDQM, MoI, and SeCoP
Dallas, TX, USA, April 16–19, 2016
Proceedings

 Springer

Editors
Hong Gao
Harbin Institute of Technology
Harbin
China

Yasushi Sakurai
Kumamoto University
Kumamoto-shi
Japan

Jinho Kim
Kangwon National University
Kangwon
Korea (Republic of)

ISSN 0302-9743   ISSN 1611-3349 (electronic)
Lecture Notes in Computer Science
ISBN 978-3-319-32054-0   ISBN 978-3-319-32055-7 (eBook)
DOI 10.1007/978-3-319-32055-7

Library of Congress Control Number: 2016934671

LNCS Sublibrary: SL3 – Information Systems and Applications, incl. Internet/Web, and HCI

Printed on acid-free paper

This Springer imprint is published by Springer Nature
The registered company is Springer International Publishing AG Switzerland

Preface

Along with the main conference, the DASFAA 2016 workshops provide an international forum for researchers to discuss research results. This year, we had four workshops held in conjunction with DASFAA 2016:

- The Third International Workshop on Semantic Computing and Personalization (SeCoP 2016)
- The Third International Workshop on Big Data Management and Service (BDMS 2016)
- The First International Workshop on Big Data Quality Management (BDQM 2016)
- The Second International Workshop on Mobile of Internet (MoI 2016)

All the workshops were selected after a public call-for-proposals process, and each of them focuses on a specific area that contributes to the main themes of DASFAA 2016. In total, 32 papers were accepted, including 15 papers for SeCoP 2016, seven papers for BDMS 2016, eight papers for BDQM 2016, and two papers for MoI 2016. We would like to thank the workshop organizers and Program Committee members for their tremendous effort in making the DASFAA 2016 workshops a success. In addition, we are grateful to the main conference organizers for their generous support.

April 2016

Hong Gao
Jinho Kim
Yasushi Sakurai

Organization

The Third International Workshop on Semantic Computing and Personalization (SeCoP 2016)

Workshop Chairs

Haoran Xie	Caritas Institute of Higher Education, Hong Kong, SAR China
Fu Lee Wang	Caritas Institute of Higher Education, Hong Kong, SAR China
Yi Cai	South China University of Technology, China
Wei Chen	Chinese Academy of Agricultural Sciences, China
Yanghui Rao	Sun Yat-Sen University, China

Program Committee

Di Zou	The Hong Kong Polytechnic University, Hong Kong, SAR China
Jianfeng Si	Institute for Infocomm Research, Singapore
Xudong Mao	Alibaba, China
Raymong Y.K. Lau	City University of Hong Kong, Hong Kong, SAR China
Rong Pan	Sun Yat-Sen University, China
Yunjun Gao	Zhejiang University, China
Shaojie Qiao	Southwest Jiaotong University, China
Jianke Zhu	Zhejiang University, China
Neil Y. Yen	University of Aizu, Japan
Derong Shen	Northeastern University, USA
Jing Yang	Research Center on Fictitious Economy and Data Science CAS, China
Yuqing Sun	Shangdong University, China
Raymong Wong	Hong Kong University of Science and Technology, Hong Kong, SAR China
Jie Tang	Tsinghua University, China
Jian Chen	South China University of Technology, China
Wong Tak Lam	The Hong Kong Institute of Education, Hong Kong, SAR China
Xiaodong Zhu	University of Shanghai for Science and Technology, China
Zhiwen Yu	South China University of Technology, China
Wenjuan Cui	China Academy of Sciences, China
Shaoxu Song	Tsinghua University, China
Guangliang Chen	Delft University of Technology, The Netherlands
Tao Wang	Southampton University, UK

The Third International Workshop on Big Data Management and Service (BDMS 2016)

Workshop Chairs

Xiaoling Wang	East China Normal University, China
Kai Zheng	The University of Queensland, Australia
An Liu	Soochow University, China

Program Committee

Muhammad Aamir Cheema	Monash University, Australia
Cheqing Jin	East China Normal University, China
Qizhi Liu	Nanjing University, China
Bin Mu	Tongji University, China
Xuequn Shang	Northwestern Polytechnical University, China
Weiwei Sun	Fudan University, China
Yan Wang	Macquarie University, Australia
Lizhen Xu	Southeast University, China
Xiaochun Yang	Northeastern University, China
Kun Yue	Yunnan University, China
Dell Zhang	University of London, UK
Xiao Zhang	Renmin University of China, China
Fuzheng Zhang	Microsoft Research Asia, China
Defu Lian	University of Electronic Science and Technology of China, China
Zhuoming Xu	Hohai University, China

The First Workshop on Big Data Quality Management (BDQM 2016)

Workshop Chairs

Hongzhi Wang	Harbin Institute of Technology, China
Jing Gao	University at Buffalo, State University of New York, USA

Program Committee

Xiaochun Yang	Northeastern University, China
Yueguo Chen	Renmin University, China
Nan Tang	Qatar Computing Research Institute, Qatar
Jiannan Wang	Simon Fraser University, Canada
Xianmin Liu	Harbin Institute of Technology, China
Zhijing Qin	Pinterest, USA
Guoliang Li	Tsinghua University, China
Cheqing Jin	East China Normal University, China

Wenjie Zhang	University of New South Wales, Australia
Shuai Ma	Beihang University, China
Zhaonian Zou	Harbin Institute of Technology, China

The Second International Workshop on Mobile of Internet (MoI 2016)

Workshop Chairs

Li Wang	TaiYuan University of Technology, China
Huawei Shen	Chinese Academy of Science, China
Lidong Wu	University of Texas at Tyler, USA
Bruno Gonçalves	Aix-Marseille Université, France

Program Committee

Weili Wu	University of Texas at Dallas, USA
XiuFang Feng	TaiYuan University of Technology, China
HongWei Xie	TaiYuan University of Technology, China
Changyou Zhang	Institute of Software, Chinese Academy of Sciences, China
Donghyun Kim	North Carolina Central University, USA
Yansha Guo	Tianjin University of Technology and Education, China
Tong Wang	Harbin Institute of Technology, China
Shuyuan Mary Ho	Florida State University, USA
Xin Lu	College of Information System and Management, National University of Defense Technology, China
Yanyan Lan	Chinese Academy of Science, China
Jiafeng Guo	Chinese Academy of Science, China
Xiaolong Jin	Chinese Academy of Science, China

Contents

SeCoP 2016

Weibo Mood Towards Stock Market 3
   Wen Hao Chen, Yi Cai, and Kin Keung Lai

Improving Diversity of User-Based Two-Step Recommendation Algorithm
with Popularity Normalization 15
   Xiangyu Zhao, Wei Chen, Feng Yang, and Zhongqiang Liu

Followee Recommendation in Event-Based Social Networks 27
   Shuchen Li, Xiang Cheng, Sen Su, and Le Jiang

SBTM: Topic Modeling over Short Texts 43
   Jianhui Pang, Xiangsheng Li, Haoran Xie, and Yanghui Rao

Similarity-Based Classification for Big Non-Structured
and Semi-Structured Recipe Data 57
   Wei Chen and Xiangyu Zhao

An Approach of Fuzzy Relation Equation and Fuzzy-Rough Set
for Multi-label Emotion Intensity Analysis 65
   Chu Wang, Daling Wang, Shi Feng, and Yifei Zhang

Features Extraction Based on Neural Network for Cross-Domain Sentiment
Classification 81
   Endong Zhu, Guoyan Huang, Biyun Mo, and Qingyuan Wu

Personalized Medical Reading Recommendation: Deep Semantic Approach . . . 89
   Tatiana Erekhinskaya, Mithun Balakrishna, Marta Tatu,
   and Dan Moldovan

A Highly Effective Hybrid Model for Sentence Categorization 98
   Zhenhong Chen, Kai Yang, Yi Cai, Dongping Huang,
   and Ho-fung Leung

Improved Automatic Keyword Extraction Given More Semantic
Knowledge 112
   Kai Yang, Zhenhong Chen, Yi Cai, DongPing Huang,
   and Ho-fung Leung

Generating Computational Taxonomy for Business Models of the Digital
Economy 126
   Chao Wu, Yi Cai, Mei Zhao, Songping Huang, and Yike Guo

How to Use the Social Media Data in Assisting Restaurant
Recommendation   134
    Wenjuan Cui, Pengfei Wang, Xin Chen, Yi Du, Danhuai Guo,
    Yuanchun Zhou, and Jianhui Li

A Combined Collaborative Filtering Model for Social Influence Prediction
in Event-Based Social Networks   142
    Xiao Li, Xiang Cheng, Sen Su, Shuchen Li, and Jianyu Yang

Learning Manifold Representation from Multimodal Data for Event
Detection in Flickr-Like Social Media   160
    Zhenguo Yang, Qing Li, Wenyin Liu, and Yun Ma

Deep Neural Network for Short-Text Sentiment Classification   168
    Xiangsheng Li, Jianhui Pang, Biyun Mo, Yanghui Rao,
    and Fu Lee Wang

BDMS 2016

VMPSP: Efficient Skyline Computation Using VMP-Based Space
Partitioning   179
    Kaiqi Zhang, Donghua Yang, Hong Gao, Jianzhong Li, Hongzhi Wang,
    and Zhipeng Cai

Real-Time Event Detection with Water Sensor Networks Using
a Spatio-Temporal Model   194
    Yingchi Mao, Xiaoli Chen, and Zhuoming Xu

Bayesian Network Structure Learning from Big Data: A Reservoir
Sampling Based Ensemble Method   209
    Yan Tang, Zhuoming Xu, and Yuanhang Zhuang

Correlation Feature of Big Data in Smart Cities   223
    Yi Zhang, Xiaolan Tang, Bowen Du, Weilin Liu, Juhua Pu,
    and Yujun Chen

Nearly Optimal Probabilistic Coverage for Roadside Data Dissemination
in Urban VANETs   238
    Yawei Hu, Mingjun Xiao, An Liu, Ruhong Cheng, and Hualin Mao

OCC: Opportunistic Crowd Computing in Mobile Social Networks   254
    Hualin Mao, Mingjun Xiao, An Liu, Jianbo Li, and Yawei Hu

Forest of Distributed B+Tree Based on Key-Value Store for Big-Set
Problem   268
    Thanh Trung Nguyen and Minh Hieu Nguyen

BDQM 2016

An Efficient Schema Matching Approach Using Previous Mapping
Result Set   285
  Hongjie Fan, Junfei Liu, Wenfeng Luo, and Kejun Deng

A Distributed Load Balance Algorithm of MapReduce for Data Quality
Detection   294
  Yitong Gao, Yan Zhang, Hongzhi Wang, Jianzhong Li, and Hong Gao

A Formal Taxonomy to Improve Data Defect Description   307
  João Marcelo Borovina Josko, Marcio Katsumi Oikawa,
  and João Eduardo Ferreira

ISSA: Efficient Skyline Computation for Incomplete Data   321
  Kaiqi Zhang, Hong Gao, Hongzhi Wang, and Jianzhong Li

Join Query Processing in Data Quality Management   329
  Mingliang Yue, Hong Gao, Shengfei Shi, and Hongzhi Wang

Similarity Search on Massive Data Based on FPGA   343
  Yanzheng Wang, Hong Gao, Shengfei Shi, and Hongzhi Wang

Skyline Join Query Processing over Multiple Relations   353
  Jinchao Zhang, Zheng Lin, Bo Li, Weiping Wang, and Dan Meng

Detect Redundant RDF Data by Rules.   362
  Tao Guang, Jinguang Gu, and Li Huang

MoI 2016

Behavior-Based Twitter Overlapping Community Detection   371
  Lixiang Guo, Zhaoyun Ding, and Hui Wang

Versatile Safe-Region Generation Method for Continuous Monitoring
of Moving Objects in the Road Network Distance.   377
  Yutaka Ohsawa and Htoo Htoo

Author Index   393

SeCoP 2016

Weibo Mood Towards Stock Market

Wen Hao Chen[1]([✉]), Yi Cai[2], and Kin Keung Lai[1]

[1] Department of Management Science, City University of Hong Kong,
Kowloon Tong, Hong Kong
wenhachen2-c@my.cityu.edu.hk, mskklai@cityu.edu.hk
[2] School of Software Engineering, South China University of Technology, Guangzhou, China
ycai@scut.edu.cn

Abstract. Behavioral economics and behavioral finance believe that public mood is correlated with economic indicators and financial decisions are significantly driven by emotions. A growing body of research has examined the correlation between stock market and social media public mood state. However most research is conducted on English social media websites, the number of research on how public mood states in Chinese social media websites affect the stock market in China is limited. This paper first summarizes the previous research on text mining and social media sentiment analysis. After that, we investigate whether measurements of collective public mood states derived from Weibo which is a social media website similar as Twitter but most posts are written in Chinese are correlated to the stock market price in China. We use a novel Chinese mood extracting method using two NLP (Natural Language Processing) tools: Jieba and Chinese Emotion Words Ontology to analyze the text content of daily Weibo posts. A Granger Causality analysis is then used to investigate the hypothesis that the extracted public mood or emotion states are predictive of the stock price movement in China. Our experimental results indicate that some public mood dimensions such as "Happiness" and "Disgust" are highly correlated with the change of stock price and we can use them to forecast the price movement.

Keywords: Sentiment analysis · Text mining · Behavioral finance · Twitter · Weibo chinese emotion words ontology

1 Introduction

In recent years, social media websites, such as Facebook, Twitter and Weibo become more and more popular. Large volume of information is generated by the online users including comments, news and discussions in these websites. The user number of social medial websites is enormous and there are different user communities with diversified natures in these websites. The data collected from these websites, to some extent, represents the public mood or opinions. An increasing number of researches have attempted to integrate sentiment analysis result of the online text into different models. However, as Chinese is quite different with English and it is more complicated in terms of recognition, segmentation and analysis, the number of researches on sentiment analysis of Chinese social media websites is limited.

© Springer International Publishing Switzerland 2016
H. Gao et al. (Eds.): DASFAA 2016 Workshops, LNCS 9645, pp. 3–14, 2016.
DOI: 10.1007/978-3-319-32055-7_1

1.1   Public Mood in Social Media Websites

Social media is known as a computer-mediated tool which allows people to create, share or exchange information. It has many different forms including blogs, micro-blogs, photo sharing, video sharing, forum, tagging system and social network. The increasing popularity of social media websites and Web 2.0 has led to exponential growth of user-generated content, especially text content on Internet. Abbasi and Chen [1] have demonstrated that the information retrieval and automated analysis technique are very useful for understanding the online content such as Web pages or forum posts and social interactions in online communities. Through analyzing the blog content, Liang et al. [2] indicated that a company can get the first hand knowledge or feedback from his clients. And it can also help to understand how the online customer networks appear and evolve [3]. This kind of information extracted from blogs enable organizations to make better decision on critical business area which is important for business intelligence [4]. In terms of integration of social media sentiment to applications in different industries, researches are conducted to use the sentiment from twitter to forecast spikes in book sales [5] and the revenues of box-office for movies in North America [6]. Another research area for using text sentiment analysis result is the recommendation system. A lot of researches have been done to integrate sentiment analysis results of online text into recommendation systems [7].

1.2   Behavioral Economic and Finance

Behavioral economic and finance study the impact of psychological or emotion factors on the related decision making area and the affect on stock price, returns and the risk of the market. Psychological research has already approved that emotions, in addition to information has a significant affect in human decision-making [8]. Behavioral finance has further demonstrated that investment decision of investors is more likely driven by their emotions [9]. And the momentum generated from the public emotions with other factors such as economic factors determine prices of the stock market.

Early research on stock market prediction focus on building models with random walk theory and Efficient Market Hypothesis [10]. However many researches show that the stock market price does not follow a random walk and some researchers suggest that some indicators of the price can be extracted from the online social media. Schumaker and Chen [11] applied machine learning methods to financial news articles and find that the sentiment in news articles has an immediately impact on the market price. The prediction model has the best performance adding the news article factors. Based on the community sentiment retrieved from the posts on the Yahoo Finance Forum through an expert classification system, Liu et al. [12] has indicated the correlation between the sentiment and the stock price. Gibert [13] using the LiveJournal as a source, has extracted the anxiety, worry and fear mood from the posts in that websites. He found that the increases on expression of anxiety have indicated that the S&P 500 Index will move downward soon. Using Granger Causality, Bollen et al. [14] has investigated the correlation between the collective mood states from large-scale twitter feeds and the value of the Dow Jones Industrial Average (DJIA) over time. He proved that there is a positive correlation between the "calm" mood states collected and the Index price. Using twitter posts as well, Zhang et al. [15] found that the sentiment or emotions can be used to predict NASDAQ and S&P 500 index as well.

Over the past 10 years significant progress has been achieved in extracting states of public mood directly from social media websites such as blog content and large scale Twitter post. However, the research on Chinese social media websites is rare. As a result, we will try to discuss how to extract public mood states from Chinese social media website such as Weibo and whether the public mood extracted can be used to predict the stock price in China.

2   Background

Through the literature review, we found out that the nature of social media website makes it a valuable source for mining public emotions or mood states. First, a lot of users are posting their status and opinions on the websites which represent their mood in some extent. Second, these posts are highly time related. Active users in these websites post their opinions every day and even every hour. Third, the online communities are based on the friendship in reality. Users are discussing different topics in the website as they have done in the real community. Some social media applications such as Weibo and Wechat even have an approval mechanism for the users which makes sure each user in these systems is a real. Fourth, some organizations and companies are using social media such as Facebook, Twitter, Weibo and Wechat as their formal channel to publish their opinion and messages. As a result the information collected from these social media websites can help us to understand the public mood at certain period.

The number of previous research on Chinese social media mining is limited and the number of research on how to use the textual sentiment analysis towards Chinese financial market is even lower. However China has already been the second largest economies and the stock market in China has a great impact to the global stock market in the recent years. The intraday trading volume on Shanghai Stock Exchange (SSE) has been more than 1 thousand billion RMB on 20th April 2015 which is more than any stock exchange intraday trading volume in the history. How the China stock market behavior according to the public mood change is important for economists, traders and socialist. As a result, we present a novel method to extract public mood states from Chinese social medium website Weibo and demonstrate that public mood states extracted are correlated with the price movement in the stock market. Proposing a model to use the sentiment analysis result of Weibo post to predict the stock price is not the target of this paper, we are only interested in discussing whether the public mood states extracted from Weibo can represent the real opinion of publics towards the stock market and how the public mood states impact the stock market. Although researches on behavioral finance and economic have already demonstrate the correlation between public emotion and the stock price, we will discuss whether Weibo is a valid source to monitor the public mood correlated with the stock market and demonstrate the causality between the Weibo public mood and the stock price.

3 Methodology

3.1 Data and System Framework Overview

Researchers have done some research on using the overall public mood from Twitter to predict the stock market price [14]. However there is a lot of noise in the public mood states generated from the social media. If we want to discuss the emotion impact on the financial industry, we need to find out the emotion text which is highly related to the stock market. For example, even most users have bad comments on a movie in twitter that should not have a big impact on the stock price of an oil company. As a result, to filter the unnecessary noise, this paper will discuss how users' comments or emotions related to some special topics which affect the real stock market. One of the data source discussed in this paper is the text content from a Chinese social media website called Weibo. It is the largest microblog platform and the counterparty of Twitter in China. It was launched in August 2009. As of mid 2014, there are 167 million monthly active users, more than 25 million posts each day.

We obtain a collection of public posts that was recorded from 1st Jan 2015 to 31th July 2015. For each post these records provide an identifier, the data-time of the submission and the text content. After that we classified all these post by date and remove the stop-words and punctuation for each post. As we are discussing the public mood related to China stock market, we choose 3 topics for discussion which are: "Shanghai-Hong Kong Stock Connect (SH-HK connect) and Shenzhen-Hong Kong Stock Connect (SZ-HK connect)", "Interest rate and reserve rate cut in China", "Greece Government-Debt Crisis". The reason why we choose these 3 topics is that from 1st Jan 2015 to 31th 2015, they are most discussed topics related to the stock market in the website. SH-HK connect has been launched on November 2014 and SZ-HK connect is planned to launched in October 2015. It allows the individual and institution traders in Shanghai or Shenzhen to trade the stock market in HK through the Connect. And the traders in HK also can use it to trade the stock market in Shanghai or Shenzhen. It provides different choices and an arbitrage chance for the traders in Hong Kong and Mainland China. It is considered as a great milestone in the development of China stock market and people believe that it will have a positive impact on China stock market. Greece Government-Debt Crisis is the debt crisis that happens in European Union (EU) in 2015, people worry that if Greece cannot pay the debt on time, it will cause the financial crisis in EU and the international crisis will be triggered again. Interest rate and reserve rate cut means the central bank in China plan to cut the bank lending interest rate and the amount of cash banks must keep in reserve. It will increase the flow of money in the economy and reduce the cost of financing to promote and support the investment and developments of the real economy. As it will help to boost the growth of economy, it has a positive impact on the stock market. The second source of data is the daily end price of stocks in Hong Kong and Shanghai stock Exchanges which could be downloaded from public available websites such as Yahoo! Finance.

Our experiments can be summarized as follows. In the first phase, we will filter the posts and only extract the posts containing the keywords related to the 3 topics. Posts are grouped under the same topic. In the second phase, we use two tools to measure the sentiment score for different emotions in the posts. The first tool, Jieba, is used to analyze the Chinese text and segment the text into meaningful Chinese words. The second tool, Chinese Emotion Word Ontology (CEWO), is a Chinese sentiment word dictionary which classifies the Chinese emotion words into 7 different emotion categories. We propose a novel method to generate the sentiment score or mood states score for each post based on the CEWO. After that we combine the sentiment score for all posts published on the same day under the same topic. As a result, the daily sentiment score for different emotions are generated for each topic. The sentiment score or the score of public mood states across the observation time can be regarded as time series. In the third phase, we investigate the hypothesis that public mood states or emotions measured by our mechanism are predictive of future stock price using Granger causality analysis. As the public mood states generated from different topics are related with different stock markets, we will discuss their impact on different kinds of stocks. Proposing an optimal stock price prediction model is not the target of this paper, we are only interested in discussing the correlation between the stock price and the public mood information.

3.2 Data Processing

As mentioned above, the raw text data is restored in database first. And then the Chinese segmentation tool Jieba is used to segment the text into different words. The Jieba tool use the dynamic programming to find out the most probable combination based on the word frequency. The segmentation result will be restored in the database as well. The next step is to generate the sentiment score or mood states from each post. For this part, we will use the Chinese Emotion Word Ontology constructed by the IR lab in Dalian University of Technology[1]. This ontology is constructed based on Ekman's theory of 6 basic emotions. And one more emotion "good" is added to make it more comfortable for Chinese language analysis. It is widely used in Chinese text mining research. Table 1 shows the mapping from English to Chinese for the 7 categories of emotion words. These categories are also known as the dimensions of public mood. Each emotion word in the Ontology has its own emotion category tag such as "Happiness", intensity value and polarity value recorded.

Table 1. The 7 Emotion Categories in the Chinese Emotion Word Ontology.

English	Chinese
Happiness	乐
Good	好
Sadness	怒
Surprise	哀
Fear	惧
Disgust	恶
Anger	惊

---

[1] http://ir.dlut.edu.cn/.

The proposed method to generate the sentiment score or public mood states including 4 steps. First, Jieba is used to segment the post and translate the post to a list of meaningful words including different lexical class such noun, verb, adv and adj. Secondly, we will process the word list and search each word of the list in the Chinese Emotion Word Ontology based on its lexical class. If a word i is found in the Ontology, the intensity value of the word in the Ontology will be recorded as s_i. Thirdly, the sentiment score of the post will be calculated as follows:

$$y = \max_{j=1,\ldots,7}\left(\sum_{i \in I(j)} s_i\right) \tag{1}$$

Where $I(j)$ is a set including all words found in jth emotion category of the Ontology for a post, $I(1)$ means the words found in the first emotion category of the Ontology which is the category of "Happiness". We assume that each post only presents one kind of emotion which is the one that has the overall highest intensity value or sentiment score. As a result each post recorded in the database will be attached an emotion tag which identifies the emotion category it belongs to. In the fourth steps, the frequency of the posts belong to the same emotion category is calculated for each day. The value will be represented as $x_{i,t}$ where i belongs to one of the 7 emotion categories and t means date t in the observation period. Then the time series for emotion i can be represented as X_i.

After processing the raw text data of Weibo posts, for each topic discussed, we generate 7 time series. Each time series represents the movement of one dimension of the public mood state or emotion such as "Happiness". Another data source is the time series of the stock price. The time series of the stock close price is downloaded from Yahoo! Finance and other public websites. As the emotion time series are associated with different stocks. We download the close price from 1st Jan to 31th July 2015 for 4 different stocks including the Hang Seng Index (HSI), Shanghai Stock Exchange Composite Index (SSECI), Hong Kong Exchanges and Clearing Limited corporation stock (stock code is 388) and Shanghai Connect Index (stock code is 000159).

3.3 Public Mood Validation

To validate the ability of the proposed method to capture various aspects of public moods, we apply it to the Weibo posts published from 1st Jan to 31th July. The distribution of the posts belonging to different dimensions of public mood or emotions indicates that the main part of the positive topic posts such as "interest rate cute" are the posts with the positive emotion such as "Happiness" and "Good" and vice versa. The distribution of the posts is shown in Fig. 1.

As we mentioned, the interest rate and reserve rate cut will booth the economy and the stock market. Actually, the China government has cut the interest rate or reserve rate three times during 1st Jan to 31 July 2015 (on 28th Feb, 10th May, 27th June). As a result, most posts about the interest rate and reserve rate cut locate in the positive emotions such as "Happiness" and "Good" which is 85 % of all related posts. The distribution of the posts related to SH-HK and SZ-HK connect is similar. However the distribution of the posts related to Greece government-debt crisis is different. As a lot of

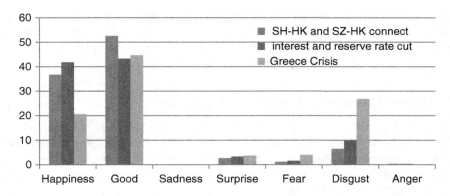

Fig. 1. The percentage distribution of the posts with different emotions under 3 topics.

people is worried about the future of the EU financial system and whether the Greece crisis will cause an international government debt crisis, there is almost 27 % posts is classified as "Disgust" emotion. There are also a lot of posts locating in the "Good" category as the Greece government-debt problem is solved in July. As the posts are mainly remain in the three categories: "Happiness", "Good", "Disgust", we will discuss the correlation between the stock market and these three emotions.

Figure 2 shows the two resulting public mood time series after the data processing which are "Happiness" time series and "Good" time series for the topic "interest rate and reserve rate cut". On 10th May, China central bank announced a new round of interest rate and reserve rate cut which is an event that may have a unique, significant and complex effect on the public mood. The result demonstrates that the time series generated by our method successfully identifies the public's emotional response to the announcement of interest rate cut. It has a significant but short-lived uptick in positive sentiment "Happiness" and "Good" to that day. Actually after the announcement, both emotions'

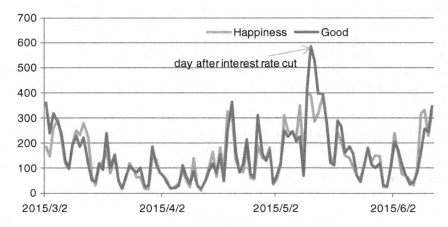

Fig. 2. Tracking public mood states from posts between Mar to June 2015 shows public responses to the announcement of interest rate cut.

sentiment score increase a lot sharply. In addition, we can find that before the announcement the sentiment score starts to climb up as there are some rumors appearing in the market which mentioned that the government will cut the interest rate soon and the expectation of the new round interest rate cut becomes higher.

3.4   Statistical Modeling: Granger Causality Analysis

Many economic time series are unit root process [16]. Therefore, we applied the Augmented Dickey-Fuller (ADF) test to each time series for identifying the unit root process. As the generated time series are proved to be unit root process, first differencing method is used to minimize the chance of spurious relationships. The time series that is not covariance stationary is converted to D_i,

$$D_{i,t} = \Delta x_{i,t} = x_{i,t} - x_{i,t-1} \tag{2}$$

To avoid the effect of various magnitudes, the time series D_i is normalized by computing the z-scores:

$$Z_{i,t} = \frac{D_{i,t} - \bar{D}_i}{\sigma_i^2} \tag{3}$$

Where $\bar{D}_i$ is the average of D_i and σ_i is the standard deviation of D_i over time period t. In our experiments, as we will discuss the 10 days moving average of stock price, we defined t equal to 10 and $\bar{D}_i$ is the 10 days moving average of the sentiment score change.

Our time series of stock prices is defined to reflect the daily changes in stock market value, i.e., the delta value between the close price of a stock in day t and day $t - 1$. Z-score computing is also applied to the delta value which helps to standardize the stock price change. $P_{i,t}$ equals to the z-score value of the delta. After establishing that our public mood time series responds to some special events such as the interest rate cut, we are concerned with the question whether other variations of the public mood state affect the price in the stock market. To answer this question, we applied the Granger causality analysis to the time series produced by our method, Z_i, and the stock price time series, P_i. Similar as Bollen's research [14], we try to demonstrate that the hypothesis that the lagged value of Z_i has a significant correlation with the price movement P_i cannot be rejected and the public mood time series can be used to forecast the movement of stock price.

We perform the Granger causality analysis according to the linear model as shown in Eq. (4) for the period of time between 1st Jan to 31st July 2015. As we have three topics to discuss and these topics are related to different stocks, different linear regressions between specific stock price time series and the public mood time series will be conducted based on the following formulas:

$$P_t = \alpha + \sum_{i=1}^{n} \beta_i P_{t-i} + \sum_{i=1}^{n} \gamma_i Z_{t-i} + \varepsilon_t \tag{4}$$

Where P_t is the z-score of the stock day close price difference between day t and $t - 1$, Z_t is the z-score of the sentiment score change of a specific public mood dimension or emotion category. The value of n is set to 7, as previous researches [14] demonstrate that public mood on day $t - 3$ has the highest correlation with stock price on day t.

First we will discuss the topic "SH-HK and SZ-HK connect". Table 2 presents the linear regression result between the time series for the close price of Shanghai Connect Index (stock code: 000159) and the time series of the emotion "Happiness" related to the topic "SH-HK and SZ-HK connect". Shanghai Connect Index is the composite index including all stocks that can be traded through the SH-HK connect by international traders in Hong Kong. As a result, the price movement of the Shanghai Connect Index is directly affected by traders' opinions or emotions about SH-HK connect. The result shows that the lag value of Z_{t-3} has a P-value equal to 0.028515 which is lower than the significance level 0.05. Based on the result of the Granger causality, we cannot reject the null hypothesis that the mood series do not predict the price movement of Shanghai Connect Index, i.e. $\gamma_{\{1,2,3,...,7\}} \neq 0$. However we find out that only the lag with 3 days has significant causal relations with price movement of the Index which is similar as Bollen's research [14] on Twitter post. In his research, he mentioned changes of public mood match shifts in the stock price that occur 3–4 day later. This behavior can be explain as that it may take some time for the public mood in social media websites to gain a momentum than will affect investors decision in the real market. Other 6 emotions time series are not significant correlated with the time series of Shanghai Connect Index according to the p-value.

Table 2. The linear regression analysis result between the "Happiness" time series and the close price movement of Shanghai Connect Index.

Lag	Coefficient	Std error	t stat	P-value
1 Day	−0.062263	0.096308	−0.646502	0.519301
2 Days	−0.128086	0.098495	−1.300428	0.196172
3 Days	−0.224765	0.101275	−2.219358	**0.028515**
4 Days	−0.025998	0.103179	−0.251972	0.801533
5 Days	−0.022974	0.101925	−0.225403	0.822084
6 Days	0.094800	0.098601	0.961458	0.338431
7 Days	0.115507	0.093337	1.237520	0.218529

To further demonstrate the forecast ability of the "Happiness" time series to other stocks, the p-value of the Granger causality correlation with 2 more stocks are calculated as well. The 2 stocks are HSI and the Hong Kong Exchanges and Clearing Limited corporation stock (stock code is 388). They are highly related to the topic "SH-HK and SZ-HK connect". HSI is the composite Index of Hong Kong stocks which will be affected by the SH-HK connect as more capitals can be invested on Hong Kong market. Hong Kong Exchanges and Clearing Limited Company is the company that provides the service of SH-HK connect. More trader are investigating through the SH-HK connect, more commissions it can receive. As it can gain more profits, the valuation of the stock is changed and the price will be increased as more investors will be attracted.

The results are shown in Table 3. According to the p-value 0.045231, the lag value of Z_{t-3} has significant correlation with the price movement of HSI. Similar as HSI, the price movement of 388 is significantly correlated with the lag value of Z_{t-4}. The p-value is 0.010188 which is much lower than the significance level 0.05. The null hypothesis cannot be rejected and the "Happiness" time series is predictive of the stock price movement including Shanghai Connect Index, HSI and the stock of Hong Kong Exchanges and Clearing Limited.

Table 3. The p-value of linear regress analysis result between "Happiness" time series and 3 different stocks.

Lag	000159	388	HSI
1 Day	0.519301	**0.096123**	0.538607
2 Days	0.196172	0.546172	0.428106
3 Days	**0.028515**	0.458417	**0.045231**
4 Days	0.801533	**0.010188**	0.155256
5 Days	0.822084	0.130202	0.198888
6 Days	0.338431	0.220351	0.500008
7 Days	0.218529	0.934022	0.301673

To demonstrate the predictive ability of public mood states with different topics, we also apply the Granger causality analysis to other topics. For the topic "Greece Government-Debt Crisis", we find out that the public mood time series other than "Disgust" is not significant correlated with the price movement of HSI. Only the lag value Z_{t-2} of the time series of "Disgust" has significant causal relation with the price movement of HSI considering the significance level 0.1. The experiment result is shown in Table 4.

Table 4. The linear regress analysis result between "Disgust" time series and the close price of HSI.

Lag	Coefficient	Std error	t stat	P-value
1 Day	−0.07973	0.131315	−0.60719	0.546991
2 Days	−0.27357	0.146389	−1.86881	**0.068634**
3 Days	0.108656	0.1709	0.635789	0.528363
4 Days	0.261093	0.176888	1.476039	0.147392
5 Days	0.220684	0.17898	1.233007	0.22443
6 Days	0.004027	0.166583	0.024174	0.980829
7 Days	−0.03112	0.146273	−0.21275	0.832549

For the topic "interest rate and reserve rate cut", as it is a method that China government used to booth the China stock market, we will discuss the correlation between the public mood time series and the close price of SSECI. The results are shown in Table 5. According to the p-value, the lag value of Z_{t-3} for emotion "Happiness" is highly correlated with the price movement of SSECI considering the significance level 0.05. And for the "Good" time series, if we consider the significance level 0.1, the lag value of Z_{t-3} is also correlated. In addition, the lag value of t-2 and t-3 of the "Disgust" time series have significant causal relationship with the SSECI value as well. We are not testing the actual causation but whether one time series has predictive information about the price movement or not. The result shows that the hypothesis that the stock price movement is correlated with the change of public mood extracted from Weibo posts cannot be rejected and the time series of "Happiness", "Good" and "Disgust" could be used to predict the price movement of SSECI.

Table 5. The linear regress analysis result between public mood states related to "interest rate cut" and the close price of SSECI.

Lag	Happiness	Good	Disgust
1 Day	0.254616	0.515584	0.395321
2 Days	0.111199	0.455075	**0.041898**
3 Days	**0.007864**	**0.0676**	**0.008356**
4 Days	0.944177	0.201313	0.959366
5 Days	0.858916	0.925238	0.995641
6 Days	0.8697	0.555839	0.193032
7 Days	0.355159	0.069943	0.384485

4   Conclusion

As the stock market in China has already been one of the largest stock market in the world, it is important to understand its behavior according to the public mood movements. In this paper, we investigate whether public mood states derived from large-scale collection of Weibo posts on weibo.com is correlated or predictive of different stock price in China market. Our results show that the movement of the public mood state can be indeed tracked through using our proposed sentiment score generation method on the large-scale Weibo posts. In addition, the Granger causality analysis results demonstrate that the null hypothesis that the public mood is correlated with the stock price cannot be rejected and the time series of public mood states is predictive of the price movement of different stocks. Among the 7 observed emotions, only some are Granger causative of the stock price movement. The emotions "Happiness" and "Disgust" have significant causative relationships with different stock prices. Changes of these two emotions are correlated with the price movement that occurs 3–4 days. These 3 days time lag is also discussed in other researches of using public mood in Twitter to predict stock price [14]. In the future, more experiments will be conducted to understand why the time lag exists and how to avoid and use it for real time algorithm trading. Other important factors will be examined in future research such as the geo-location effect and spam detection. In addition, in this paper, we only discuss the relationship between the close price of stocks and the daily movement of public mood. How the intraday public mood change affects the real-time stock price movement will be discussed in future research as well.

References

1. Abbasi, A., Chen, H.: CyberGate: a design framework and system for text analysis of computer-mediated communication. MIS Q. **32**, 811–837 (2008)
2. Liang, H., Tsai, F.S., Kwee, A.T.: Detecting novel business blogs. In: Proceedings of the 7th International Conference on Information, Communications and Signal Processing (2009)
3. Chau, M., Xu, J.: Mining communities and their relationships in blogs: a study of hate groups. Int. J. Hum Comput Stud. **65**, 57–70 (2007)
4. O'Leary, D.E.: blog mining-review and extensions: 'from each according to his opinion'. Decis. Support Syst. **54**, 821–830 (2011)
5. Gruhl, D., Guha, R., Kumar, R., Novak, J., Tomkins A.: The predictive power of online chatter. In: Proceedings of the Eleventh ACM SIGKDD International Conference on Knowledge Discovery in Data Mining, pp. 78–87 (2005)
6. Mishne, G., Glance N.: Predicting movie sales from blogger sentiment. In: Proceedings of AAAI-CAAW 2006, The Spring Symposia on Computational Approaches to Analyzing Weblogs (2006)
7. Sun, J., Wang, G., Cheng, X., Fu, Y.: Mining affective text to improve social media item recommendation. Inf. Process. Manage. **51**, 444–457 (2015)
8. Dolan, R.J.: Emotion, cognition, and behavior. Science **298**(5596), 1191–1194 (2002)
9. Nofsinger, J.R.: Social mood and financial economics. J. Behav. Finan. **6**(3), 144–160 (2005)
10. Fama, E.F.: The behavior of stock-market prices. J. Bus. **38**(1), 34–105 (1965)
11. Schumaker, R.P., Chen, H.: Textual analysis of stock market prediction using breaking financial news: the AZFin text system. ACM Trans. Inf. Syst. **27**(2), 1–19 (2009)
12. Liu, A.Y., Gu, B., Konana, P., Ghosh, J.: Predicting stock price from financial message boards with a mixture of experts framework. In: Intelligent Data Exploration & Analysis Laboratory, pp. 1–14 (2006)
13. Gilbert, E., Karahalio, E.: Widespread worry and the stock market. In: Proceedings of the International AAAI Conference on Weblogs and Social Media (2010)
14. Bollen, J., Mao, H., Zeng, X.: Twitter mood predicts the stock market. J. Comput. Sci. **2**(1), 1–8 (2011)
15. Zhang, X., Fuehres, H., Gloor, P.A.: Predicting stock market indicator through twitter "I hope it is not as bad as I fear". Procedia-Soc. Behav. Sci. **26**, 55–62 (2011)
16. Nelson, C.R., Plosser, C.R.: Trends and random walks in macroeconmic time series: some evidence and implications. J. Monetary Econ. **10**, 139–162 (1982)

Improving Diversity of User-Based Two-Step Recommendation Algorithm with Popularity Normalization

Xiangyu Zhao[1,3], Wei Chen[2,4(✉)], Feng Yang[3,5], and Zhongqiang Liu[1,6]

[1] Beijing Research Center for Information Technology in Agriculture, Beijing, China
[2] Agricultural Information Institute, Chinese Academy of Agricultural Sciences,
Beijing, China
chenwei@caas.cn
[3] National Engineering Research Center for Information Technology in Agriculture,
Beijing, China
{zhaoxy,yangf}@nercita.org.cn
[4] Key Laboratory of Agri-information Service Technology,
Ministry of Agriculture, Beijing, China
[5] Key Laboratory of Agri-informatics, Ministry of Agriculture, Beijing, China
[6] Beijing Engineering Research Center of Agricultural Internet of Things,
Beijing, China
liuzq@nercita.org.cn

Abstract. Recommender systems become increasingly significant in solving the information overload problem. Beyond conventional rating prediction and ranking prediction recommendation technologies, two-step recommendation algorithms have been demonstrated that they have outstanding accuracy performance in top-N recommendation tasks. However, their recommendation lists are biased towards popular items. In this paper, we propose a popularity normalization method to improve the diversity of user-based two-step recommendation algorithms. Experiment results show that our proposed approach improves the diversity performance significantly while maintaining the advantage of two-step recommendation approaches on accuracy metrics.

Keywords: Recommender system · Collaborative filtering · Diversity · Two-step recommendation · Popularity normalization

1 Introduction

As the development of Internet and Mobile Internet, massive user-generated data give an opportunity to offer personalized information service with semantic computing [18,25,26]. Recommendation is an important kind of personalization technology which may help people in retrieving potentially useful information in a huge set of choices, especially in the current age of information overload. Collaborative filtering (CF) is a leading approach to build recommender systems which has gained much popularity recently [5,6,20,21]. CF is based on the

© Springer International Publishing Switzerland 2016
H. Gao et al. (Eds.): DASFAA 2016 Workshops, LNCS 9645, pp. 15–26, 2016.
DOI: 10.1007/978-3-319-32055-7_2

analysis of past interactions between users and items, and hence can be readily applied in a variety of domains, without requiring external information about the traits of the recommended items.

Conventional CF approaches are based on users' rating values, for example from 1 to 5, and consider the recommendation problem as a rating prediction problem. These approaches estimate the ratings of items that have not been rated by the target user based on the rating history with a heuristic method [7,9] or a learned model [4,16,31], and recommend top-N items with the highest predicted ratings. Therefore, many researchers focus on improving the prediction accuracy of unknown ratings. They consider that high quality of rating predictions directly indicates good recommendations [16,31]. However, what people want from recommender systems is not whether the system can predict rating values accurately, but recommendations that match their interests [8]. Some researchers demonstrate that there is no trivial relationship between rating prediction accuracy and recommendation quality, as the rating prediction accuracy is not always consistent with ranking effectiveness [8,19,28]. Therefore, different from these rating prediction approaches, some researchers directly consider the recommendation problem as a ranking problem [17,19,24]. They propose models for ranking predictions by directly modeling user preferences with respect to a set of items rather than the rating scores on individual items.

We agree that the recommendation problem is a ranking problem. However, directly optimizing ranking targets may loss the semantic information of the recommendation scenario. Different from these ranking prediction CF, we proposed a two-step recommendation framework to generate recommendations by simulating the steps of users generating their rating data in previous work [28,30]. Experiments show that our proposed approaches based on this framework gain better accuracy than conventional ones. However, beyond accuracy, other quality factors, such as diversity and novelty, are also important for recommendation technology [1,3,10,12,15,23,27,32]. Some studies argue that one of the goals of recommender systems is to provide a user with highly idiosyncratic or personalized items, and more diverse recommendations result in more opportunities for users to get recommended such items [2]. In 2014, the ACM conference on Recommender Systems holded an independent session "Diversity, Novelty and Serendipity"[1]. Unfortunately, diversity and novelty of these two-step recommendation approaches are not very good. It appears that some recommendations of these approaches are biased towards well-known items, which may have been known by users. In this circumstance the recommendations are accurate, but not that valuable as the lack of novelty.

Therefore, the goal of this paper is to improve the diversity and novelty performance of two-step recommendation approaches while maintaining their advantages of accuracy. An improved user-based two-step recommendation algorithm with popularity normalization (UTSP) is proposed by adjusting the importance of items according to their popularity.

---

[1] http://recsys.acm.org/recsys14/session-diversity-novelty-serendipity/.

The remainder of the paper is organized as follows. Section 2 introduces two-step recommendation algorithm. UTSP is proposed in Sect. 3. Experiments are carried out on the MovieLens dataset in Sect. 4 to compare the proposed approach with some benchmark ones. We review related literature in Sect. 5. Finally, Sect. 6 concludes the paper.

2 Two-Step Recommendation Algorithm

A typical Collaborative Filtering recommendation task is based on the rating data generated by users. These data contain two layers of user behaviors. The first one is that the current user selects an item to rate. The second one is rating it with a value. However, conventional recommendation approaches directly use rating values or rating ordinal relation to build rating prediction or ranking prediction algorithms and generate recommendations with no concern on the behaviors that users select items to rate. The effectiveness of these approaches is based on a condition that if a user rates an item, he/she may rate it with a value which is predicted by recommendation algorithms. Unfortunately, the prerequisite may not be satisfied. A user may not tend to rate an item, as it does not match his/her interest.

Therefore, we have proposed two-step recommendation algorithms to solve the recommendation problem in our previous work [28,30]. In a two-step recommendation algorithm, the unknown user behaviors can be predicted as simulation of user generating ratings. That is predicting the probability $\hat{P}(u, i)$ that user u rates item i (in the first step), and then predicting the value $\hat{r}(u, i)$ which u may rate i with (in the second step). After that, the ranking score can be computed as:

$$ranking(u, i) = \hat{P}(u, i)\hat{r}(u, i) \tag{1}$$

This ranking score can be interpreted by the generation steps with a probability semantic. In addition, for a certain $\langle u, i \rangle$, the probability that the user may rate the item is $\hat{P}(u, i)$. Therefore, the probability that the user will not rate the item is $1 - \hat{P}(u, i)$. In recommender systems, typical values for the rated item are in 1–5 or 1–10 scale. In order to model rating values and rating behaviors in a unique model, the items that a user does not want to rate can be considered as being rated with value 0^2. In this way, the ranking score can be viewed as the mathematical expectation of users' rating on the items. This can be considered as another interpretation of the ranking score:

$$\begin{aligned}
&ranking(u, i) \\
&= \hat{P}(u, i)\hat{r}(u, i) \\
&= \hat{P}(u, i)\hat{r}(u, i) + (1 - \hat{P}(u, i)) \cdot 0 \\
&= E[r(u, i)]
\end{aligned} \tag{2}$$

---

[2] 0 is a typical value out of the range of rating scale, which can be used to distinguish the rating value and the rating behavior.

The goal of the first step is to predict the rating behaviors. Intuitively, historical rating behaviors are relevant to it, whereas rating values are not. Therefore, the probability is predicted using only rating behaviors in the first step of our proposed framework. In the second step, all users' historical rating data (both rating behaviors and rating values) are used to predict unknown ratings. This is a classical rating prediction problem. Existing techniques focus on rating prediction can be used in this step. After the two-step calculation, the ranking score can be computed with (1). The recommendation results can be generated based on the rankings.

3   UTSP Recommendation Algorithm

Recommender systems are explored to solve information overload problem for users. It means that the purpose of recommendation is inherently linked to a notion of discovery, as recommendation makes most sense when it exposes the user to a relevant experience that he/she would not have found by himself/herself. However, it is found that recommender systems actually can reduce the aggregate diversity, which has been described as "Harry Potter Problem[3]" [11,29]. Harry Potter is a runaway bestseller, which always appears in customers' recommendation list whatever books they are browsing. That is, recommended items are biased towards popular, well-known items. This can be explained by the fact that the idiosyncratic items often have limited historical data and, thus, are more difficult to be recommended to users; in contrast, popular items typically have more ratings and, therefore, can be recommended to more users. This phenomenon exists and is even worse in two-step recommendation algorithms, hence a UTSP algorithm is proposed to improve the diversity and novelty performance in this section.

3.1   The First Step

The target of the first step is to predict the probability that a user rates an item with users' historical rating behaviors. The rating behaviors are binary data, hence a user can be described as an n-dimensional vector in which 1 represents rated items and 0 represents unrated ones, which can be written as:

$$V_U(u) = (v_1, v_2, \cdots, v_n) \tag{3}$$

$$v_i = \begin{cases} 1, i \in I(u) \\ 0, i \notin I(u) \end{cases} (i \in [1, n]) \tag{4}$$

where $I(u)$ represents the item set rated by user u.

Conventional user-based two-step recommendation algorithm (UTS) directly use this model to predict the probability that a user rates an item. If ignoring the effect of user similarity, the probability can be calculated as:

$$\hat{P}(u, i) = \frac{1}{|N(u)|} \sum_{a \in N(u)} V_U(a)[i] \tag{5}$$

---

[3] http://glinden.blogspot.com/2006/03/early-amazon-similarities.html.

where $V_U(a)[i]$ is the ith element of the binary user model for user a, $N(u)$ represents the neighbor set of user u, which contains the most similar users to user u.

This is the probability that an item rated by the neighbors for a given user. Intuitively, it is biased towards popular items as they have more ratings overall. Let's take the book domain for a motivate example. Harry Potter, as a bestseller, is bought by about 20 % of users, while Data Mining is a professional book for computer science researchers, only bought by no more than 0.3 % of total users. But in the neighbor set (with 50 neighbors) of a specific user a, there are 10 users who have bought Harry Potter, and 5 users who have bought Data Mining. If using (5) to generate recommendation directly, Harry Potter will be recommend to a. However, Data Mining might be a much better recommendation as the purchase rate in a's neighbor set is far larger than the overall rate, the user may be a computer science researcher. It means that the increment of the purchase rate in a user's neighborhood is a more important measure than the value of purchase rate itself. It can be calculated as:

$$\hat{P}(u,i) = \frac{\sum_{a\in N(u)} V_U(a)[i]/|N(u)|}{|U(i)|/|U|} \tag{6}$$

where $U(i)$ represents the set of users who have rated item i, U represents the set of all the users. The value of the increased rate may be larger than 1, which is not suitable for the definition of probability. Moreover, since the main target of recommender systems is to recommend items for given users, and $|N(u)|$ and $|U|$ are constants for a given user, (6) can be simplified as:

$$\hat{P}(u,i) = \frac{\sum_{a\in N(u)} V_U(a)[i]}{|U(i)|} \tag{7}$$

As $U(i)$ is the popularity of item i, (7) can be considered as a normalization function with popularity. Furthermore, the attitude from a more similar user is always considered as more important information. Therefore, by including the similarity information, the probability that a user rates an item can be calculated as:

$$\hat{P}(u,i) = \frac{\sum_{a\in N(u)} sim(u,a) \cdot V_U(a)[i]}{|U(i)| \cdot \sum_{a\in N(u)} sim(u,a)} \tag{8}$$

where $sim(u,a)$ is the similarity between user u and user a.

In theory, this is an effective method to predict the probability that a user rates an item. However, according to our empirical study, the recommendations from (8) are biased towards long tail items. Let's get back to the book domain example. If only one neighbor has bought Data Mining because of his/her individual interest, the book still will be recommended to the user as its popularity is much less than Harry Potter. This means that the recommended items will be biased towards individual neighbor's long tail interest rather than the common interest of the neighbor set. Therefore, we can decrease the degree of popularity normalization in order to reduce the bias towards long tail items. The revised function is written as:

$$\hat{P}(u, i) = \frac{\sum_{a \in N(u)} sim(u, a) \cdot V_U(a)[i]}{\beta \cdot \sqrt{|U(i)|} \cdot \sum_{a \in N(u)} sim(u, a)} \tag{9}$$

where β is a small constant to make sure the probability is between 0 and 1.

3.2   The Second Step

The second step is considered as a classical rating prediction problem. It can be done by making use of existing techniques. In UTSP, we use SVD++ [16] in the second step.

SVD++ is a matrix factorization approach, which is demonstrated to yield superior accuracy by considering implicit feedbacks[4] as complement of explicit feedbacks (rating values), and using them together to build recommendation models by minimizing prediction errors. The prediction model of SVD++ is as follows:

$$\hat{r}(u, i) = \mu + b_u + b_i + q_i^T (p_u + |N(u)|^{-\frac{1}{2}} \sum_{j \in Iu} y_j) \tag{10}$$

where μ is the average rating value of the known data. b_u and b_i indicate the observed deviations of user u and item i, respectively, from the average. p_u and q_i are the factorized user and item factor, respectively. y_j is an item factor which is computed according to the impact of implicit feedbacks.

SVD++ learns the values of involved parameters with a stochastic gradient descent technique by minimizing the regularized squared error function [16] associated with:

$$\min_{p*, q*, b*, y*} \sum_{<u,i>} (r_{ui} - \mu - b_u - b_i - q_i^T (p_u + $$
$$|N(u)|^{-\frac{1}{2}} \sum_{j \in N(u)} y_i))^2 + \lambda_6 (b_u^2 + b_i^2) \tag{11}$$
$$+ \lambda_7 (\|q_i\|^2 + \|p_u\|^2 + \sum_{j \in N(u)} \|y_i\|^2)$$

where r_{ui} is the actual rating value for item i rated by user u, λ_6 and λ_7 are two regularization parameters. The predicted ratings can be calculated by (10) using the learned parameters.

Based on the above models, UTSP can predict $\hat{P}(u, i)$ according to (9), predict $\hat{r}(u, i)$ according to (10), compute the rankings of the unrated items for users according to (1), and then generate recommendations.

---

[4] Types of implicit feedback include rating behaviors, purchase history, browsing history, and search patterns.

4   Experiment

4.1   Experiment Setup

In this paper, we focus on both accuracy and diversity performance in top-N recommendation task, and use 4 metrics to evaluate our proposed approach. 1-call [22] and the Normalized Discounted Cumulative Gain (NDCG) [14] are used as accuracy metrics, whereas Coverage (COV) is used for evaluate the diversity of recommendations, and coverage in long tail (CIL) is mainly for evaluating novelty [30].

The proposed recommendation approach is evaluated on the MovieLens dataset, which consists of 100,000 ratings which are assigned by 943 users on 1682 movies. Collected ratings are in a 1-to-5 star scale. We use 5-fold cross validation for the evaluation. Starting from the initial data set, five distinct splits of training and test data are generated. For each data split, 80 % of the original set is included in the training data and 20 % of it is included in the test data. Users' rating history in the training set is used to generate recommendations according to different algorithms. The test set is then used to evaluate the recommendation results.

The proposed approach is compared with some benchmark ones for both rating prediction and ranking prediction approaches. For rating prediction approaches, UserCF [9] and SVD++ [16] are used for comparison. UserCF is a user-based CF with Jaccard as its similarity function. SVD++ is a state-of-the-art rating prediction approach. For ranking prediction approaches, pLPA [19] is used for comparison, which is a probabilistic latent preference analysis model which directly optimizes ranking target based on a pairwise ordinal model.

In addition, some approaches need user-specific parameters. The details of parameter assignments for different approaches are as follows: the size of nearest neighbors for UserCF is 50; SVD++ has 50 features and 25 iteration steps with $\lambda_6 = \lambda_7 = 0.05$, and $\gamma_1 = \gamma_2 = 0.002$; pLPA has 6 latent preferences and 30 iterations [19]. The first step of UTS and UTSP has the same neighbor size as UserCF, and the second step of UTS and UTSP has the same parameters as SVD++. Therefore, the effectiveness of the proposed approaches is irrelevant to the impact of these parameters.

With these parameters, all of the above mentioned approaches are evaluated by 1-call, NDCG, COV and CIL.

4.2   Experiment Results

In this subsection, UTSP will be compared with some benchmark recommendation approaches, including UserCF, SVD++, pLPA, and UTS, to demonstrate its effectiveness. For each approach, we report the NDCG values at the $1st$, $3rd$ and $5th$ positions in the recommendation list, and 1-call, COV and CIL at the $5th$ position. Table 1 illustrates the results, where the top 2 best performed approaches for each metric have been highlighted.

Table 1. Performance of two-step recommendation approaches.

	NDCG			1-call	COV	CIL
	1	3	5			
UserCF	0.0101	0.0100	0.0131	0.0880	**0.2949**	**0.1700**
SVD++	0.0372	0.0448	0.0468	0.1994	0.0386	0.0065
pLPA	0.1506	0.1326	0.1211	0.3213	0.0428	0.0000
UTS	**0.2082**	**0.1855**	**0.1750**	**0.4369**	0.1034	0.0036
UTSP	**0.1877**	**0.1690**	**0.1587**	**0.4358**	**0.2592**	**0.0910**

As can be seen from the results, the two rating prediction approaches, UserCF and SVD++, get the worst accuracy. It means that there is no trivial relationship between the accuracy of rating prediction and quality of top-N recommendation. pLPA, a ranking prediction recommendation approach, can improve top-N recommendation accuracy from rating prediction approaches, which indicates that the recommendation problem is a ranking problem. For UTS and UTSP, these two-step recommendation approaches can further increase the recommendation accuracy. This shows that two-step recommendation is more suitable for top-N recommendation task.

Focusing on diversity metrics, UTS almost gets the worst performance. However, by using popularity normalization, UTSP improves its diversity performance significantly and achieves the 2nd best among all the approaches. If considering many recommended items from UserCF are irrelevant as its poor accuracy, UTSP recommends most effective items in terms of diversity. It means that UTSP outperforms all the benchmark recommendation approaches if considering both accuracy and diversity performance comprehensively.

5   Related Work

In this section, the review of literatures is divided into three parts. The first one is about conventional rating prediction recommendation algorithms. The second one includes some studies on ranking prediction recommendation approaches. The last one focuses on the two-step recommendation approaches.

5.1   Rating Prediction Approaches

Recommendation techniques have been studied for several years. Conventional recommendation approaches are based on rating prediction. In these approaches, the past interactions between users and items are analyzed by collaborative filtering. Algorithms of collaborative filtering can be divided into two classes: memory-based and model-based [2].

Memory-based algorithms are heuristic methods that make rating predictions based on the entire collection of items previously rated by users [7,9]. They are based on a basic assumption that people who agreed in the past tend to agree again in the future. The level of agreement can be measured by similarity. Based on the similarity calculation, recommender systems predict ratings for unknown items using adjusted weighted sum of known ratings and recommend items with high predicted values [9].

Model-based CF is another kind of typical CF methods. Model-based algorithms use the collection of ratings to learn a model, typically using some statistical machine-learning methods, which are then used to make rating prediction. These approaches always design appropriate loss functions and optimization procedures to learn their models by minimizing the error between predicted ratings and actual ones. Examples of such techniques include Bayesian clustering [4], matrix factorization [16,31], and probabilistic Latent Semantic Analysis [13].

These conventional approaches are based on users' rating values, their optimization goals are minimizing prediction errors. Though they cannot generate top-N recommendation effectively, these techniques can be applied in the second step of two-step recommendation approaches.

5.2   Ranking Prediction Approaches

Differently from those rating prediction approaches, some researches directly consider the recommendation problem as a ranking prediction problem. They propose models for ranking prediction by directly modeling user preferences with respect to a set of items rather than the rating scores on individual items.

Weimer et al. [24] present a method (CofiRank) which uses Maximum Margin Matrix Factorization and considers maximum NDCG as the optimizing target. The approach is adaptable to different scores. Since the optimizing target of Cofirank is a listwise one, the approach scales well on collaborative filtering tasks.

Liu et al. [19] propose a probabilistic latent preference analysis (pLPA) model to make ranking predictions. From a user's observed ratings, they extract his/her preferences in the form of pairwise comparisons of items which are modeled by a mixture distribution based on Bradley-Terry model. An EM algorithm for fitting the corresponding latent class model as well as a method for predicting the optimal ranking is described.

Koren et al. [17] propose a collaborative filtering recommendation framework, which is based on the technique that considers user feedback on products as ordinal, rather than the more common numerical point of view. Their approach is based on a pointwise ordinal model, which allows it to linearly scale with data size. In addition, the approach can predict a full probability distribution of the expected item ratings, rather than only a single score for an item, and estimate the confidence level in each individual prediction.

It is demonstrated that these ranking prediction approaches can get better ranking results than rating prediction ones. However, experiments show that good performance on ranking prediction does not necessarily indicate good quality of top-N recommendation, which is the main purpose of recommender systems.

5.3   Two-Step Recommendation Approaches

Typical recommendation task is based on the rating data which contain two layers of user behaviors. The first one is that the current user selects an item to rate. The second one is rating it with a value. In this circumstance, simply using either rating prediction or ranking prediction idea to generate recommendations is ineffective since their basement condition that if a user rates an item, he/she may rate it with a value which is predicted by recommendation algorithms may not be satisfied. Therefore, two-step recommendation approaches try to solve recommendation problem in a different way.

Hofmann [13] decomposes the recommendation problem into the prediction of selected items and the prediction of the rating conditioned on the selected items. This mimics a scenario in which the user is free to select an item of his/her choice and also provides a rating for it. [28] finds that whether a user rates an item can be considered as a measure of interest no matter whether the value is high or low, and the rating values themselves represent the attitude to the quality of the target item, especially in the information overloaded age. Therefore, used-based and item-based two-step recommendation approaches are proposed by recommending items matching users' interests first, and then finding high quality items which users will like from the interested item set. [30] further proposes a two-step recommendation framework by simulating user generating ratings. That is predicting the probability that a user rates an item in the first step, and then predicting the value which the user may rate the item with in the second step. After that, the ranking score, which is used for generating recommendations, can be computed as the product of the probability and the value. Based on the framework, a hybrid approach of topic model and matrix factorization is proposed.

All the above two-step recommendation approaches gains good performance on accuracy in top-N recommendation task. The main difference between them is that Hofmann's approach is a intra two-step recommendation approach, which learns a unified model containing both steps, whereas others are inter two-step recommendation approaches, which combine two models, each of which processes in one step, respectively.

6   Conclusions

User-based two-step recommendation approach directly uses the probability that an item is rated by the neighbors to predict given user's possible rating behavior. It may cause recommendations to bias toward popular items.

By analyzing this problem, we propose a popularity normalization approach to improve UTS, which leads to significant diversity improvement while maintaining the good performance on accuracy. Experiment results show that our proposed approach outperforms the benchmark, including UserCF, SVD++, pLPA, and UTS while considering both accuracy and diversity performance comprehensively.

Acknowledgements. This work is supported by the National Key Technology R&D Program of China (project no. 2014BAD10B08).

References

1. Adamopoulos, P., Tuzhilin, A.: On over-specialization and concentration bias of recommendations: Probabilistic neighborhood selection in collaborative filtering systems.In: Proceedings of the 8th ACM Conference on Recommender Systems, pp. 153–160. ACM (2014)
2. Adomavicius, G., Kwon, Y.O.: Improving aggregate recommendation diversity using ranking-based techniques. IEEE Trans. Knowl. Data Eng. **24**(5), 896–911 (2012)
3. Bradley, K., Smyth, B.: Improving recommendation diversity. In: Proceedings of the Twelfth Irish Conference on Artificial Intelligence and Cognitive Science, Maynooth, Ireland, pp. 85–94. Citeseer (2001)
4. Breese, J.S., Heckerman, D., Kadie, C.: Empirical analysis of predictive algorithms for collaborative filtering. In: Proceedings of the Fourteenth Conference on Uncertainty in Artificial Intelligence, pp. 43–52. Morgan Kaufmann Publishers Inc. (1998)
5. Cai, Y., Lau, R.Y.K., Liao, S.S.Y., Li, C., Leung, H.-F., Ma, L.C.K.: Object typicality for effective web of things recommendations. Decis. Support Syst. **63**, 52–63 (2014)
6. Cai, Y., Leung, H., Li, Q., Min, H., Tang, J., Li, J.: Typicality-based collaborative filtering recommendation. IEEE Trans. Knowl. Data Eng. **26**(3), 766–779 (2014)
7. Chen, W., Niu, Z., Zhao, X., Li, Y.: A hybrid recommendation algorithm adapted in e-learning environments. World Wide Web **17**(2), 271–284 (2014)
8. Cremonesi, P., Koren, Y., Turrin, R.: Performance of recommender algorithms on top-n recommendation tasks. In: Proceedings of the fourth ACM conference on Recommender systems, pp. 39–46. ACM (2010)
9. Delgado, J., Ishii, N.: Memory-based weighted majority prediction. In: ACM SIGIR 1999 Workshop on Recommender Systems. Citeseer (1999)
10. Ekstrand, M.D., Harper, F.M., Willemsen, M.C., Konstan, J.A.: User perception of differences in recommender algorithms. In: Proceedings of the 8th ACM Conference on Recommender systems, pp. 161–168. ACM (2014)
11. Fleder, D., Hosanagar, K.: Blockbuster culture's next rise or fall: The impact of recommender systems on sales diversity. Manage. Sci. **55**(5), 697–712 (2009)
12. Garcin, F., Faltings, B., Donatsch, O., Alazzawi, A., Bruttin, C., Huber, A.: Offline and online evaluation of news recommender systems at swissinfo.ch. In: Proceedings of the 8th ACM Conference on Recommender systems, pp. 169–176. ACM (2014)
13. Hofmann, T.: Latent semantic models for collaborative filtering. ACM Trans. Inf. Syst. (TOIS) **22**(1), 89–115 (2004)
14. Järvelin, K., Kekäläinen, J.: Cumulated gain-based evaluation of IR techniques. ACM Trans. Inf. Syst. (TOIS) **20**(4), 422–446 (2002)
15. Kapoor, K., Kumar, V., Terveen, L., Konstan, J.A., Schrater, P.: I like to explore sometimes: Adapting to dynamic user novelty preferences. In: Proceedings of the 9th ACM Conference on Recommender Systems, pp. 19–26. ACM (2015)
16. Koren, Y.: Factorization meets the neighborhood: a multifaceted collaborative-filtering model. In: Proceedings of the 14th ACM SIGKDD International Conference on Knowledge Discovery and Data Mining, pp. 426–434. ACM (2008)
17. Koren, Y., Sill, J.: OrdRec: An ordinal model for predicting personalized item ratingdistributions. In: Proceedings of the Fifth ACM Conference on Recommender Systems, pp. 117–124. ACM (2011)

18. Li, X., Xie, H., Song, Y., Li, Q., Shanfeng Zhu, F., Wang, L.: Does summarization help stock prediction? News impact analysis via summarization. IEEE Intell. Syst. **30**, 26–34 (2015)
19. Liu, N.N., Zhao, M., Yang, Q.: Probabilistic latent preference analysis for collaborative filtering. In: Proceedings of the 18th ACM Conference on Information and Knowledge Management, pp. 759–766. ACM (2009)
20. Perugini, S., Gonçalves, M.A., Fox, E.A.: Recommender systems research: A connection-centric survey. J. Intell. Inf. Syst. **23**(2), 107–143 (2004)
21. Ricci, F., Shapira, B.: Recommender Systems Handbook. Springer, Heidelberg (2011)
22. Shi, Y., Karatzoglou, A., Baltrunas, L., Larson, M., Oliver, N., Hanjalic, A.: CliMF: Learning to maximize reciprocal rank with collaborativeless-is-more filtering. In: Proceedings of the Sixth ACM Conference on Recommender Systems, pp. 139–146. ACM (2012)
23. Vargas, S., Castells, P.: Improving sales diversity by recommending users to items. In: Proceedings of the 8th ACM Conference on Recommender systems, pp. 145–152. ACM (2014)
24. Weimer, M., Karatzoglou, A., Le, Q.V., Smola, A.J.: Cofi rank-maximum margin matrix factorization for collaborative ranking. In: Advances in Neural Information Processing Systems, pp. 1593–1600 (2007)
25. Xie, H.-R., Li, Q., Cai, Y.: Community-aware resource profiling for personalized search in folksonomy. J. Comput. Sci. Technol. **27**(3), 599–610 (2012)
26. Xie, H., Yu, L., Li, Q.: A hybrid semantic item model for recipe search by example. In: IEEE International Symposium on Multimedia (ISM), pp. 254–259. IEEE (2010)
27. Zhang, M., Hurley, N.: Avoiding monotony: Improving the diversity of recommendation lists. In: Proceedings of the 2008 ACM Conference on Recommender Systems, pp. 123–130. ACM (2008)
28. Zhao, X., Niu, Z., Chen, W.: Interest before liking: Two-step recommendation approaches. Knowl. Based Syst. **48**, 46–56 (2013)
29. Zhao, X., Niu, Z., Chen, W.: Opinion-based collaborative filtering to solve popularity bias in recommender systems. In: Decker, H., Lhotská, L., Link, S., Basl, J., Tjoa, A.M. (eds.) DEXA 2013, Part II. LNCS, vol. 8056, pp. 426–433. Springer, Heidelberg (2013)
30. Zhao, X., Niu, Z., Chen, W., Shi, C., Niu, K., Liu, D.: A hybrid approach of topic model and matrix factorization based on two-step recommendation framework. J. Intell. Inf. Syst. **44**(3), 335–353 (2014)
31. Zhou, Y., Wilkinson, D., Schreiber, R., Pan, R.: Large-scale parallel collaborative filtering for the netflix prize. In: Fleischer, R., Xu, J. (eds.) AAIM 2008. LNCS, vol. 5034, pp. 337–348. Springer, Heidelberg (2008)
32. Ziegler, C.-N., McNee, S.M., Konstan, J.A., Lausen, G.: Improving recommendation lists through topic diversification. In: Proceedings of the 14th International Conference on World Wide Web, pp. 22–32. ACM (2005)

Followee Recommendation
in Event-Based Social Networks

Shuchen Li, Xiang Cheng$^{(\boxtimes)}$, Sen Su, and Le Jiang

State Key Laboratory of Networking and Switching Technology,
Beijing University of Posts and Telecommunications, Beijing, China
{lsc,chengxiang,susen,jiangle}@bupt.edu.cn

Abstract. Recent years have witnessed the rapid growth of event-based social networks (EBSNs) such as Plancast and DoubanEvent. In these EBSNs, followee recommendation which recommends new users to follow can bring great benefits to both users and service providers. In this paper, we focus on the problem of followee recommendation in EBSNs. However, the sparsity and imbalance of the social relations in EBSNs make this problem very challenging. Therefore, by exploiting the heterogeneous nature of EBSNs, we propose a new method called Heterogenous Network based Followee Recommendation (HNFR) for our problem. In the HNFR method, to relieve the problem of data sparsity, we combine the explicit and latent features captured from both the online social network and the offline event participation network of an EBSN. Moreover, to overcome the problem of data imbalance, we propose a Bayesian optimization framework which adopts pairwise user preference on both the social relations and the events, and aims to optimize the area under ROC curve (AUC). The experiments on real-world data demonstrate the effectiveness of our method.

Keywords: Followee recommendation · Event-based social networks · Heterogenous network

1 Introduction

In the past few years, event-based social networks (EBSNs), such as Plancast[1] and DoubanEvent[2], have grown rapidly and attracted millions of users. These EBSNs provide online platforms for users to establish, manage and join social events. In these EBSNs, followee recommendation can bring great benefits to both users and service providers. On one hand, users could find like-minded people or their friends in real life to follow, and thus are able to enjoy better user experiences through effective followee recommendation. On the other hand, service providers can exploit followee recommendation to drive users' engagement and loyalty. In this paper, we focus on the problem of followee recommendation in EBSNs.

---

[1] http://www.plancast.com/.
[2] http://beijing.douban.com/.

© Springer International Publishing Switzerland 2016
H. Gao et al. (Eds.): DASFAA 2016 Workshops, LNCS 9645, pp. 27–42, 2016.
DOI: 10.1007/978-3-319-32055-7_3

To make a study, we collect real data from Plancast and DoubanEvent. Based on our analysis, the social relations in Plancast and DoubanEvent are extremely sparse and imbalanced. In particular, the distributions of the number of followees and followers in both the datasets almost follow a power-law distribution. The problem of data sparsity and imbalance can significantly degrade the recommendation performance, due to the lack of enough data and the huge amount of negative samples. However, since an EBSN is a heterogeneous network which consists of an online social network and an offline event participation network, in addition to the social relations, there is an unprecedented source which can be utilized for our problem: the event participation records. As users attend events mainly based on their interests, the events attended by a user can reflect the user's interest. According to the social theory of homophily, users with similar interests are more likely to establish social relations. Besides, since the events are held at physical places, users who have attended the same event may have a chance to meet each other and develop new social links between them. Therefore, we can exploit the event participation records to improve the effectiveness of followee recommendation.

To relieve the problem of data sparsity, we utilize both the social relations and event participation records for followee recommendation. In particular, we extract two kinds of explicit features for our problem: (1) social features, which are extracted from the online social network and (2) event-based features, which are captured from the offline event participation network. Moreover, to take advantage of the latent factor model, we employ matrix factorization model in the online social network and the offline event participation network to capture the latent features in both the networks. More importantly, to derive users's latent features better and more comprehensively, we assume that these two networks share the same latent user feature in the matrix factorization model for each user. Finally, we combine all the explicit and latent features into a unified recommendation model.

To overcome the problem of data imbalance, a common way is to consider the area under ROC curve (AUC) as the optimization object, which is not influenced by the distribution of classes. Since our problem can be regarded as a ranking problem with implicit feedback, inspired by the Bayesian Personal Ranking (BPR) [16], we propose a Bayesian optimization framework which aims to optimize the AUC. By exploiting the heterogeneous network, our framework adopts pairwise user preference on both the social relations and the events, and optimizes for ranking pairs correctly.

To sum up, the primary contributions of our research are listed as follows.

- To the best of our knowledge, we are the first to study the problem of followee recommendation in EBSNs.
- We extract several social and event-based explicit features from the online social network and the offline event participation network. Moreover, we combine all the explicit and latent features into a unified recommendation model.
- We propose a new Bayesian optimization framework which adopts pairwise user preference on both the social relations and the events.

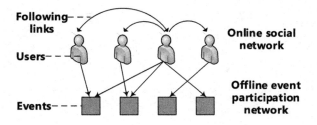

Fig. 1. An illustration of EBSNs.

– We evaluate the performance of our method using real-world data collected from Plancast and DoubanEvent. Experimental results show that our method is superior to alternatives and methods that consider only part of the factors exploited in this paper.

The rest of this paper is organized as follows. In Sect. 2, we give an overview of EBSNs, formally define the problem, and make some analyses about our data. In Sect. 3, we show the details of our followee recommendation model. We report the experimental results in Sect. 4 and review related works in Sect. 5. Finally, we make the conclusion in Sect. 6.

2   Preliminaries

2.1   Event-Based Social Network

A graph representation of EBSNs such as Plancast and DoubanEvent is shown in Fig. 1. We can observe that users and events are the two main entities in an EBSN. In particular, an event is an activity that is held in a physical venue, e.g., a drama held in a theater. Moreover, users are the participants of events, who can express their willingness to join an event by RSVP ('Yes' or 'Maybe'). The RSVP('Yes' or 'Maybe') indicates that a user wants to attend or is interested in an event. Besides, they can establish social relations by following other users. From Fig. 1, we can also find that the network structure of an EBSN is heterogeneous, which consists of an online social network and an offline event participation network.

2.2   Problem Definition

In an EBSN, we have two types of entities: $\{U(\text{user}) \text{ and } E(\text{event})\}$, and two kinds of networks: $\{G^{on}(\text{online social network}) \text{ and } G^{off} (\text{offline event participation network})\}$. Let $U = \{u_1, u_2, ..., u_n\}$ denote the set of users and $E = \{e_1, e_2, ..., e_m\}$ denote the set of events, respectively. For each user $u \in U$, it has a set of followees $\mathcal{F}_u^+ \subseteq U$, a set of followers $\mathcal{F}_u^- \subseteq U$, and a set of attended events $E_u \subseteq E$. For convenience, we use a matrix $R \in \mathbb{R}^{n \times n}$ to represent the online social network G^{on}, where $r_{ij} = 1$ indicates that user u_i is a follower of user u_j and $r_{ij} = 0$ means

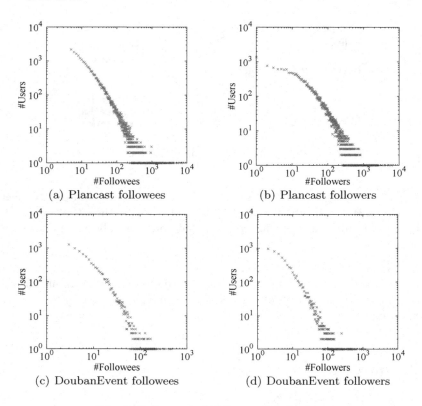

(a) Plancast followees   (b) Plancast followers

(c) DoubanEvent followees   (d) DoubanEvent followers

Fig. 2. Followee and Follower distributions

that user u_i has not yet followed user u_j. Similarly, let matrix $V \in \mathbb{R}^{n \times m}$ represent the offline event participation network G^{off}, where v_{ij} is equal to 1 if user u_i has attended event e_j and equal to 0 otherwise. Then, the problem of followee recommendation can be formally defined as ranking all users in the candidate set $C_u = U - (\mathcal{F}_u^+ \cup u)$ for each user $u \in U$, according to user u's preference to other users, and recommending top-k users for user u.

2.3   Data Analysis

To make a study of the problem, we collect real data from Plancast and Douban-Event. The Plancast dataset was published by the work [13] and the Douban-Event dataset was obtained by querying Douban API[3]. To make data sufficient for evaluation, for the Plancast dataset, we remove the inactive users who have less than 5 followees and have not attended any event; for the DoubanEvent dataset, we remove the inactive users who have less than 3 followees and have not joined any event. After preprocessing, some statistics of the two datasets

---

[3] http://developers.douban.com/

Table 1. Statistics of the pre-processed datasets

	Plancast	DoubanEvent
Number of users	28,060	8,299
Number of events	237,994	104,591
Number of follower-followee links	1,228,714	103,604
Number of use-event links	591,870	531,113
Max number of followees	6,943	465
Max number of followers	6,149	1,148

are shown in Table 1. In summary, we obtain the density of the online social network (i.e., the social relations) as 1.56×10^{-3} for Plancast dataset and 1.5×10^{-3} for DoubanEvent dataset, which indicates the high sparsity of the social relations.

To further understand the social relations in Plancast and DoubanEvent, we analyze the distributions of the followee and follower numbers in both the datasets. As shown in Fig. 2, the distributions in both datasets almost follow a power-law distribution, which means that while most of users have a small amount of followees and followers, there exist a few users who have a large number of followees and followers. On average, each user of Plancast has 44 followees and a user in DoubanEvent has 12 followees, which indicates the extreme imbalance of the social relations, i.e., the number of a user's followees is very small with regard to the total number of the users.

To sum up, the social relations in both the datasets are extremely sparse and imbalanced, which makes the problem of followee recommendation very challenging.

3   Followee Recommendation Modeling

In this section, we introduce the details of our model. We first present the explicit features and the latent features in Sects. 3.1 and 3.2, respectively. Then, we introduce our recommendation model in Sect. 3.3. Finally, we describe the parameter learning method in Sect. 3.4.

3.1   Explicit Features

Explicit features for a pair of users are usually used to measure the similarity of two users from different points of view. According to the social theory of homophily, two users are likely to develop new social links between them if they are similar to each other. In this work, we generate two kinds of explicit features: social features and event-based features. In the following, we describe each explicit feature in detail.

Social Features. Online social relations intuitively play an important role in helping create new social links. In the following, we define several social features based on node neighborhoods in the online social network of the EBSN.

Number of common neighborhoods. Given a user pair $\langle u_i, u_j \rangle$, this feature calculates how many user u_i's followees have followed user u_j and is defined as follows:

$$common\_neighbor(u_i, u_j) = \left| \mathcal{F}_i^+ \cap \mathcal{F}_j^- \right|.$$

Ratio of overlapped neighborhoods. This feature measures the Jaccard similarity between user u_i's followees and user u_j's followers, which is defined as follows:

$$overlap\_neighbor(u_i, u_j) = \frac{\left| \mathcal{F}_i^+ \cap \mathcal{F}_j^- \right|}{\left| \mathcal{F}_i^+ \cup \mathcal{F}_j^- \right|}.$$

Adamic/Adar. This feature measures the Adaminc/Adar [1] score of two users, which sums up the reciprocal value of a logarithmic function with the number of each common neighbor's followees, and is defined as follows:

$$aa\_neighbor(u_i, u_j) = \sum_{c \in \mathcal{F}_i^+ \cap \mathcal{F}_j^-} \frac{1}{\log(|\mathcal{F}_c^+|)}.$$

Event-Based Features. Offline event participation is a unique characteristic of EBSNs, compared with conventional social networks. In this work, we define several event-based features for a user pair $\langle u_i, u_j \rangle$ as follows.

Number of common events. This feature captures the number of common events attended by two users and is formally defined as follows:

$$common\_event(u_i, u_j) = |E_i \cap E_j|.$$

Ratio of overlapped events. This feature measures the overlap ratio of two user's event sets using the Jaccard similarity, which is defined as follows:

$$overlap\_event(u_i, u_j) = \frac{|E_i \cap E_j|}{|E_i \cup E_j|}.$$

Adamic/Adar with event entropy. Distinct events may have different impacts on social link creation and, intuitively, events with a few attendees have a larger probability of creating new social links among the attendees than those with a large number of attendees, e.g., the attendees of a small home party are more likely to be good friends, while large events are usually public events such as concerts and exhibitions. To estimate the weights of the events, we introduce an entropy-based measure based on the information theory. Let N_{e_k} denote the number of attendees of event e_k and we use a discrete uniform distribution

$p_i = 1/N_{e_k}$ to describe the proportion of a certain user among the attendees. Let EN_{e_k} denote the entropy of event e_k, then we define EN_{e_k} as follows:

$$EN_{e_k} = -\sum_i p_i \log p_i = \log N_{e_k}.$$

As shown, the event entropy has a positive correlation with the number of attendees of an event. Inspired by the Adamic/Adar measure, we define a feature *ent_event* as the sum of the reciprocal value of each event entropy in common events between two users, i.e.,

$$en\_event(u_i, u_j) = \sum_{e_k \in E_i \cap E_j} \frac{1}{EN_{e_k}}.$$

Weighted common events. This feature takes the weights of events into consideration. In particular, we assign an event participation vector $\mu_i = (\mu_{i1}, \mu_{i2}, ..., \mu_{im})$ for each user $u_i \in U$, where if the user has attended event e_j then μ_{ij} is the reciprocal value of the entropy of event e_j, i.e., $\mu_{ij} = 1/EN_{e_j}$, otherwise, $\mu_{ij} = 0$. Then, we define a feature *w_common_event* as the inner product of two users' event participation vectors, i.e.,

$$w\_common\_event(u_i, u_j) = \mu_i \cdot \mu_j.$$

Weighted overlapped events. Similar to weighted common events, this feature also takes the weights of events into consideration but focuses on the overlapped events. Given two users, we define a feature *w_overlap_event* as the cosine similarity of their event participation vectors, i.e.,

$$w\_overlap\_event(u_i, u_j) = \frac{\mu_i \cdot \mu_j}{\|\mu_i\| \times \|\mu_j\|}.$$

Since each aforementioned explicit feature has a clear meaning and the feature space is small, to derive a user's preference towards another user based on the explicit features, we adopt the linear model to combine all the explicit features, which is simple but effective enough. Let z_{ij} denote the explicit feature vector of the user pair $\langle u_i, u_j \rangle$ and $r_f^u(u_i, u_j)$ denote the user u_i's preference towards user u_j based on the explicit features. Then, we define $r_f^u(u_i, u_j)$ as follows:

$$r_f^u(u_i, u_j) = w^T z_{ij}, \tag{1}$$

where w is the weight coefficient vector.

3.2   Latent Features

In this work, we employ matrix factorization model to capture the latent features of users and events, which is widely employed in recommendation system. The main idea of matrix factorization model is to seek a low-dimension latent feature vector to represent each entity and the rating scores can be approximated by a function of these low-dimension feature vectors. In our problem, there are two kinds of entities: users and events, correspondingly, we use latent feature vectors $x_i \in \mathbb{R}^{d \times 1}$ and $y_j \in \mathbb{R}^{d \times 1}$ with $d \ll |U| \wedge d \ll |E|$ to represent each user $u_i \in U$ and each event $e_j \in E$. Let $r_m^u(u_i, u_j)$ and $r_m^e(u_i, e_k)$ denote the user u_i's preference towards user u_j and event e_k based on the latent features, respectively. Then, we can define $r_m^u(u_i, u_j)$ and $r_m^e(u_i, e_k)$ as follows:

$$r_m^u(u_i, u_j) = bu_i + bu_j + x_i^T H x_j,$$
$$r_m^e(u_i, e_k) = bu_i + be_k + x_i^T y_k, \tag{2}$$

where the matrix $H \in \mathbb{R}^{d \times d}$ represents the correlations among users' latent features. Besides, bu_i, bu_j, and be_k are the bias items of user u_i, user u_j, and event e_k, respectively. Notice that, these two preferences share the same latent user feature for each user.

3.3   Recommendation Model

In this work, we combine the explicit and latent features to obtain a user's overall preference towards another user. Let $r^u(u_i, u_j)$ denote the user u_i's overall preference towards user u_j. Then, we define $r^u(u_i, u_j)$ as follows:

$$\begin{aligned} r^u(u_i, u_j) &= r_f^u(u_i, u_j) + r_m^u(u_i, u_j) \\ &= w^T z_{ij} + bu_i + bu_j + x_i^T H x_j. \end{aligned} \tag{3}$$

Moreover, we derive a user's overall preference towards an event only based on the latent features. Let $r^e(u_i, e_k)$ denote the user u_i's overall preference towards event e_j. Then, we define $r^e(u_i, e_k)$ as follows:

$$r^e(u_i, e_k) = r_m^e(u_i, e_k) = bu_i + be_k + x_i^T y_k. \tag{4}$$

Since our problem can be regarded as a ranking problem with implicit feedback, inspired by the BPR, we propose a Bayesian optimization framework which aims to optimize the AUC. In particular, for each user $u_i \in U$, it has two user sets: one is his/her followees (i.e., the positive user set), denoted as P_{u_i}, the other one is the remaining users that user u_i has not followed (i.e., the negative user set), denoted as N_{u_i}. Besides, it also has two event sets: one consists of the attended events (i.e., the positive event set), denoted as EP_{u_i}, the other one includes the remaining events (i.e., the negative event set), denoted as EN_{u_i}. Our model assumes that a user prefers the items in the positive set to these in the negative set, which is equal to the following formulas:

$$r^u(u_i, u_j) > r^u(u_i, u_k) \ \forall u_i \in U, u_j \in P_{u_i}, u_k \in N_{u_i},$$
$$r^e(u_i, e_j) > r^e(u_i, e_k) \ \forall u_i \in U, e_j \in EP_{u_i}, e_k \in EN_{u_i}.$$

Let $p(r^u(u_i, u_j) > r^u(u_i, u_k))$ and $p(r^e(u_i, e_j) > r^e(u_i, e_k))$ denote the probability that user u_i prefers user u_j to user u_k and the probability that user u_i

prefers event e_j to event u_k. Then, we define $p(r^u(u_i, u_j) > r^u(u_i, u_k))$ and $p(r^e(u_i, e_j) > r^e(u_i, e_k))$ as follows:

$$p(r^u(u_i, u_j) > r^u(u_i, u_k)) := \varepsilon(r^u_{ijk}),$$
$$p(r^e(u_i, e_j) > r^e(u_i, e_k)) := \varepsilon(r^e_{ijk}),$$

where

$$\varepsilon(x) := \frac{1}{1 + e^{-x}},$$
$$r^u_{ijk} := r^u(u_i, u_j) - r^u(u_i, u_k)$$
$$= w^T(z_{ij} - z_{ik}) + bu_j - bu_k + x_i^T H(x_j - x_k),$$
$$r^e_{ijk} := r^e(u_i, e_j) - r^e(u_i, e_k)$$
$$= be_j - be_k + x_i^T(y_j - y_k).$$

Let $\Theta = (X, H, Y, bu, be, w)$ denote the parameters of our model, where $X \in \mathbb{R}^{d \times n}$ and $Y \in \mathbb{R}^{d \times m}$ denote the latent user feature matrix and the latent event feature matrix, and $\Phi = (\sigma_x^2, \sigma_h^2, \sigma_w^2, \sigma_{bu}^2, \sigma_{be}^2, \sigma_y^2)$ denote the prior parameters of Θ. By assuming that users are independent to each other and each training sample is also independent, we aim to maximize the following posterior probability:

$$p(\Theta|R, V, \Phi) \propto$$
$$\prod_{u_i \in U, u_j \in P_{u_i}, u_k \in N_{u_i}} p(r^u(u_i, u_j) > r^u(u_i, u_k)|\Theta) \cdot$$
$$\prod_{u_i \in U, e_j \in EP_{u_i}, e_k \in EN_{u_i}} p(r^e(u_i, e_j) > r^e(u_i, e_k)|\Theta) \cdot p(\Theta|\Phi), \tag{5}$$

where R and V denote the online social network and the offline event participation network, respectively. To avoid over-fitting, we introduce Gaussian priors with zero-mean on the parameters Θ. After applying logarithmic function on Eq. (5), maximizing the posterior probability can be equivalent to minimizing the following objective function:

$$E = - \sum_{u_i \in U} \sum_{u_j \in P_{u_i}} \sum_{u_k \in N_{u_i}} \ln\varepsilon(r^u_{ijk})$$
$$- \sum_{u_i \in U} \sum_{e_j \in EP_{u_i}} \sum_{e_k \in EN_{u_i}} \ln\varepsilon(r^e_{ijk}) + \frac{\lambda_x}{2} \|X\|_F^2 + \frac{\lambda_y}{2} \|Y\|_F^2$$
$$+ \frac{\lambda_h}{2} \|H\|_F^2 + \frac{\lambda_{bu}}{2} \|bu\|_F^2 + \frac{\lambda_{be}}{2} \|be\|_F^2 + \frac{\lambda_w}{2} \|w\|_F^2, \tag{6}$$

where $\lambda_x = 1/\sigma_x^2$, $\lambda_y = 1/\sigma_y^2$, $\lambda_h = 1/\sigma_h^2$, $\lambda_{bu} = 1/\sigma_{bu}^2$, $\lambda_{be} = 1/\sigma_{be}^2$, $\lambda_w = 1/\sigma_w^2$, and $\|\cdot\|_F$ is the Frobenius norm.

3.4 Parameter Learning

To learn the parameters of our model, we use the stochastic gradient descent (SGD) algorithm to minimize the Eq. (6), since it provides fast convergence to the local optimums and has good expandability. In employing SGD, we randomly choose an instance from the training samples. For each selected training instance, we calculate its partial derivative and update the parameters Θ as follows:

$$\Theta \leftarrow \Theta - \alpha * \frac{\partial E}{\partial \Theta},$$

where α is the learning rate. In updating the parameters iteratively, each parameter moves along the descending gradient direction until it converges or the maximum number of iterations is reached.

4   Experiments

4.1   Experimental Setup

Data Allocation. We use the same datasets used in our data analysis for performance evaluation. The details of these two datasets have been shown in Sect. 2.3. For both the datasets, we use s-fold cross validation to evaluate the performance. In detail, we randomly divide each user u_i's positive user set P_{u_i} into s subsets and repeat the holdout method s times. Each time, one of the s subsets is used as the testing set, denoted as $S_{u_i}^{test}$, and the other s-1 subsets form a positive training user set, denoted as $S_{P_{u_i}}^{train}$. We combine $S_{u_i}^{test}$ and its original negative user set N_{u_i} into a new negative training user set, denoted as $S_{N_{u_i}}^{train}$. Our model is trained based on each user's training sets $S_{P_{u_i}}^{train}$ and $S_{N_{u_i}}^{train}$, and the recommendation performance is evaluated on each user's testing set $S_{u_i}^{test}$. Finally, we take the average performance results of all the trials. In this work, we use 5-fold cross validation.

Evaluation Metrics. To evaluate the recommendation performance, we adopt three widely used evaluation metrics: AUC, Precision@k, and MAP@n.

AUC is especially suited for measuring the overall performance in highly imbalanced dataset, as in our case where the number of a user's followees is usually very small with regard to the total number of the users. In our problem, the AUC is defined as follows:

$$AUC = \frac{\sum\limits_{u_i \in U} \sum\limits_{u_j \in S_{u_i}^{test}} \sum\limits_{u_k \in N_{u_i}} \delta(r^u(u_i, u_j) > r^u(u_i, u_k))}{\sum\limits_{u_i \in U} |S_{u_i}^{test}| \cdot |N_{u_i}|},$$

where $\delta(x)$ is an indicator function which is equal to 1 if x is true and equal to 0 otherwise.

Precision@k and MAP@n are mainly used to evaluate the performance of top-k recommendation. In our problem, the Precision@k measures the ratio of users in the top-k recommendation list that are corresponding to the followees in users' testing sets. MAP@n is the mean of all the users' Average Precision (AP) scores at position n. The AP@n score of each user is defined as follows:

$$AP_{u_i}@n = \frac{\sum\limits_{k=1}^{n} \text{Precision}@k \times I(k)}{|S_{u_i}^{test}|},$$

where $I(k)$ is an indicator function which is equal to 1 if the user at rank k is in user u_i's testing set $S_{u_i}^{test}$ and equal to 0 otherwise. To measure the overall results of recommendation, we set n to the maximum value for MAP@n. For convenience, in the following, we omit the argument n in MAP@n and replace Precision@k with P@k.

Comparison Methods. Since our proposed method, denoted as HNFR, combines all the explicit features (social features and event-based features) and latent features, we devise methods that incorporate these factors individually. Beside, we compare our method with some existing works which are designed for followee recommendation in traditional social networks. In summary, we compare HNFR with the following methods:

- **FoF** [5]: For each user u, this method ranks the candidates according to the number of user u's followees who have followed the candidate.
- **CB-MF** [21]: This method first employs a LDA-based method on the social relations to discover communities and then applies WRMF [9] on each discovered community.
- **BPR-SF:** This method only uses the social features described in Sect. 3.1 and employs BPR for optimization.
- **BPR-EF:** Similar to BPR-SF, this method only considers the event-based features introduced in Sect. 3.1.
- **BPR-AF:** This method combines all the social features and event-based features, which can be used to verify the effectiveness of integrating all the explicit features.
- **BPR-MF:** This is the basic matrix factorization model for followee recommendation discussed in Sect. 3.2, which only considers the latent features in the online social network and applies BPR for optimization.
- **BPR-MAF:** This method uses the explicit and latent features in the online social network, however, it does not consider the latent features in the offline event participation network.

In our experiment, all the methods are implemented with the LibRec [6] library using JAVA.

Parameter Settings. Empirically, for the regularization parameters λ_x, λ_y, λ_h, λ_{bu}, and λ_{be}, we set all of them to 10^{-2}. Specially, we set the regularization parameter λ_w to 10^{-3}. Besides, we empirically set the dimension of latent features to 20. In both the datasets, we set the initial learning rate α to 10^{-2} for the methods containing the latent features and 10^{-4} for the methods that only use the explicit features. Note that, the LibRec library will adjust the learning rate automatically during the training process.

4.2   Experimental Results

Performance Comparison Under the AUC Metric. In this section, we evaluate the performance of all the methods under the AUC metric. The AUC metric can reflect the overall performance under pairwise ranking.

The AUC results of all the methods in both datasets are shown in Table 2. We can clearly observe that our proposed model, HNFR, always outperforms all the baseline methods in both the datasets significantly. Moreover, the performance of HNFR is better than BPR-MAF, which demonstrates the effectiveness of considering the latent features in the offline event participation network. In both the datasets, BPR-AF achieves better performance than BPR-SF and BPR-EF, which indicates the strength of combining social features and event-based features. We also find that matrix factorization based methods such as BPR-MF and BPR-MAF perform much better than the methods which only consider explicit features like BPR-SF, BPR-EF, and BPR-AF. Besides, we observe that the performance of most methods in Plancast dataset is much better than those in DoubanEvent dataset, which might be due to that the density of the online social network in Plancast dataset is denser than that in DoubanEvent dataset.

Performance Comparison in Top-k Recommendation. In this section, we evaluate the overall performance of all the methods in top-k recommendation.

Table 2. The AUC results

	FoF	CB-MF	BPR-SF	BPR-EF	BPR-AF	BPR-MF	BPR-MAF	HNFR
Plancast	0.84236	0.89496	0.84455	0.60117	0.86309	0.92363	0.92959	**0.94738**
DoubanEvent	0.75785	0.78461	0.75888	0.72511	0.80788	0.88330	0.88560	**0.91735**

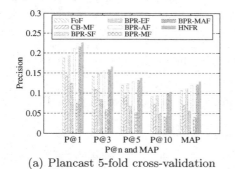
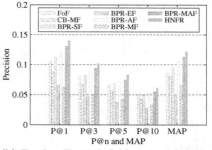

(a) Plancast 5-fold cross-validation   (b) DoubanEvent 5-fold cross-validation

Fig. 3. The P@k and MAP performance

As users mainly focus on the top of the recommendation results, we evaluate the P@k performance under $k = 1, 3, 5, 10$. Meanwhile, we use MAP to measure the overall recommendation ranking list.

The P@k and MAP performance of all the methods in both datasets are shown in Fig. 3. As shown, our method HNFR also achieves the best performance in all the tests, which verifies the effectiveness of our proposed method in followee recommendation in EBSNs. Moreover, we observe that BPR-MF performs much worse than the methods that only consider explicit features, which is contrary to the performance under AUC metric. This may be caused by the following reasons: (1) explicit features play an important role in top-k recommendation; (2) pairwise ranking learning strategy does not focus on the top-k recommendation results. Analogously, we also observe that most methods in Plancast dataset perform better than the corresponding methods in DoubanEvent dataset.

5   Related Work

In this section, we briefly review the related works, including followee/friend recommendation and collaborative filtering.

Followee/Friend Recommendation. The problem of followee/friend recommendation in social networks, which can also be regarded as a special kind of link prediction problem that foucuses on predicting the social links among users, has been widely studies for many years [2,5,7,8,12,14,18–21]. Liben-Nowell et al. [12] first study the link prediction problem in social networks. They develop and compare several unsupervised methods based node proximity in a social network, such as methods based on node neighborhoods and methods based on the ensemble of all paths. Zhao et al. [21] propose a community-based followee recommendation in Twitter-style social networks. They first employ an LDA-based community discovery method based on the followers/folloewee relations and then apply matrix factorization method on the discovered communities to recommend followees.

In addition to using the network structure information, there are many works exploiting other types of information such as user-generated content, user profile, and locations. For example, Chen et al. [5] study the people recommendation problem in an enterprise social networking and propose four recommendation methods based on social network structure or user-generated contents. They find that methods based on social network structure are good at recommending known contacts for users while methods based on user-generated content are stronger in discovering new friends. Moreover, Hannon et al. [8] propose to recommend followees in Twitter based on user profiling. They evaluate different profiling strategies based on the tweets or relations of users' Twitter social graphs and find that the collaborative strategies perform better than the content strategies. In addition, Yuan et al. [20] study how to exploit sentiment homophily for link prediction. They propose a topic-sentiment affiliation based graphical model which incorporates the sentiment features extracted from tweets, structural features based on social graph, and topical features based on the topical affiliation of two users.

Recently, there exist some studies focusing on different friend recommendation problem to satisfy different needs. For instance, Barbieri et al. [2] study the problem of link prediction with explanations for user recommendation. They propose a stochastic topic model over directed and nodes-attributed graphs which can produce different types of explanation for different kinds of links (a topical link or a social link). In addition, Wan et al. [19] study the problem of informational friend recommendation which aims to recommend friends according to users' informational needs. They first employ collaborative filtering method to predict a user's rating for each post and then rank the candidate users based on their informational utilities.

In this work, we propose to exploit the offline event participation information for followee recommendation, which is a unique and important characteristic of EBSNs, and no previous works in followee recommendation have considered such informantion. Moreover, we design a novel recommendation model utilizing all the latent and explicit features of both the social relations and the offline event participation records, which is also different from previous works.

Collaborative Filtering. Collaborative filtering is a main recommendation method, which has been widely employed in recommendation systems. [3,4,9–11,15–17]. Matrix factorization plays an important role in collaborative filtering techniques. The basic concept of matrix factorization is to seek latent representations for both items and users, which are usually low-dimensional vectors of factors in the latent space. In these works, Koren et al. [10] propose to incorporate bias factors, temporal dynamics, or confidence levels into matrix factorization. Salakhutdinov et al. [17] propose probabilistic algorithms for matrix factorization which scale linearly with the number of observations. Lee et al. [11] propose a matrix factorization algorithm with non-negativity constraints which produces a parts-based representation of the original matrix.

There are some works focusing on the matrix factorization based recommendation with implicit feedbacks [9,15,16]. For example, Hu et al. [9] proposed a factor model which treats the implicit feedbacks as indication of positive and negative preference associated with vastly varying confidence levels. Rendle et al. [16] propose a Bayesian optimization criterion named BPR for personal ranking from implicit feedbacks and apply it to matrix factorization. Unlike previous works, they adopted pairwise user preference towards items, which have the underlying assumption that a user prefers item viewed by the user to all other non-observed items.

In this work, inspired by BPR, we propose a new Bayesian optimization framework which adopts pairwise user preference on both the online social network and the offline event participation network.

6   Conclusion

In this paper, we study the problem of followee recommendation in EBSNs. We propose a new followee recommendation method called HNFR, which exploits the heterogeneous nature of EBSNs. In our method, we combine all the explicit and latent features which are captured from the online social network and the offline event participation network. Moreover, we propose a Bayesian optimization framework which adopts pairwise user preference on both the social relations and the events. The experimental results on real-world data demonstrate the effectiveness of our method.

Acknowledgements. The work was supported by National Natural Science Foundation of China under Grant 61502047.

References

1. Adamic, L.A., Adar, E.: Friends and neighbors on the web. Soc. Netw. **25**(3), 211–230 (2003)
2. Barbieri, N., Bonchi, F., Manco, G.: Who to follow and why: link prediction with explanations. In: SIGKDD, pp. 1266–1275 (2014)
3. Cai, Y., Lau, R.Y.K., Liao, S.S.Y., Li, C., Leung, H., Ma, L.C.K.: Object typicality for effective web of things recommendations. Decis. Support Syst. **63**, 52–63 (2014)
4. Cai, Y., Leung, H., Li, Q., Min, H., Tang, J., Li, J.: Typicality-based collaborative filtering recommendation. IEEE Trans. Knowl. Data Eng. **26**(3), 766–779 (2014)
5. Chen, J., Geyer, W., Dugan, C., Muller, M.J., Guy, I.: Make new friends, but keep the old: recommending people on social networking sites. In: CHI, pp. 201–210 (2009)
6. Guo, G., Zhang, J., Sun, Z., Yorke-Smith, N.: Librec: a java library for recommender systems. In: Posters, Demos, Late-breaking Results and Workshop Proceedings of User Modeling, Adaptation, and Personalization (UMAP 2015) (2015)
7. Guy, I., Ronen, I., Wilcox, E.: Do you know?: recommending people to invite into your social network. In: IUI, pp. 77–86 (2009)
8. Hannon, J., Bennett, M., Smyth, B.: Recommending twitter users to follow using content and collaborative filtering approaches. In: RecSys, pp. 199–206 (2010)
9. Hu, Y., Koren, Y., Volinsky, C.: Collaborative filtering for implicit feedback datasets. In: ICDM, pp. 263–272 (2008)
10. Koren, Y., Bell, R.M., Volinsky, C.: Matrix factorization techniques for recommender systems. IEEE Comput. **42**(8), 30–37 (2009)
11. Lee, D.D., Seung, H.S.: Algorithms for non-negative matrix factorization. In: NIPS, pp. 556–562 (2000)
12. Liben-Nowell, D., Kleinberg, J.M.: The link prediction problem for social networks. In: CIKM, pp. 556–559 (2003)
13. Liu, X., He, Q., Tian, Y., Lee, W., McPherson, J., Han, J.: Event-based social networks: linking the online and offline socialworlds. In: SIGKDD, pp. 1032–1040 (2012)
14. Menon, A.K., Elkan, C.: Link prediction via matrix factorization. In: Gunopulos, D., Hofmann, T., Malerba, D., Vazirgiannis, M. (eds.) ECML PKDD 2011, Part II. LNCS, vol. 6912, pp. 437–452. Springer, Heidelberg (2011)

15. Pan, R., Zhou, Y., Cao, B., Liu, N.N., Lukose, R.M., Scholz, M., Yang, Q.: One-class collaborative filtering. In: ICDM, pp. 502–511 (2008)
16. Rendle, S., Freudenthaler, C., Gantner, Z., Schmidt-Thieme, L.: BPR: bayesian personalized ranking from implicit feedback. In: UAI, pp. 452–461 (2009)
17. Salakhutdinov, R., Mnih, A.: Probabilistic matrix factorization. In: NIPS, pp. 1257–1264 (2007)
18. Scellato, S., Noulas, A., Mascolo, C.: Exploiting place features in link prediction on location-based social networks. In: SIGKDD, pp. 1046–1054 (2011)
19. Wan, S., Lan, Y., Guo, J., Fan, C., Cheng, X.: Informational friend recommendation in social media. In: SIGIR, pp. 1045–1048 (2013)
20. Yuan, G., Murukannaiah, P.K., Zhang, Z., Singh, M.P.: Exploiting sentiment homophily for link prediction. In: RecSys, pp. 17–24 (2014)
21. Zhao, G., Lee, M., Hsu, W., Chen, W., Hu, H.: Community-based user recommendation in uni-directional social networks. In: CIKM, pp. 189–198 (2013)

SBTM: Topic Modeling over Short Texts

Jianhui Pang[1], Xiangsheng Li[1], Haoran Xie[2], and Yanghui Rao[1(✉)]

[1] Sun Yat-sen University, Guangzhou, China
{pangjh3,lixsh6}@mail2.sysu.edu.cn, raoyangh@mail.sysu.edu.cn
[2] Caritas Institute of Higher Education, New Territories, Hong Kong
hrxie2@gmail.com

Abstract. With the rapid development of social media services such as Twitter, Sina Weibo and so forth, short texts are becoming more and more prevalent. However, inferring topics from short texts is always full of challenges for many content analysis tasks because of the sparsity of word co-occurrence patterns in short texts. In this paper, we propose a classification model named sentimental biterm topic model (SBTM), which is applied to sentiment classification over short texts. To alleviate the problem of sparsity in short texts, the similarity between words and documents are firstly estimated by singular value decomposition. Then, the most similar words are added to each short document in the corpus. Extensive evaluations on sentiment detection of short text validate the effectiveness of the proposed method.

Keywords: Short text classification · Sentiment detection · Topic-based similarity · Biterm topic model

1 Introduction

With the broad availability of portable devices such as tablets, online users can now conveniently express their opinions through various channels, which generates large-scale sentimental short texts and accelerates the sentiment detection research. Many works on sentiment detection have focused on classifying the sentiments via latent topics [2,14], since a topic represents the real-world event that indicates the subject or context of the sentiment [19]. However, short texts, as indicated by the name, typically only include a few words, causing the sparsity of word co-occurrence patterns. Thus, traditional topic models such as probabilistic latent semantic analysis [8] and latent Dirichlet allocation [3], will suffer from this severe data sparsity problem when inferring the latent topics. Based on the idea that two frequently co-occurred words are more likely to belong to a same topic, Cheng et al. [4] proposed a biterm topic model (BTM), which extracts biterms for each document and alleviates the problem of data sparsity in generating topics.

In light of these considerations, we propose a method named sentimental biterm topic model (SBTM) for sentiment classification over short texts. To tackle the issue of sparse content, we first extract the most similar words for each

© Springer International Publishing Switzerland 2016
H. Gao et al. (Eds.): DASFAA 2016 Workshops, LNCS 9645, pp. 43–56, 2016.
DOI: 10.1007/978-3-319-32055-7_4

short document by singular value decomposition and incorporate these words into the related texts. Then, we extract biterms from each text and employ BTM to infer the topic probability distribution. Finally, the similarity of short texts is estimated by the topic relevancy rather than common words. Thus, our method reduces the negative influence of short texts, in which, the topical association of different words improves the performance of short-text sentiment classification. Experiments using a sensibly small and unbalanced sentimental dataset validate the effectiveness of the proposed SBTM.

The remainder of this paper is organized as follows. In Sect. 2, we introduce the related studies in sentiment detection, short text classification, latent semantic analysis and topic modeling over short texts. In Sect. 3, we propose a novel sentimental BTM method and our classification algorithm. The experimental evaluation is described in Sect. 4. In Sect. 5, we draw our conclusions.

2   Related Work

2.1   Sentiment Detection

Sentiment detection aims to identify and extract the attitudes of a subject (i.e., an opinion holder, a commentator and so forth), towards either a topic or the overall tore of a document [6]. Traditional methods of sentiment detection were mainly based on lexicons, supervised and unsupervised learning algorithms. Lexicon-based methods [14–16] constructed word-level or topic-level emotional dictionaries to tag sentiments. Models based on supervised learning used traditional categorization algorithms, which include naïve Bayes [11], maximum entropy and support vector machines [12]. Unsupervised learning methods detected the corresponding emotional orientation by counting the occurrence of positive and negative terms and estimating the overall polarity of documents [21]. However, the methods aforementioned mainly target on the long text, and perform not well on short texts.

2.2   Short Text Classification

Short text brings new challenges to the classification task. Although short texts may not share any common words, they could relate to each other semantically. However, most traditional classifiers were based on term frequencies and common words shared between paired texts, such as naïve Bayes [11] or support vector machine [22]. Therefore, these algorithms suffer the problem of feature sparsity in short texts. To overcome the above limitation, Sahami and Heilman exploited the external texts from the web to enrich the short text [17]. The key idea of enriching the texts is valuable, but the enriched texts may not always be connected to the original text. Similarly, Banerjee et al. [1] applied the existing knowledge bases such as WordNet or Wikipedia to identify the semantic association between words. However, if a short document contains words which do not exist in the knowledge base, the performance was poor.

2.3   Latent Semantic Analysis

To estimate the similarity between two short texts, we should not only focus on the common words they both have, but also consider the latent semantic of words. On early stage of research, the synonymous words were extracted by singular value decomposition (SVD), which projects documents into a lower dimensional space called latent semantic space. The similar model is named latent semantic analysis (LSA) [5], which enables us to calculate the similarity between two texts even if they have not any common words. Probabilistic latent semantic analysis (PLSA) and latent Dirichlet allocation (LDA) are two other popular models. PLSA [8] projects the word-by-document matrix to a lower rank space, i.e., the document-topic distribution θ and each topic can be represented by the topic-word distribution ϕ. Extended from PLSA, LDA [3] further adds Dirichlet priors for the document-specific topic mixtures, making it possible to generate unseen documents. Both PLSA and LDA are proposed to capture the latent semantic of documents, and can be used to mine the latent user community [25], community-aware resource profiling [23] and user profiling [24].

2.4   Topic Modeling over Short Texts

With the prevalent of tweets, questions, instant-messages and news headlines, topic models have been extensively studied to overcome the limitations of short texts. Early studies mainly focused on exploiting the external knowledge to enrich the representation of short texts [9,13]. However, these methods only achieve a good performance when the auxiliary data are closely related to the original data. More importantly, finding such auxiliary data may be expensive or even impossible. To overcome these shortages, Cheng et al. [4] proposed a biterm topic model (BTM) based on the key idea that if two words co-occurred more frequently, they are more likely to belong to a same topic. For each short text, BTM extracts biterms by exploiting its contextual information. For example, a document with three distinct words will generate three biterms:

$$(\omega_1, \omega_2, \omega_3) \Rightarrow \{(\omega_1, \omega_2), (\omega_1, \omega_3), (\omega_2, \omega_3)\}$$

where $(.,.)$ is unordered. After the pre-processing, each short text is represented as a biterm set. Each two words in a biterm are drawn independently from a topic. Unlike most topic models, BTM learns the latent topic components in a corpus by modeling the generation of biterms. However, if there are less than 3 words in a document, BTM can not expand it by extracting biterms. Therefore, we aim to overcome this shortage in our research.

3   Sentimental Biterm Topic Model

In this section, we detail the proposed sentimental biterm topic model (SBTM). Firstly, we give the process of our method. Secondly, we introduce the algorithm to calculate the similarity between a word and a short document. Thirdly, we give the detail of BTM that is employed. Finally, we describe the method of sentiment detection by the topic-based similarity.

3.1   Process of SBTM

Given a corpus with N_D short documents, the process of our sentiment detection model is listed as follows:

1. Generate a term-document matrix M using TF-IDF values (or the one-hot representation), where $Weight_{\omega,d}$ represents the weight of the ω-th word in the d-th document.
2. Employ SVD to process matrix M, from which we get matrix U (left-singular vectors of M), matrix V (right-singular vectors of M), and matrix Σ (singular values matrix of M). Since each row u_ω of matrix U represents the feature of each word ω in the corpus and each row v_d of matrix V represent the feature of each document in the corpus, we can calculate the cosine similarity c between each u_ω and v_d. The dimension of u_ω and v_d could be chosen according the matrix Σ. After calculation, we get a matrix S, which consists of W rows and N_D column. Each entry $S_{\omega,d}$ represents the similarity of the ω-th word and the d-th document.
3. After each row of matrix S being sorted, for each document, we can get n most similar words. Then we add these n words into the related documents to expand the short text. In our work, we add 3 words to each document. Finally, a new expanding corpus C_{new} is generated.
4. We extract biterms from C_{new}. Since there are more than 3 words in each document, we can always extract more biterms than the original document. Then, we employ BTM to detect the topic probability distribution $P(z|d)$ for each document.
5. We extract the similarity between two documents by calculating the correlation coefficient $Corr$ of their topic probability distribution. Then, we set a threshold λ for sentiment classification. If $Corr$ between text A and text B is larger than λ, we consider that their sentiments are similar. Finally, we predict the sentiment of unlabeled text d based on a training set with sentimental labels.

3.2   Detect Similarities Between Words and Texts

Given a term-document matrix, we can get a term-feature matrix and a document-feature matrix after applying SVD. Early research made use of the document-feature matrix to calculate the similarity between two different documents to extract the latent semantic [5].

In our model, we use both term-feature and document-feature matrices to compute the similarity between words and documents. The purpose is to find the most related words from other documents in the corpus, which are added to enrich the contextual information of each short document. To estimate the similarities between words and documents, we calculate the cosine similarity between word-vector and document-vector. Besides, the dimension of the vector can be reduced according to matrix Σ.

The process of expanding each short document is described in Algorithm 1, from which a new expanding corpus C_{new} is generated.

Algorithm 1. Expand short documents

Input:
  1: M: term-document matrix;
  2: C: a collection of short documents;
  3: n_{new_word}: the number of most similar words added to related documents;
  4: U: term-feature matrix of M;
  5: V^T: transpose of document-feature matrix;
  6: Σ: singular values matrix of M;
Output:
  7: C_{new}: a new corpus after expanding documents;
  8: **procedure** EXPAND SHORT DOCUMENTS
  9:     Apply SVD to the term-document matrix: $U\Sigma V^T = \text{SVD } (M)$;
 10:     Compute the cosine similarity between each word and each document;
 11:     **for all** $V_d \in V$ **do**
 12:         **for all** $U_\omega \in U$ **do**

$$Similarity_{d,\omega} = \cos\left(\frac{V_d * U_\omega}{|V_d| * |U_\omega|}\right)$$

 13:         **end for**
 14:     Sort($Similarity_d$);
 15:     Select n_{new_word} most related words and add them to document d;
 16:     Append new document d into new corpus C_{new};
 17:     **end for**
 18: **end procedure**

3.3 Topic Extraction via BTM

After generating the new corpus mentioned earlier, we employ BTM to extract the latent topics for each document.

According to the statement of BTM [4], we assume that the new corpus contains N_B biterms $\mathbf{B} = \{b_i\}_{i=1}^{N_B}$ and K topics express over W unique words in the vocabulary. Let $z \in [1, K]$ be a topic indicator variable, the prevalence of topics in the corpus, i.e., $P(z)$ can be represented by a K-dimensional multinomial distribution $\theta = \{\theta_k\}_{k=1}^K$ with $\theta_k = P(z = k)$ and $\sum_{k=1}^K \theta_k = 1$. A $K \times W$ matrix Φ represents the word distribution of topics, i.e., $P(\omega|z)$, where the k-th row ϕ_k is a W-dimensional multinomial distribution with entry $\phi_{k,\omega} = P(\omega|z = k)$ and $\sum_{\omega=1}^W \phi_{k,\omega} = 1$. Then, BTM uses symmetric Dirichlet priors for θ and ϕ_k with single-valued hyper-parameters α and β. Formally, the generative process of BTM is described as follows:

1. Draw $\theta \sim$ Dirichlet (α);
2. For each topic $k \in [1, K]$, draw $\phi_k \sim$ Dirichlet (β);
3. For each biterm $b_i \in \mathbf{B}$, draw $z_i \sim$ Multinomial(θ),
    and draw $\omega_{i,1}, \omega_{i,2} \sim$ Multinomial(ϕ_{z_i}).

The graphical representation of BTM is shown in Fig. 1 [4]. Each node in the graph denotes a random variable, where shading represents an observed variable. A plate denotes the replication of the model within it, and the number of replicates is given at the bottom right corner of the plate.

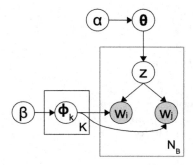

Fig. 1. The graphic representation of BTM

Given the parameters θ and Φ, the conditional probability of biterm b_i is estimated as follows:

$$
\begin{aligned}
P(b_i|\theta, \Phi) &= \sum_{k=1}^{K} P(\omega_{i,1}, \omega_{i,2}, z_i = k|\theta, \Phi) \\
&= \sum_{k=1}^{K} P(z_i = k|\theta_k) P(\omega_{i,1}|z_i = k, \phi_{k,\omega_{i,1}}) P(\omega_{i,2}|z_i = k, \phi_{k,\omega_{i,2}}) \\
&= \sum_{k=1}^{K} \theta_k \phi_{k,\omega_{i,1}} \phi_{k,\omega_{i,2}}.
\end{aligned}
\tag{1}
$$

For all biterms, BTM maximizes the following likelihood function for the whole corpus:

$$
P(\mathbf{B}|\theta, \Phi) = \prod_{i=1}^{N_B} \sum_{k=1}^{K} \theta_k \phi_{k,\omega_{i,1}} \phi_{k,\omega_{i,2}}.
\tag{2}
$$

To conduct approximate inference for θ and Φ, BTM estimates the parameters using Gibbs sampling [7], which uses samples drawn from the posterior distributions of latent variables sequentially conditioned to the current values of all other variables. For each biterm b_i, BTM samples topic z_i according to the following conditional distribution:

$$
P(z_i = k|z_{-i}, \mathbf{B}) \propto (n_{-i,k} + \alpha) \frac{(n_{-i,\omega_{i,1}|k} + \beta)(n_{-i,\omega_{i,2}|k} + \beta)}{(n_{-i,\cdot|k} + W\beta + 1)(n_{-i,\cdot|k} + W\beta)},
\tag{3}
$$

where z_{-i} is the topic assignments for all biterms excluding b_i, $n_{-i,k}$ is the number of biterms assigned to topic k excluding b_i, $n_{-i,\omega|k}$ is the number of times that word ω assigned to topic k excluding b_i, and $n_{-i,\cdot|k} = \sum_{\omega=1}^{W} n_{-i,\omega|k}$.

After a sufficient number of iterations, BTM counts the number n_k in each topic k, and the number of times $n_{\omega|k}$ that each word ω assigned to topic k. Finally, Φ and θ are estimated by

$$\phi_{k,\omega} = \frac{n_{\omega|k} + \beta}{n_{\cdot|k} + W\beta}, \tag{4}$$

$$\theta_k = \frac{n_k + \alpha}{N_B + K\alpha}. \tag{5}$$

Based on the latent topic of each biterm, BTM can extract the topic proportion for each document by assuming that each biterm's topic z is conditionally independent of document d, as follows:

$$P(z|d) = \sum_{i=1}^{N_d} P(z|b_i^{(d)})P(b_i^{(d)}), \tag{6}$$

where $P(z = k|b_i^{(d)})$ can be calculated via Bayes' formula based on the parameters learned in BTM, and $P(b_i^{(d)})$ can be estimated by the frequency of each biterm in documents, as follows:

$$P(z = k|b_i^{(d)}) = \frac{\theta_k \phi_{k,\omega_{i,1}^{(d)}} \phi_{k,\omega_{i,2}^{(d)}}}{\sum_{k'} \theta_{k'} \phi_{k',\omega_{i,1}^{(d)}} \phi_{k',\omega_{i,2}^{(d)}}}, \tag{7}$$

$$P(b_i^{(d)}) = \frac{n(b_i^{(d)})}{\sum_{i=1}^{N_d} n(b_i^{(d)})}. \tag{8}$$

According to Eq. 3, the topic assignment for biterms generated from our new corpus is shown in Algorithm 2. Then, we estimate the topic probability distribution $P(z|d)$ for each document by Eq. 6.

3.4   Sentiment Detection by Topic-Based Similarity

To detect the sentiment of unlabeled documents, we first estimate the similarity between labeled document d_1 and unlabeled document d_2 based on the topic probability distribution as follows:

$$Similarity(d_1, d_2) = \frac{cov(P(z|d_1), P(z|d_2))}{\sqrt{Var(P(z|d_1))}\sqrt{Var(P(z|d_2))}}, \tag{9}$$

where $cov(P(z|d_1), P(z|d_2))$ is the covariance between $P(z|d_1)$ and $P(z|d_2)$, $Var(P(z|d))$ is the variance of $P(z|d)$.

Then, a threshold λ is used to determine whether the labeled documents' sentiments are related to the sentiment of each unlabeled document. Finally, we estimate the sentimental score of the unlabeled document by calculating the mean value of all sentimental scores of related labeled documents, and assign the unlabeled document to the sentiment with the largest value. The process of short text sentiment detection is shown in Algorithm 3.

Algorithm 2. Topic assignment for biterms

Input:
  1: K: the number of topic;
  2: α: single-valued hyper-parameter for θ;
  3: β: single-valued hyper-parameter for ϕ_k;
  4: $\mathbf{B}$: the biterm set;
Output:
  5: Φ: the topic proportion of a document;
  6: **procedure** TOPIC ASSIGNMENT FOR BITERMS
  7:      Randomly initialize the topic assignments for all biterms;
  8:      **repeat**
  9:          **for all** $b_i = (\omega_{i,1}, \omega_{i,2}) \in \mathbf{B}$ **do**
  10:             Draw topic k from $P(z_i | z_{-i}, \mathbf{B})$ by Eq. 3.
  11:             Update n_k, $n_{\omega_{i,1}|k}$ and $n_{\omega_{i,2}|k}$
  12:         **end for**
  13:     **until** N_{iter} times
  14:     Compute Φ by Eq. 4 and θ by Eq. 5.
  15: **end procedure**

Algorithm 3. Short text sentiment detection

Input:
  1: *labeled_d*: the collection of labeled documents
  2: *unlabeled_d*: the new unlabeled documents
  3: $P(z|d)$: the topic probability distribution for each document
  4: λ: the threshold value
Output:
  5: *result*: Sentimental category of *unlabeled_d*;
  6: **procedure** SHORT-TEXT-SENTIMENT-DETECTION
  7:      **for all** $d_i \in unlabeled\_d$ **do**
  8:          **for all** $d_j \in labeled\_d$ **do**
  9:              $similarityList[i][j] = \text{Similarity}(d_i, d_j)$ (refer to Eq. 9)
  10:             **if** $similarityList[i][j] > \lambda$ **then**
  11:                 $Sentiment_{d_i} + = Sentiment_{d_j}$
  12:             **end if**
  13:         **end for**
  14:         $Sentiment_{d_i} = avg(Sentiment_{d_i})$
  15:     **end for**
  16:     $result = $ the category with the largest averaged sentimental value;
  17: **end procedure**

---

4   Experiments

In this section, we evaluate our proposed method by designing experiments to observe the influence of parameters λ and K on the performance of short text sentiment classification.

4.1 Dataset

SemEval is an English dataset used in the 14th task of the 4th International Workshop on Semantic Evaluations (SemEval-2007). This dataset includes news headlines and user scores over emotions of anger, disgust, fear, joy, sad and surprise. There are 1,246 valid news headlines with the total score of the 6 emotions larger than 0.

Table 1. Statistics of *Semeval*

Dataset	Emotion label	# of articles	# of ratings
SemEval	anger	87	12,042
	disgust	42	7,634
	fear	194	20,306
	joy	441	23,613
	sad	265	24,039
	surprise	217	21,495

The statistics of *SemEval* is summarized in Table 1. The number of articles of each emotional label represents the sum of documents having the most ratings over that emotion. For instance, there are 441 news headlines which have the most user ratings over the largest emotional category of "joy". The smallest emotional category is "disgust", which only has 42 documents. Among the 1,246 pieces of short text, the first 246 and the remaining 1,000 of them are used for training and testing, respectively.

4.2 Experimental Setup

To make an appropriate comparison with other baseline models, the accuracy at top 1 is employed as the indicator of performance [2]. According to the evaluation metric, a predicted ranked list of emotion labels is correct if the list's first item is identical to the actual ranked list's first item. If two emotion labels in the actual ranked list have the same number of votes, then their positions are interchangeable.

The parameters of SBTM are shown in Table 2. To find the best values of parameters α, β and n_{new_word}, we tuned them using 5-fold cross-validation.

4.3 Comparison with BTM

In this subsection, we evaluate the influence of parameters K and λ on the performance of SBTM and BTM. The accuracy with different parameters is shown in Table 3.

Table 3(a) presents the accuracies of SBTM and BTM with different values of K given $\lambda = 0.001$. On one hand, the accuracy of SBTM_K has a maximal

Table 2. Parameters of SBTM

Parameter	Value
n_{new_word}	3
N_{iter}	500
α	$50/K$
β	0.01

Table 3. Accuracies with different parameter values

(a) Influence of K ($\lambda = 0.001$)

K	30	40	50	60	70	80	90	100	110	Averaged value
BTM_K	36.3	35.0	35.4	36.3	35.8	37.8	36.1	37.3	37.6	36.4
SBTM_K	36.1	36.8	36.1	36.6	38.4	35.8	34.4	36.3	36.7	36.6

(b) Influence of λ ($K = 70$)

λ	0.0001	0.0004	0.0007	0.001	0.004	0.007	0.01	0.04	0.07	0.1	Averaged value
BTM_λ	35.8	36.3	36.8	35.8	35.2	35.2	34.6	32.0	31.1	30.7	34.4
SBTM_λ	36.0	37.1	38.1	38.4	37.9	38.1	36.4	33.3	33.5	30.6	35.9

value (38.4 %) when $K = 70$, and the averaged value is 36.6 %. On the other hand, the best value of accuracy for BTM_K is 37.8 % when $K = 80$, and the averaged value is 36.4 %. The results indicate that the performance of SBTM is nearly the same with that of BTM when using varied scales of K.

Table 3(b) presents the accuracies of SBTM and BTM with different values of λ given $K = 70$. The results indicate that SBTM_λ outperforms BTM_λ for almost all λ values, and the averaged accuracy of SBTM_λ improves 1.5 % than BTM_λ. Compared to the performance with different values of K, we can observe that the value of λ has a larger impact on the accuracy of both SBTM and BTM. Such an observation is also validated by Figs. 2 and 3, respectively.

4.4   Comparison with Baselines

In this subsection, we implement the following baseline models for comparison with our SBTM:

1. SWAT system (SWAT). This is one of the top-performing systems on the "affective text" task in SemEval-2007 [18]. SWAT made use of the emotional content of news headlines and scored the emotions of each word w with the unigram model. The predicted labels depends on the average score of emotions of every news headline that contains w [10,20].

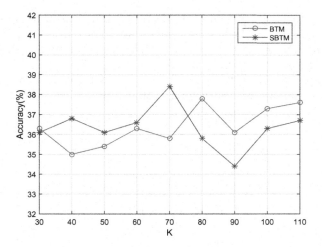

Fig. 2. Accuracies with different values of K ($\lambda = 0.001$)

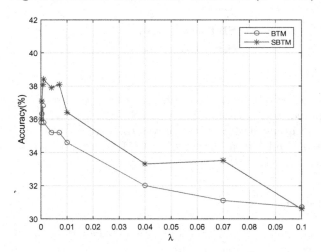

Fig. 3. Accuracies with different values of λ ($K = 70$)

2. Emotion term method (ET). This is a straightforward method to model the word-emotion associations [2]. ET used the naïve Bayes (NB) method by assuming that words are independently generated from emotion labels in two sampling steps. Different from NB, ET considered emotion ratings when estimating the values of $P(e)$ and $P(w|e)$.
3. Emotion topic model (ETM). This model introduced an additional emotion layer into ET and LDA, and exploited the emotional distribution to guide the topic generation [2]. The parameters of ETM were set according to the description in [2].

4. Multi-label supervised topic model (MSTM) and sentiment latent topic model (SLTM) [16]. MSTM began by generating topics from words, and then sampled emotions from each topic. SLTM, on the other hand, generated topics directly from social emotions.
5. Affective topic model (ATM) [15]. The exponential distribution was employed by ATM to generate user ratings for each emotion label.

Table 4. Accuracies of different models

Model	Accuracy(%)
SWAT	31.40
ET	31.00
ETM	24.65
MSTM	20.80
SLTM	20.85
ATM	32.45
SBTM_λ	**35.94**
SBTM_K	**36.60**

The accuracies of our model and other baselines on *SemEval* are shown in Table 4. Compared to the baseline models of SWAT, ET, ETM, MSTM, SLTM, and ATM, the accuracy of SBTM_λ improves 14.47 %, 15.94 %, 45.80 %, 72.79 %, 72.37 %, 10.76 %, and the accuracy of SBTM_K improves 16.56 %, 18.06 %, 48.48 %, 75.96 %, 75.54 %, 12.79 %, respectively.

5   Conclusion

In this paper, we proposed a method of sentiment detection for short text effectively. In this method, we first expand each short document by adding relative words from the corpus. Then, we employ the biterm topic model to estimate the topic probability distribution of biterms and documents. Finally, we estimate the similarity between two documents by calculating topic-level correlation coefficient values, and detect the sentiment of unlabeled documents by ranking the mean sentimental scores of the related labeled documents.

In the future, we plan to improve the method of estimating the similarity between words and documents, in addition to extend our approach to other applications, such as the emotionally aware recommendation of events in social media materials.

Acknowledgements. The authors are thankful to the anonymous reviewers for their constructive comments and suggestions on an earlier version of this paper. This research has been supported by the National Natural Science Foundation of China (61502545, 61472453, U1401256, U1501252), a grant from the Research Grants Council of the Hong Kong Special Administrative Region, China (UGC/FDS11/E06/14), and the Fundamental Research Funds for the Central Universities.

References

1. Banerjee, S., Ramanathan, K., Gupta, A.: Clustering short texts using wikipedia. In: Proceedings of the 30th Annual International ACM SIGIR Conference on Research and Development in Information Retrieval (SIGIR), pp. 787–788. ACM (2007)
2. Bao, S., Xu, S., Zhang, L., Yan, R., Su, Z., Han, D., Yu, Y.: Mining social emotions from affective text. IEEE Trans. Knowl. Data Eng. **24**(9), 1658–1670 (2012)
3. Blei, D.M., Ng, A.Y., Jordan, M.I.: Latent dirichlet allocation. J. Mach. Learn. Res. **3**, 993–1022 (2003)
4. Cheng, X., Lan, Y., Guo, J., Yan, X.: Btm: Topic modeling over short texts. IEEE Trans. Knowl. Data Eng. **26**(12), 2928–2941 (2014)
5. Deerwester, S., Dumais, S.T., Furnas, G.W., Landauer, T.K., Harshman, R.: Indexing by latent semantic analysis. J. Am. Soc. Inf. Sci. **41**(6), 391–407 (1990)
6. Gangemi, A., Presutti, V., Reforgiato Recupero, D.: Frame-based detection of opinion holders and topics: a model and a tool. IEEE Comput. Intell. Mag. **9**(1), 20–30 (2014)
7. Geman, S., Geman, D.: Stochastic relaxation, gibbs distributions, and the bayesian restoration of images. IEEE Trans. Pattern Anal. Mach. Intell. **6**(6), 721–741 (1984)
8. Hofmann, T.: Probabilistic latent semantic indexing. In: Proceedings of the 22nd Annual International ACM SIGIR Conference on Research and Development in Information Retrieval (SIGIR), pp. 50–57 (1999)
9. Jin, O., Liu, N.N., Zhao, K., Yu, Y., Yang, Q.: Transferring topical knowledge from auxiliary long texts for short text clustering. In: Proceedings of the 20th ACM International Conference on Information and Knowledge Management (CIKM), pp. 775–784. ACM (2011)
10. Katz, P., Singleton, M., Wicentowski, R.: Swat-mp: The semeval-2007 systems for task 5 and task 14. In: Proceedings of the 4th International Workshop on Semantic Evaluations (SemEval), pp. 308–313. Association for Computational Linguistics (2007)
11. Kim, S.B., Han, K.S., Rim, H.C., Myaeng, S.H.: Some effective techniques for naive bayes text classification. IEEE Trans. Knowl. Data Eng. **18**(11), 1457–1466 (2006)
12. Pang, B., Lee, L., Vaithyanathan, S.: Thumbs up? sentiment classification using machine learning techniques. In: Proceedings of the Conference on Empirical Methods in Natural Language Processing (EMNLP), pp. 79–86 (2002)
13. Phan, X.H., Nguyen, L.M., Horiguchi, S.: Learning to classify short and sparse text & web with hidden topics from large-scale data collections. In: Proceedings of the 17th International Conference on World Wide Web (WWW), pp. 91–100. ACM (2008)
14. Rao, Y., Lei, J., Liu, W., Li, Q., Chen, M.: Building emotional dictionary for sentiment analysis of online news. World Wide Web J. **17**, 723–742 (2014)
15. Rao, Y., Li, Q., Liu, W., Wu, Q., Quan, X.: Affective topic model for social emotion detection. Neural Netw. **58**, 29–37 (2014)
16. Rao, Y., Li, Q., Mao, X., Liu, W.: Sentiment topic models for social emotion mining. Inf. Sci. **266**, 90–100 (2014)

17. Sahami, M., Heilman, T.D.: A web-based kernel function for measuring the simi-
    larity of short text snippets. In: Proceedings of the 15th International Conference
    on World Wide Web (WWW), pp. 377–386(2006)
18. Snow, R., O'Connor, B., Jurafsky, D., Ng, A.Y.: Cheap and fast–but is it good?:
    evaluating non-expert annotations for natural language tasks. In: Proceedings of
    the Conference on Empirical Methods in Natural Language Processing (EMNLP),
    pp. 254–263. Association for Computational Linguistics (2008)
19. Stoyanov, V., Cardie, C.: Annotating topics of opinions. In: Proceedings of the
    6th International Conference on Language Resources and Evaluation (LREC),
    pp. 3213–3217 (2008)
20. Strapparava, C., Mihalcea, R.: Semeval-2007 task 14: affective text. In: Proceedings
    of the 4th International Workshop on Semantic Evaluations (SemEval), pp. 70–74.
    Association for Computational Linguistics (2007)
21. Turney, P.D.: Thumbs up or thumbs down? semantic orientation applied to unsu-
    pervised classification of reviews. In: Proceedings of Annual Meeting of the Asso-
    ciation for Computational Linguistics (ACL). pp. 417–424 (2002)
22. Wang, J., Yao, Y., Liu, Z.: A new text classification method based on hmm-svm.
    In: International Symposium on Communications and Information Technologies
    (ISCIT). pp. 1516–1519 (2007)
23. Xie, H., Li, Q., Cai, Y.: Community-aware resource profiling for personalized search
    in folksonomy. J. Comput. Sci. Technol. **27**(3), 599–610 (2012)
24. Xie, H., Li, Q., Mao, X., Li, X., Cai, Y., Rao, Y.: Community-aware user profile
    enrichment in folksonomy. Neural Netw. **58**, 111–121 (2014)
25. Xie, H., Li, Q., Mao, X., Li, X., Cai, Y., Zheng, Q.: Mining latent user commu-
    nity for tag-based and content-based search in social media. Comput. J. **57**(9),
    1415–1430 (2014)

Similarity-Based Classification for Big Non-Structured and Semi-Structured Recipe Data

Wei Chen[1,3] and Xiangyu Zhao[2,4(✉)]

[1] Agricultural Information Institute,
Chinese Academy of Agricultural Sciences, Beijing, China
[2] Beijing Research Center for Information Technology in Agriculture, Beijing, China
[3] Key Laboratory of Agri-information Service Technology,
Ministry of Agriculture, Beijing, China
chenwei@caas.cn
[4] National Engineering Research Center for Information Technology
in Agriculture, Beijing, China
zhaoxy@nercita.org.cn

Abstract. In current big data era, there has been an explosive growth of various data. Most of these large volume of data are non-structured or semi-structured (e.g., tweets, weibos or blogs), which are difficult to be managed and organized. Therefore, an effective and efficient classification algorithm for such data is essential and critical. In this article, we focus on a specific kind of non-structured/semi-structured data in our daily life: recipe data. Furthermore, we propose the document model and similarity-based classification algorithm for big non-structured and semi-structured recipe data. By adopting the proposed algorithm and system, we conduct the experimental study on a real-world dataset. The results of experiment study verify the effectiveness of the proposed approach and framework.

Keywords: Recipe data · Classification · User-generated contents · Semi-structured data · Non-structured data

1 Introduction

In current big data era, there has been an explosive growth of various data. Most of these large volume of data are non-structured or semi-structured (e.g., tweets, weibos or blogs). These data has the following distinct characteristics, which cannot be handled by the conventional database management systems.

- **Huge Volumed.** The user-generated data is large scale as users are easily to produce and share data with various kinds of devices.
- **Explosive Grown.** Every day, 2.5 quintillion bytes of data are produced and 90 % of the data in the world today were created within the past two years [20].

© Springer International Publishing Switzerland 2016
H. Gao et al. (Eds.): DASFAA 2016 Workshops, LNCS 9645, pp. 57–64, 2016.
DOI: 10.1007/978-3-319-32055-7_5

- **Non-/Semi-Structured.** The structure of the user-generated data are normally non-structured and semi-structured, since a piece of data (e.g., a tweet or a blog) may contain videos, images, texts and so on.

Therefore, it is very difficult to manage and organize such data by using the conventional approaches. To find the underline behaviors and patterns of the data, it is quite important to classify data into categories. To achieve this goal, an effective and efficient classification algorithm for the data is essential and critical. In this article, we focus on a specific kind of non-structured/semi-structured data in our daily life: recipe data. As the recipe data is a useful source to help people cook dishes, it is quite necessary to categorize these data to help them find favorite dishes of their preference. The main contribution of this paper are listed as follows.

- We present a document and similarity model for the big non-structured and semi-structured recipe data.
- We propose a similarity-based classification algorithm based on the document and similarity model.
- We conduct the experimental study on a real-world dataset, and verify the effectiveness of the proposed algorithm and framework.

The remaining parts of this article are organized as follows. Section 2 reviews the related work to our research. The proposed document and similarity modeling approach and the classification algorithm for big non-structured and semi-structured recipe data is introduced in Sect. 3. In Sect. 4, we report the experimental results by performing the algorithm and baseline in a real-world dataset. Section 5 summarizes the findings and some potential future directions of this research.

2   Related Work

In this section, we mainly review the research on the user-generated contents (e.g., twitters) in the Web 2.0 era. Golder and Huberman investigated the usage patterns and user behaviors of the social media contents and annotations [7]. Bischoff et al. did some statistical analysis on some social tagging data sets to gain valuable tags for search [2]. Manish et al. investigated various distinct characteristics of the user-generated tags [9]. Furthermore, the social tags were considered as an important semantic sources for modeling the items for recommendations [17,29], semantic retrieval [25] and personalized search [3,22]. More recently, researchers attempted to organize and categorize these data from the perspective of users. For example, the community-based modeling approaches [19,21,23,24] were adopted to achieve this goal. Another example is to identify the underlying patterns from the structure of social data [1,11]. Some researches also try to index the data through a cognitive approach, which identify sentiments [14,15], emotions [18], pre-knowledge [6,30], role in the group [5], preferred patterns [31] and personality [8]. There have been several classifying and organizing approaches for non-structured/semi-structured data. A classifier for the semi-structured documents was proposed by using the a structured vector model [28]. Lesbegueries et al. presented models take into account characteristics of heterogeneous human expression modes: written language and captures of drawings, maps, pictures, etc., and semantic treatments have been built to automatically manage spatial and temporal information from non-structured data [13]. The techniques and relevant issues of keyword search were discussed in [4]. Mansmann et al. proposed an approach based on introducing a data enrichment layer responsible for detecting new structural elements in the data using data mining and other techniques [16]. EsdRank, which treats vocabularies, terms and entities from external data as objects connecting query and documents, was a new technique for improving ranking using external semi-structured data such as controlled vocabularies and knowledge bases [26].

3 Methodology

The overall workflow of classification algorithm is shown in Fig. 1. Firstly, the non-/semi-structured data will be pre-processed. The detail steps include Chinese segmentation, summarization and weighted value assignment. Secondly, the similarity among different documents will be calculated. Finally, we exploit the document similarity for classification, and obtain the categorized data.

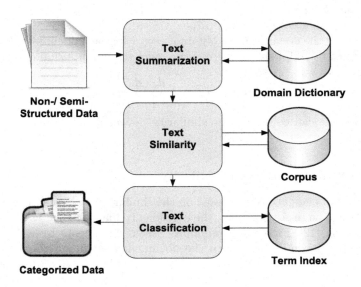

Fig. 1. The workflow of classification algorithm

3.1 Pre-process on Non/Semi-Structured Data

As the dataset is Chinese, we firstly need to segment Chinese texts into terms. To address this issue, we employ the Hidden Markov Model (HMM) to perform the Chinese segmentation. The detail of our segmentation approach is introduced in [27]. Furthermore, we build a domain dictionary to extract the summary of the texts, filter trivial terms and assign the weight to these terms. The construction of domain dictionary follows the approach introduced in [10]. After the pre-processed, we then attempt to construct the relationship among documents.

3.2   Document and Similarity Modeling

By pre-processing the texts, we obtain a set of documents to be categorized. Formally, we defined a document as

$$d_i = < t_1 : w_i(t_1); t_2 : w_i(t_2); ...t_n : w_i(t_n) > \tag{1}$$

where t_x is a term of the document, and $w_i(t_x)$ is the weight to be assigned to the term. $w_i(t_x)$ is calculated based on the hybrid paradigm [22] as follows.

$$w_i(t_x) = f_i(t_x) \times \frac{log^2 N}{m_x^2} \times \frac{f_i(t_x) \cdot (k+1)}{f_i(t_x) + k \cdot (1 - b + b \cdot \frac{l_i}{l_a})} \tag{2}$$

where $f_i(t_x)$ is the term frequency of t_x in the document, N is the total number of documents, m_x is the number of document containing t_x, l_i is the length of the document, l_a is the average length of all documents, and k and b are two parameters, which are set as 2 and 0.75.

By modeling the document, we obtain a vector for each document. It is quite straightforward to adopt the cosine measurement to compute similarity between documents.

$$Sim(d_i, d_j) = \frac{d_i \cdot d_j}{||d_i|| ||d_j||} \tag{3}$$

Therefore, we can construct the similarity between each pair of documents in the corpus.

3.3   Classification Based on Similarity

The algorithm of classification is based on the similarity we have obtained in the last step. As shown in Algorithm 1, the core idea is adapted from the prototype-based clustering [12]. Firstly, we randomly assign documents to each category as a prototype. Secondly, we assign the document has the maximal similarity with the prototype to the category. Finally, the class labels are outputs when all documents in the corpus has been assigned.

4   Experiment

The experiment is based on a dataset collected from a folksonomy-based multi-media recipe dataset in [3]. The dataset contains 500 recipes (documents) in five main kinds of dishes in China (i.e., Cantonese dishes, Sichuan dishes and so on)

Data: A set of documents D $(d_i \in D)$; The set of categories C $(c_x \in C)$
Result: Class label c_x for d_i
for *each $c_x \in D$* **do**
　　Randomly assign $|C|$ documents to each c_x;
　　Set these documents d_x as the prototype of c_x;
　　$D \leftarrow D - \{d_x\}$
end
for *each document $d_i \in D$* **do**
　　Compute $sim(d_i, d_x)$ by Equation (3);
　　Find thed_i, d_x with maximal $sim(d_i, d_j)$;
　　$c_x \leftarrow d_i$;
　　$d_x \leftarrow Mean(c_x$;
　　$D \leftarrow D - \{d_i\}$;
end
Output class label c_x for d_i;

Algorithm 1. The Similarity-based Classification

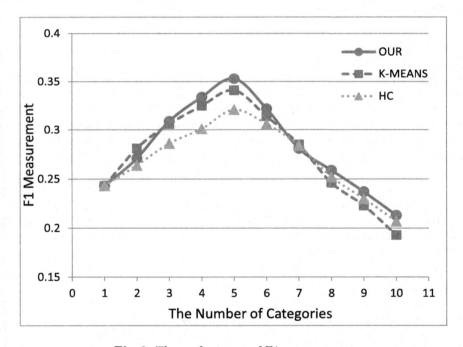

Fig. 2. The performance of F1 measurement

associated with the general descriptions on the characteristics, ingredients, main taste and so on. Furthermore, 203 users have annotated 7, 889 tags on these recipes. Averagely, each user has annotated 16.7 recipes. For the purpose of evaluating the classification algorithm on non/semi-structured documents, we combine the user-generated tags for each recipe with the corresponding recipe descriptions to generate a non/semi-structured document.

To evaluate the effectiveness of the proposed framework and algorithm, we use the F1 measurement, which is the harmonic mean of precision and recall. As the proposed method is an unsupervised method for classification, we adopt a voting strategy to determine the predicted classification label in a category. That is, the final label of a category is determined by the label which have the most members in the category. The accuracy and recall are therefore evaluated by matching the final labels and the ground-truth labels. As shown in Fig. 2, the performance of the proposed methods (OUR), K-Means (K-MEAN) and Hierarchial method (HC) are illustrated. It is also quite clear that the F1 measurement achieve by our method has the best performance ($F1 = 0.535$, $c = 5$). The effectiveness of our proposed method is verified. It is worth to point out that all methods have the best performance in the number of categories ($c = 5$). A reasonable explanation for this observation is that the dataset contains five kinds of dishes as we mentioned. The classification is overfitting the dataset when the number of categories is greater than 5, while it is underfitting when the number of categories is less than 5.

5   Conclusion

In this paper, we present a document and similarity model for the big non-structured and semi-structured user-generated data. Based on the model, we propose a similarity-based classification algorithm for processing the user-generated data. Furthermore, We conduct the experimental study on a real-world dataset, and verify the effectiveness of the proposed algorithm and framework. In future research, we plan to build user profiles to further improve the effectiveness of the system.

Acknowledgement. This work is supported by Fundamental Research Funds of Agricultural Information Institute, Chinese Academy of Agricultural Sciences (No. 2014-J-011), and Project of Ministry of Agriculture of China "Agricultural information monitoring and early-warning".

References

1. Armstrong, T.G., Ponnekanti, V., Borthakur, D., Callaghan, M.: Linkbench: A database benchmark based on the facebook social graph. In: Proceedings of the ACM SIGMOD International Conference on Management of Data, pp. 1185–1196. ACM (2013)
2. Bischoff, K., Firan, C.S., Nejdl, W., Paiu, R.: Can all tagsbe used for search? In: Proceedings of CIKM 08, Napa Valley, California, USA, October 26-30, pp. 193–202. ACM, New York, NY, USA (2008)
3. Cai, Y., Li, Q., Xie, H., Yu, L.: Personalized resource search by tag-based user profile and resource profile. In: Chen, L., Triantafillou, P., Suel, T. (eds.) WISE 2010. LNCS, vol. 6488, pp. 510–523. Springer, Heidelberg (2010)

4. Chen, Y., Wang, W., Liu, Z., Lin, X.: Keyword search on structured and semi-structured data. In: Proceedings of the ACM SIGMOD International Conference on Management of Data, pp. 1005–1010. ACM (2009)
5. Feng, X., Peng, Y., Xie, H., Yan, Z.: Role-based learning path discovery for collaborative business environment. In: International Conference on Control, Automation and Systems Engineering (CASE), pp. 1–4. IEEE (2011)
6. Feng, X., Xie, H., Peng, Y., Chen, W., Sun, H.: Groupized learning path discovery based on member profile. In: Luo, X., Cao, Y., Yang, B., Liu, J., Ye, F. (eds.) ICWL 2010. LNCS, vol. 6537, pp. 301–310. Springer, Heidelberg (2011)
7. Golder, S.A., Huberman, B.A.: Usage patterns of collaborative tagging systems. J. Inf. Sci. **32**, 198–208 (2006)
8. Gou, L., Zhou, M.X., Yang, H., Knowme, S.: Understanding automatically discovered personality traits from social media and user sharing preferences. In: Proceedings of the SIGCHI Conference on Human Factors in Computing Systems, pp. 955–964. ACM (2014)
9. Gupta, M., Li, R., Yin, Z., Han, J.: Survey on social tagging techniques. SIGKDD Explor. Newsl. **12**, 58–72 (2010)
10. Islam, A., Inkpen, D.: Semantic text similarity using corpus-based word similarity and string similarity. ACM Trans. Knowl. Disc. Data (TKDD) **2**(2), 10 (2008)
11. Jin, T., Xie, H., Lei, J., Li, Q., Li, X., Mao, X., Rao, Y.: Finding dominating set from verbal contextual graph for personalized search in folksonomy. In: IEEE/WIC/ACM International Joint Conferences on Web Intelligence (WI) and Intelligent Agent Technologies (IAT), vol. 1, pp. 367–372. IEEE (2013)
12. Kuncheva, L., Bezdek, J.C., et al.: Nearest prototype classification: Clustering, genetic algorithms, or random search? IEEE Trans. Syst. Man Cybern., Part C: Appl. Rev. **28**(1), 160–164 (1998)
13. Lesbegueries, J., Gaio, M., Loustau, P.: Geographical information access for non-structured data. In: Proceedings of the ACM Symposium on Applied Computing, pp. 83–89. ACM (2006)
14. Li, X., Xie, H., Chen, L., Wang, J., Deng, X.: News impact on stock price return via sentiment analysis. Knowl. Based Syst. **69**, 14–23 (2014)
15. Li, X., Xie, H., Song, Y., Li, Q., Shanfeng Zhu, F., Wang, L.: Does summarization help stock prediction? News impact analysis via summarization. IEEE Intell. Syst. **30**, 26–34 (2015)
16. Mansmann, S., Rehman, N.U., Weiler, A., Scholl, M.H.: Discovering olap dimensions in semi-structured data. Inf. Syst. **44**, 120–133 (2014)
17. Mao, X., Li, Q., Xie, H., Rao, Y.: Popularity tendency analysis of ranking-oriented collaborative filtering from the perspective of loss function. In: Bhowmick, S.S., Dyreson, C.E., Jensen, C.S., Lee, M.L., Muliantara, A., Thalheim, B. (eds.) DAS-FAA 2014, Part I. LNCS, vol. 8421, pp. 451–465. Springer, Heidelberg (2014)
18. Rao, Y., Lei, J., Wenyin, L., Li, Q., Chen, M.: Building emotional dictionary for sentiment analysis of online news. World Wide Web **17**(4), 723–742 (2014)
19. Tang, J., Chang, Y., Liu, H.: Mining social media with social theories: A survey. ACM SIGKDD Explorations Newsletter **15**(2), 20–29 (2014)
20. Xindong, W., Zhu, X., Gong-Qing, W., Ding, W.: Data mining with big data. IEEE Trans. Knowl. Data Eng. **26**(1), 97–107 (2014)
21. Xie, H.-R., Li, Q., Cai, Y.: Community-aware resource profiling for personalized search in folksonomy. J. Comput. Sci. Technol. **27**(3), 599–610 (2012)
22. Xie, H., Li, Q., Mao, X.: Context-aware personalized search based on user and resource profiles in folksonomies. In: Sheng, Q.Z., Wang, G., Jensen, C.S., Xu, G. (eds.) APWeb 2012. LNCS, vol. 7235, pp. 97–108. Springer, Heidelberg (2012)

23. Xie, H., Li, Q., Mao, X., Li, X., Cai, Y., Rao, Y.: Community-aware user profile enrichment in folksonomy. Neural Netw. **58**, 111–121 (2014)
24. Xie, H., Li, Q., Mao, X., Li, X., Cai, Y., Zheng, Q.: Mining latent user community for tag-based and content-based search in social media. Comput. J. **57**(9), 1415–1430 (2014)
25. Xie, H., Yu, L., Li, Q.: A hybrid semantic item model for recipe search by example. In: IEEE International Symposium on Multimedia (ISM), pp. 254–259. IEEE (2010)
26. Xiong, C., Callan, J.: Esdrank: Connecting query and documents through external semi-structured data. In: International Conference on Information and Knowledge Management, pp. 951–960. ACM (2015)
27. Yang, W., Ren, L.-Y., Tang, R.: A dictionary mechanism for chinese word segmentation based on the finite automata. In: International Conference on Asian Language Processing (IALP), pp. 39–42. IEEE (2010)
28. Yi, J., Sundaresan, N.: A classifier for semi-structured documents. In: Proceedings of the Sixth ACM SIGKDD International Conference on Knowledge Discovery and Data Mining, pp. 340–344. ACM (2000)
29. Yu, L., Li, Q., Xie, H., Cai, Y.: Exploring folksonomy and cooking procedures to boost cooking recipe recommendation. In: Du, X., Fan, W., Wang, J., Peng, Z., Sharaf, M.A. (eds.) APWeb 2011. LNCS, vol. 6612, pp. 119–130. Springer, Heidelberg (2011)
30. Zou, D., Xie, H., Li, Q., Wang, F.L., Chen, W.: The load-based learner profile for incidental word learning task generation. In: Popescu, E., Lau, R.W.H., Pata, K., Leung, H., Laanpere, M. (eds.) ICWL 2014. LNCS, vol. 8613, pp. 190–200. Springer, Heidelberg (2014)
31. Zou, D., Xie, H., Wang, F.L., Wong, T.-L., Wu, Q.: Investigating the effectiveness of the uses of electronic and paper-based dictionaries in promoting incidental word learning. In: Cheung, S.K.S., Kwok, L.-F., Yang, H., Fong, J., Kwan, R. (eds.) ICHL 2015. LNCS, vol. 9167, pp. 59–69. Springer, Heidelberg (2015)

An Approach of Fuzzy Relation Equation and Fuzzy-Rough Set for Multi-label Emotion Intensity Analysis

Chu Wang[1], Daling Wang[1,2(✉)], Shi Feng[1,2], and Yifei Zhang[1,2]

[1] School of Computer Science and Engineering,
Northeastern University, Shenyang, China
wangchu@research.neu.edu.cn,
{wangdaling,fengshi,zhangyifei}@cse.neu.edu.cn
[2] Key Laboratory of Medical Image Computing (Northeastern University),
Ministry of Education, Shenyang 110819, People's Republic of China

Abstract. There are a large number of subjective texts which contain people's all kinds of sentiments and emotions in social media. Analyzing the sentiments and predicting the emotional expressions of human beings have been widely studied in academic communities and applied in commercial systems. However, most of the existing methods focus on single-label sentiment analysis, which means that only an exclusive sentiment orientation (negative, positive or neutral) or an emotion state (joy, hate, love, sorrow, anxiety, surprise, anger, or expect) is considered for a document. In fact, multiple emotions may be widely coexisting in one document, paragraph, or even sentence. Moreover, different words can express different emotion intensities in the text. In this paper, we propose an approach that combining fuzzy relation equation with fuzzy-rough set for solving the multi-label emotion intensity analysis problem. We first get the fuzzy emotion intensity of every sentiment word by solving a fuzzy relation equation, and then utilize an improved fuzzy-rough set method to predict emotion intensity for sentences, paragraphs, and documents. Compared with previous work, our proposed algorithm can simultaneously model the multi-labeled emotions and their corresponding intensities in social media. Experiments on a well-known blog emotion corpus show that our proposed multi-label emotion intensity analysis algorithm outperforms baseline methods by a large margin.

Keywords: Opinion mining · Fuzzy relation equation · Sentiment analysis · Multi-labeled emotion · Emotion intensity · Fuzzy-rough set

1 Introduction

With the development of Web 2.0 techniques, more and more people are willing to express their feelings and emotions via the social media platform such as blog, microblog, and online forum. Therefore, detecting and analyzing the sentiments embedded in social media has become a popular research topic for both academic communities and commercial companies.

© Springer International Publishing Switzerland 2016
H. Gao et al. (Eds.): DASFAA 2016 Workshops, LNCS 9645, pp. 65–80, 2016.
DOI: 10.1007/978-3-319-32055-7_6

For sentiment analysis, previous researches usually focused on sentiment orientation classification, i.e., classifying the subjective text into two-orientation (*positive* and *negative*) or three-orientation (*positive*, *neutral*, and *negative*). However, currently more and more researches consider the sentiment categories such as *joy*, *hate*, *love*, *sorrow*, *anxiety*, *surprise*, *anger*, *expect* [11] called as fine-grained emotion (we call this research as emotion analysis in this paper). Compared with sentiment orientation classification, the fine-grained emotion analysis could capture users' meticulous sentiments (i.e., emotions), and are more suitable for public opinion monitoring about online hot events.

In fact, in social media platform, multiple emotions may be coexisting in just one document, paragraph, or even sentence, as shown in the following example.

"As a teacher, I'm very happy. I love my work, and like my school. However, the air quality of the city is too bad, and housing prices are so expensive."

In above example, for sentiment orientation analysis task, the paragraph contains positive (expressed by sentiment word "love", "like", and "happy") and negative (expressed by sentiment word "bad" and "expensive") sentiments. For fine-grained emotion analysis, the paragraph simultaneously contains joy (expressed by sentiment word "happy"), love (expressed by sentiment word "love" and "like"), anger (expressed by sentiment word "bad"), and anxiety (expressed by sentiment word "expensive") emotions. For such analysis task, a multi-label fine-grained emotion analysis will be required. Moreover, for the same emotion love, the emotion intensity of word "love" and "like" is different.

Generally, sentiment orientation and emotion are expressed implicitly by sentence structure, semantic, and sentiment words including adjectives, verbs, and adverbs. Table 1 shows some examples. Here each post is associated with several different emotion labels and a value between $0 \sim 1$ indicates the intensity of every emotion. In Table 1, the third post has the labels of *joy*, *hate*, *love*, and the forth post has the labels *hate*, *anger*. If an author expresses stronger emotion, the corresponding post will have a higher intensity. Because the word "fantastic" expresses a more intense emotion than the word "OK", so the intensities of *joy* and *love* emotions in first post are higher than those of second post. Analyzing the emotions in social text needs to not only recognize the multi-label co-existing emotions but also calculate their corresponding intensities.

The emotion analysis problem that simultaneously considering multi-label fine-grained emotions and their corresponding intensity is really rarely studied in the previous literature. To tackle this challenge, we regard the multiple emotion intensity detection in social text as an uncertain classification problem, and propose an approach that combining fuzzy relation equation with fuzzy-rough set for solving the problem. We first calculate the fuzzy emotion intensity (expressed as range) of every sentiment word by solving a fuzzy relation equation, then utilize an improved fuzzy-rough set method to predict emotion intensity for the social subjective text at sentence, paragraph, and document level.

In our method, human emotions have eight basic kinds of categories as defined in Quan's research [11], and each one with ten levels of intensity which is annotated between 0.1 and 1. Due to the intrinsic characteristic of human language, different sentiment words may express different emotion intensity. Even the same sentiment word may

Table 1. The examples of multiple emotions with different intensities

Social text	Joy	Hate	Love	Sorrow	Anxiety	Surprise	Anger	Expect
The movie tonight is fantastic, the dinner sucks!	0.5	0.2	0.5	0	0	0	0.7	0
The movie tonight is OK, but I still looking forward to Avengers 2!	0.2	0	0.2	0	0	0	0	0.5
I like this phone's screen, but it has a short battery life.	0.1	0.3	0.3	0	0	0	0	0
This is a totally rubbish phone! It cost me 500$ but was broke in one month!	0	0.6	0	0	0	0	0.8	0

have different emotion intensity in different context. To tackle these problems, firstly, we put all the sentiment words in a fuzzy relation equation and calculate each word's emotion value that is mapped into an intensity range with upper and lower bounds. Then we model the words by our improved fuzzy-rough set method which considering both the bounds of the intensity ranges and the importance of the words. Finally the intensities of the emotions are tuned by fuzzy modifiers to determine the overall emotions embedded in the social text. Experiment results using a well annotated blog emotion dataset show that our proposed algorithm significantly outperforms other baselines.

The rest of the paper is organized as follows: Sect. 2 introduces fuzzy relation equation and fuzzy-rough set. In Sect. 3, we propose the fuzzy relation equation and fuzzy-rough set based multi-label, fine-gained emotion intensity prediction method. In Sect. 4, we show our experiment setup and results. We survey the related work with sentiment and emotion analysis in Sect. 5. Finally we conclude the paper and give the future work in Sect. 6.

2 Fuzzy Relation Equation and Fuzzy-Rough Set

In this paper, we use improved fuzzy-rough set based on fuzzy relation equation for our multi-label and fine-gained emotion intensity computing. We introduce the fuzzy relation equation and fuzzy-rough set in this section.

2.1 Rough Set

Rough set theory was proposed by Pawlak in 1982 [10], and it is useful to deal with uncertainty analysis problems. In traditional Pawlak's rough set theory [10], the pair (U, R) is called as an approximation space, where U is a universe and R is an equivalence relation on U.

Suppose R is an indiscernibility relation on U, with respect to R, equivalence class of an element x in U could be defined an as follows:

$$[x]_R = \{y|(x, y) \in R\} \tag{1}$$

The quotient set of U by the relation R is denoted by U/R, and

$$U/R = \{X_1, X_2, \ldots, X_m\} \tag{2}$$

Let $[x]_R = \{y \in U | (x, y) \in R\}$ be the equivalence class of $x \in U$, the Pawlak approximation space is defined as follows.

Let X_i ($i = 1, 2, \ldots, m$) be an equivalence class of R, and the equivalence classes of R are called elementary sets. If given an arbitrary set $X \in R$, X may be characterized by a pair of upper and lower approximations defined as follows [10, 22]:

$$\bar{R}(X) = \{x \in U, [x]_R \subseteq X\} \tag{3}$$

$$\underline{R}(X) = \{x \in U, [x]_R \cap X \neq \phi\} \tag{4}$$

2.2 Fuzzy Set and Fuzzy Relation Equation

Fuzzy set theory was proposed by Zadeh in 1965 [20]. U is a finite and non-empty set, and is called as universe. In this paper, the universe U is considered to be finite. Fuzzy set A is a mapping from U into the unit interval $[0, 1]$:

$$\mu : U \rightarrow [0, 1] \tag{5}$$

where for each $x \in U$, $\mu_A(x)$ is called as the membership degree of x in A. The fuzzy power set is denoted by $F(U)$ [20] showing the set of all fuzzy sets in the universe U.

The fuzzy relation equation was proposed by Ernest et al. [5] which is an equation of the form $H \circ X = P$, where H and P are fuzzy sets, X is a fuzzy relation, "$\circ$" is fuzzy inner product, for $h_i \in H$, $x_i \in X$, $H \circ X = \vee(H \wedge X)$, where $\wedge$ means fuzzy intersection, it defined as $h_i \wedge x_i = \min(h_i, x_i)$, and $\vee$ means fuzzy union. It is defined as: $h_i \vee x_i = \max(h_i, x_i)$. And $H \circ X = P$ stands for the composition of H with X.

For a given Fuzzy matrix $H = [h_{m \times l}] \in \mu$, $P = [p_{n \times l}] \in \mu$, we calculate fuzzy matrix $X = [x_{n \times m}] \in \mu$ to meet the formula $H \circ X = P$ as:

$$\begin{bmatrix} x_{11} & \cdots & x_{1m} \\ \vdots & \ddots & \vdots \\ x_{n1} & \cdots & x_{nm} \end{bmatrix} \cdot \begin{bmatrix} h_{11} & \cdots & h_{1l} \\ \vdots & \ddots & \vdots \\ h_{m1} & \cdots & h_{ml} \end{bmatrix} = \begin{bmatrix} p_{11} & \cdots & p_{1l} \\ \vdots & \ddots & \vdots \\ p_{n1} & \cdots & p_{nl} \end{bmatrix} \tag{6}$$

For most fuzzy relation equation, if it has solution, the solutions always have the following characteristics:

(1) The solution of the equation always appears in the form of set.
(2) A fuzzy relation equation always has more than one solution.

2.3 Fuzzy-Rough Set

Fuzzy-rough set was first proposed by Dubois and Prade in 1990 [3]. A fuzzy subset $R \in F(U \times W)$ is seemed as a fuzzy binary relation between U and W, and $R(x, y)$ is defined as the degree of relation between x and y, where $(x, y) \in U \times W$ [17, 18].

Definition 1: U and W are two finite and nonempty universes. Suppose that R is an arbitrary relation from U to W, the triple (U, W, R) is called a generalized fuzzy approximation space. For any set $A \in F(U)$, the upper and lower approximations of A, $\bar{R}(A)$ and $\underline{R}(A)$, with respect to the approximation space (U, W, R) are fuzzy sets of U whose membership functions, and for each $x \in U$, are defined respectively as:

$$\bar{R}(A) = \vee_{y \in W}[R(x, y) \wedge A(y)], \ x \in U \tag{7}$$

$$\underline{R}(A) = \vee_{y \in W}[1 - R(x, y) \vee A(y)], \ x \in U \tag{8}$$

The pair $(\bar{R}(A), \underline{R}(A))$ is referred to as a generalized fuzzy rough set, and R is referred to as upper and lower generalized fuzzy rough approximation operators.

3 Improving Fuzzy-Rough Set Based on Fuzzy Relation Equation for Multi-label Emotion Intensity Prediction

Since the emotion intensity between [0, 1] can be regarded as fuzzy degree, we consider to use fuzzy relation equation and fuzzy-rough set for emotion intensity analysis. In this paper, we propose an improved fuzzy-rough set based method that combined with fuzzy relation equation. The overall framework is shown as Fig. 1.

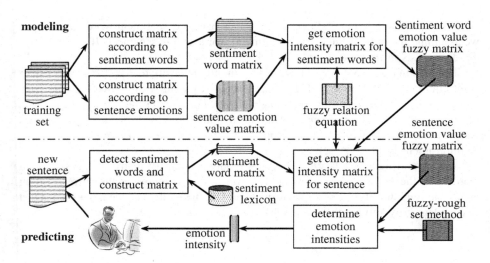

Fig. 1. Framework of multi-label emotion intensity analysis

In detail, in the modeling stage, we use an annotated blog set as training data in which all the sentiment words are labeled and the corresponding eight emotion labels and intensities are given for every sentence. Based on the training set, we can construct X matrix with sentiment words and P matrix with eight emotion intensities, and further calculate matrix H according to Formula (6) $H \circ X = P$. Here X can be regarded as eight emotion intensities of every sentiment word. As mentioned in Sect. 2.2, H is a fuzzy matrix. After modeling (obtaining matrix H), in the predicting stage, for a new sentence we can construct X by detecting sentiment word of the sentence, we further apply an improved fuzzy-rough set method for obtaining eight emotion intensities of the sentence. Moreover, the process can be extended to paragraph and document level.

The main techniques used in modeling and predicting stage include fuzzy relation equation and fuzzy-rough set (here we improve it) method, so in this section, we introduce their applications in our multi-label emotion intensity analysis.

3.1 Fuzzy Relation Equation Calculation

As mentioned above, our training set has been labeled emotion intensity at sentence level, and the embedded sentiment words are also known (even unknown, we can detect them with existing technique and sentiment lexicon). So in modeling stage, we can use a fuzzy relation equation (Formula (6)) to get emotion intensity fuzzy matrix of all sentiment words in the training set.

For clearer significance, here we rewrite fuzzy relation equation in Formula (6) and present it as Formula (9).

$$
\begin{bmatrix} vw_{11} & \cdots & vw_{1n} \\ \vdots & \ddots & \vdots \\ vw_{m1} & \cdots & vw_{mn} \end{bmatrix} \cdot \begin{bmatrix} ve_{11} & \cdots & ve_{18} \\ \vdots & \ddots & \vdots \\ ve_{n1} & \cdots & ve_{n8} \end{bmatrix} = \begin{bmatrix} vs_{11} & \cdots & vs_{18} \\ \vdots & \ddots & \vdots \\ vs_{m1} & \cdots & vs_{m8} \end{bmatrix} \tag{9}
$$

In above equation, the first item in left $VW = [vw_{mn}]$ is sentiment word matrix, here we assuming the training set T has n sentiment words and m sentences. If jth sentiment word w_j exists in ith sentence s_i, $vw_{ij} = 1$ else $vw_{ij} = 0$.

The item in right $VS = [vs_{m8}]$ is sentence emotion intensity matrix, here we consider eight emotions as the same as [11], i.e., $e_1 = joy$, $e_2 = hate$, $e_3 = love$, $e_4 = sorrow$, $e_5 = anxiety$, $e_6 = surprise$, $e_7 = anger$, $e_8 = expect$. For the ith sentence s_i, $vs_{i1}, vs_{i2}, \ldots,$ vs_{i8} represent the emotion intensity value of joy, $hate$, $love$, $sorrow$, $anxiety$, $surprise$, $anger$, $expect$, respectively. We can construct the matrix based on known multi-label emotion intensity of every sentence in T.

The second item in left $VE = [ve_{n8}]$ is emotion intensity matrix of all sentiment words. For the ith sentiment word w_i, $ve_{i1}, ve_{i2}, \ldots, ve_{i8}$ represent the corresponding eight emotion intensity values of w_i, respectively. In the modeling stage, our goal is just to calculate it by solving the fuzzy relation equation in Formula (9).

The following Algorithm 1 describes the process for achieving *VE* matrix.

Algorithm 1: Modeling Algorithm for Multi-label Emotion Intensity

---

Input: Training Set T // In T, all sentiment words have been labeled and eight emotion
                    intensities of every sentence are known
Output: emotion intensity matrix *VE* about all sentiment words
Description:
    1.   For every sentence $s \in T$;
    2.   {Construct a row of *VW* matrix with all sentiment words in s;
    3.   Construct a row of *VS* matrix with eight emotion intensity values of s;
    4.   }
    5.   Solve and Return fuzzy matrix *VE* with fuzzy relation equation in Formula (9);

3.2 Improving Fuzzy-Rough Set

According to the framework in Fig. 1, in the predicting stage, for a new sentence s, we apply *VE* matrix returned by Algorithm 1 to achieve a new emotion intensity matrix *VE'*. Based on *VE'* we further calculate eight emotion intensity values corresponding to s. Because *VE* is a fuzzy matrix, *VE'* is also fuzzy. In this case, we further process it with an improved fuzzy-rough set for determining emotion intensity values for s. Here we introduce the improved fuzzy-rough set method and its application in our work.

Definition 2: (F^{-1}, W) is considered as a relation of a fuzzy set over universe E iff F^{-1} is a mapping of U into the set of all fuzzy subsets from the set W, where F^{-1} is a mapping given by $F^{-1}: E \rightarrow F(W)$. $F(E)$ denotes W, which is regarded as all fuzzy subsets of parameter set, then $F^{-1}(e)(w) \in [0, 1]$, $\forall e \in E, w \in W$ [8].

    In this section, we will establish an improved model based on fuzzy-rough sets, which is able to calculate the weight of each attribute, i.e., the emotion intensity, for the given sentence. In Sect. 3.1, we associate each sentiment word with the eight basic emotions: *joy, hate, love, sorrow, anxiety, surprise, anger* and *expect*, and most sentences have more than one emotion words in this situation. We need a new algorithm to estimate the multi-label emotion intensity value of the whole sentence. The process can also extend to paragraph or document.

Definition 3: Let (F^{-1}, W) be a fuzzy set over E, the triple relation (E, W, F^{-1}) is called as the fuzzy approximation space. For any $A \in F(W)$, the upper and lower approximations of A, $\bar{F}(A)$ and $\underline{F}(A)$ with respect to the fuzzy approximation space (E, W, F^{-1}) are fuzzy sets of U whose membership functions, are defined as followings:

$$\bar{F}(A)(x) = \vee y \in W[(F^{-1}(x)(y)) \wedge A(y)], x \in E \tag{10}$$

$$\underline{F}(A)(x) = \wedge y \in W[(1 - F^{-1}(x)(y)) \vee A(y)], x \in E \tag{11}$$

After we find out attributes of sentiment words existing in sentence s, to make this algorithm more suitable for sentiment analysis task and human language logic, according to Definitions 1 and 4, we propose an improved fuzzy-rough set method. For any $A \in F(W)$, the upper approximation and lower approximation of A, $\bar{F}(A)$ and $\underline{F}(A)$. Under this specific situation of this paper, we defined $A = \{\bar{A}, \underline{A}\}$, because of the fuzzy relation equation the solution of VE in Formula (9) may be a range, which is described as $[F(e)(\underline{w}), F(e)(\bar{w})]$, then we defined $\bar{A}$ for $\bar{F}(A)$, $\underline{A}$ for $\underline{F}(A)$, and if $F(e)(w) = 0$ then $\bar{F}(A) = 0$ and $\underline{F}(A) = 0$, if $F(e) \neq 0$ then

$$\bar{F}(A)(e) = \vee_{w \in W}[F(e)(\bar{w}) \wedge \bar{A}(w)], e \in E \tag{12}$$

$$\underline{F}(A)(e) = \begin{cases} \wedge_{w \in W}[(1 - F(e)(\underline{w})) \vee \underline{A}(w)], & e \in E, F(e)(\underline{w}) \in [0.5, 1] \\ \wedge_{w \in W}[F(e)(\underline{w}) \wedge \underline{A}(w)], & e \in E, F(e)(\underline{w}) \in [0, 0.5) \end{cases} \tag{13}$$

In this paper, all the emotional intensities are between 0 and 1, so it is suitable for fuzzy degree. Here A is very important, and we choose the strongest emotion intensity of the solutions, which the upper bound is showed as $\bar{A}$, and lower bound as $\underline{A}$. Specifically, we will calculate the strongest emotion intensity which is belong to upper bound as the result of $\bar{A}$, and lower bound as the result of $\underline{A}$. the logic of the improving fuzzy rough algorithm is the more approximate to A, the stronger of the emotion intensity of the sentiment words is. So we can construct the decision object A on the evaluation of the words universe W.

Take one sentence for example in our training set: "我想, 人其实内心都有顽强的意志力的, 只不过有些人没释放出来而已 (*I think actually there is strong willpower in people" heart, but some people just have not release yet)*". We calculate $\bar{F}(A)$ and $\underline{F}(A)$ with Formula (12) and (13), the result is shown as Table 2.

Table 2. The examples of emotion computation based on our improved fuzzy-rough set

	顽强(strong)	意志力 (willpower)	只不过 (just)	而已 (yet)	$\bar{F}(A)$	$\underline{F}(A)$	intensity
Joy	0	0	0	0	0	0	0
Hate	0	0	0	0	0	0	0
Love	[0.4,0.7]	0.5	0	0	0.7	0.4	1.1
Sorrow	0	0	0	0	0	0	0
Anxiety	0	0	[0.4,0.7]	[0.2,0.5]	0.7	0.2	0.9
Surprise	0	0	0	0	0	0	0
Anger	0	0	0	0	0	0	0
Expect	0	0	[0.3,0.6]	[0.3,0.8]	0.8	0.3	1.1
$\bar{A}$	0.7	0.5	0.7	0.8			
$\underline{A}$	0.4	0.5	0.4	0.3			

In Table 2, four words are selected from the sentences by the sentiment lexicon, and the shadow part is the new emotion intensity value matrix, which is calculated from returned result *VE* by Algorithm 1 according to sentiment words in the sentence. In the dataset, the emotion labels and intensity values of the sentiment words are marked by people with 顽强 strong (*love* = 0.7), 意志力 willpower (*love* = 0.5), 只不过 just (*anxiety* = 0.4, *expect* = 0.3), and 而已 yet (*anxiety* = 0.3, *expect* = 0.3). Obviously, most these human annotated values are in the range of our predicted results. It can prove the effectiveness of our Algorithm 1.

Moreover, in the dataset, this sentence is annotated by the emotional intensity: *love* = 0.5, *anxiety* = 0.4 and *expect* = 0.5. By further normalization process with linear regression, our calculated values are the same order as the result annotated by human.

In summary, for a new sentence, the process of predicting its multi-label emotion intensity is shown as Algorithm 2.

Algorithm 2: Predicting Multi-label Emotion Intensity for a New Sentence

Input: a new sentence s, emotion intensity matrix VE from Algorithm 1
Output: emotion intensity values of s
Description:
    1.   Find all sentiment word $w \in s$;
    2.   Construct sentiment word matrix VW' with all above $w \in s$;
    3.   For every $w \in VW'$
    4.   Achieve emotion intensity value of w from VE';
    5.   Construct emotion intensity value fuzzy matrix W;
    6.   Compute strongest emotional intensity object A as:

$$\overline{A} = \sum_{i=1}^{W} \frac{maxF(\overline{w}_i)}{w_i}, w_i \in W, i.e., \overline{A}(w_i) = max\{F(e_j)(\overline{w}_i)\big|e_j \in E\}$$

$$\underline{A} = \sum_{i=1}^{W} \frac{maxF(\underline{w}_i)}{w_i}, w_i \in W, i.e., \underline{A}(w_i) = max\{F(e_j)(\underline{w}_i)\big|e_j \in E\}$$

    7.   Calculate the improving fuzzy-rough upper approximation $\overline{F}(A)$ and fuzzy-rough lower approximation $\underline{F}(A)$ // see formula (12) and (13);
    8.   Calculate the choice value $m = \overline{F}(A)(e_i) + \underline{F}(A)(e_i)$, $e_i \in E$;
        // E is the universe of the emotion
    9.   Normalize m into [0, 1] with Linear Regression Algorithm;
    10. Return m as emotion intensity;

---

4 Experiments

For proving the validity and advantage, in this section we give comparison experiments.

4.1 Dataset and Evaluation Metric

In this section we use Quan's [11] Chinese blog dataset to evaluate our proposed method. The corpus contains 1,487 documents, with 11,953 paragraphs, 38,051 sentences, and

971,628 Chinese words. All sentiment words are labeled, and every sentence and sentiment word are annotated with eight basic kinds of emotions with intensities. An example is shown as Fig. 2 (which is actually the sentence of Table 2).

Although Quan's annotated dataset contains emotion intensity labels for each key sentiment word in blogs, our proposed fuzzy-rough set based method does not rely on the labels at the word level to predict the emotions in sentences, paragraphs and documents. In the training stage, all we need are the emotion intensity labels for the sentences in the training dataset and a sentiment lexicon. Therefore, we ignore the emotion labels of the sentiment words and their corresponding emotion intensities. We use 5 fold cross validation for the experiments. The dataset is divided into 5 parts on average, and each time there are four training sets and one test set.


```
<sentence S="我想，人其实内心都有顽强的意志力的，只不过有些人没释放出来而已。">
  <S_no>第7段第3句标注</S_no>
  <Segmented_S> 我想/n ，/w 人/n 其实/d 内心/n 都/d 有/v 顽强/a 的/u 意志力/n 的/u ，/w 只不过/d 有些/r 人/n 没/d 释放/v 出来/v 而已/y 。</Segmented_S>
  <S_Length>32</S_Length>
  <Keywords Anger="0" Anxiety="0" Expect="0" Hate="0" Joy="0" Love="0.7" Opinionholder="0" POS="a" Sorrow="0" Surprise="0" end="-1" position="10" start="-1">顽强
    </Keywords>
  <Keywords Anger="0" Anxiety="0" Expect="0" Hate="0" Joy="0" Love="0.5" Opinionholder="0" POS="n" Sorrow="0" Surprise="0" end="-1" position="13" start="-1">意志力
    </Keywords>
  <Keywords Anger="0" Anxiety="0.4" Expect="0.3" Hate="0" Joy="0" Love="0" Opinionholder="0" POS="d" Sorrow="0.0" Surprise="0" end="-1" position="18" start="-1">只不
    过</Keywords>
  <Keywords Anger="0" Anxiety="0.3" Expect="0.3" Hate="0" Joy="0" Love="0" Opinionholder="0" POS="y" Sorrow="0" Surprise="0" end="-1" position="29" start="-1">而已
    </Keywords>
  <noword end="-1" position="24" start="-1" modifier_word_length="2" modifier_word_position="25">没</noword>
  <Opinion_Fact>opinion</Opinion_Fact>
  <Polarity>积极</Polarity>
  <Opinion_holder end="0" position="0" start="0">我想</Opinion_holder>
  <Joy>0.0</Joy>
  <Hate>0.0</Hate>
  <Love>0.5</Love>
  <Sorrow>0.0</Sorrow>
  <Anxiety>0.4</Anxiety>
  <Surprise>0.0</Surprise>
  <Anger>0.0</Anger>
  <Expect>0.5</Expect>
```


Fig. 2. An annotated blog fragment in the training set

Since not all the traditional multi-label learning metrics could meet the need of the multi-label intensity prediction problem, in this experiment, we use four evaluation metrics [21], and some are revised versions of the formulas to fit our problem.

$$\textbf{Subset Accuracy} : subsetacc_s(h) = \frac{1}{p}\sum_{i=1}^{p} [|h(x_i = Y_i)|] \quad (14)$$

The subset accuracy evaluates the fraction of correctly classified examples, i.e. the predicted label set is identical to the ground-truth label set. Intuitively, subset accuracy can be regarded as a multi-label counter part of the traditional accuracy metric, and tends to be overly strict especially when the size of label space is large.

$$\textbf{Hamming Loss} : hloss_s(h) = \frac{1}{p}\sum_{i=1}^{p} \frac{1}{q}|h(x_i)\Delta Y_i| \quad (15)$$

where Δ stands for the symmetric difference between two sets. The hamming loss evaluates the fraction of misclassified instance-label pairs, i.e., a relevant label is missed or an irrelevant is predicted. This formula is suitable for the multi-label learning problem. In this paper, because we focus on multi-label intensities of multiple emotions, we need to revise the hamming loss metric. In the formula below, we try to measure the difference between the value we predict and the ground truth one:

$$hloss_s(h) = \frac{1}{p}\sum_{i=1}^{p}\frac{1}{q}|h(x_i) - \Delta Y_i| \tag{16}$$

One−error : $One-error_s(h) = \frac{1}{p}\sum_{i=1}^{p}\frac{1}{q}[[\arg\max_{y\in Y} f(x_i, y)] \notin Y_i] \tag{17}$

The one-error evaluates the fraction of examples whose top-ranked label is not in the relevant label set. In our paper, it means that the strongest emotion of the eight we predict is wrong. As we can see, the value of the One-error is the fewer, the better.

Average precision :

$$avgprec_s(h) = \frac{1}{p}\sum_{i=1}^{p}\frac{1}{Y_i}\sum_{y\in Y_i}\frac{|\{y'|rank_f(x, y') \leq rank_f(x_i, y), y' \in Y_i\}|}{rank_f(x_i, y)} \tag{18}$$

The average precision evaluates the average fraction of relevant labels ranked higher than a particular label $y \in Y_i$. In this paper, it means whether the descending order of the value of each emotion is right. This metric is the larger, the better.

4.2 Experiment Setup

As few related researches were proposed for multi-label emotion intensity, we will compare our method with the following methods that can be divided into two categories.

(1) **Using Word Emotion Intensity Labels**. This kind of methods means that we leverage the word emotion labels and their corresponding intensity in the training blog dataset to predict the emotions in the test set. Note that this will reduce the difficulty of the learning problem. These methods include:

- Fuzzy Union (FU for short). It is defined as: $(A \cup B)(x) = \max(A(x), B(x))$ for all $x \in X$. Taking the value of joy from Fig. 2 for example: Love (Fuzzy union) = max(顽强(love), 意志力(love)) = max (0.7, 0.5) = 0.7.
- Naïve Bayes (NB for short). We assumed that every emotion is independent from each other. Taking the value of *joy* from Fig. 2 for example: Love (Naïve Bayes) = P(顽强| *love*) * P(意志力| *love*) * P(*love*) = (6/12) * (5/17) * (12/25) = 0.07058.
- Multi-label Prediction(ML for short). In this method, we have already known the emotion labels of sentiment words which are annotated by people in the corpus. So we aggregate the labels of the words to predict the binary emotion labels (0/1) of the testing text.
- Our proposed fuzzy-rough set based method (FRS for short). In this case, we do not need the fuzzy relation equation to predict the emotions of the word and we can directly use Formulas (12) and (13) to predict the emotion intensities in the test set [15].

(2) **Ignoring Word Emotion Intensity Labels**. This kind of methods means that we ignore the labels and intensities of the words in the training dataset and only utilize the labels at the sentence level to train classifiers. These methods include:

- Regression Analysis (RA for short). We assume two vector spaces: B and W, then constructing an equation: $Y = BW + C$. The object function Y is the emotional value result of the training set, W is the sentiment words of the training set, B is the coefficient need to be learned, and C is the adjustment factor. We built 8 RA models for 8 kinds of emotions.
- Our proposed fuzzy relation equation and fuzzy-rough set based method (FRE-FRS for short). We leverage Algorithm 2 to predict the emotion intensities in the test set.

4.3   Experiment Results and Discussion

In the first experiment, we evaluate the label prediction accuracy with Subset Accuracy, which depends on fuzzy relation equation. Using our method, the percent of emotions figured out in the article is showed in Table 3.

Next we compare our multi-label method with other baseline algorithms mentioned in Sect. 4.2. Our method ignores the emotions and their intensity of sentiment words when training because of its unrealistic in most datasets. In this experiment, we also give the comparison results using the emotion intensity labels of the sentiment words during the training stage. The comparison methods include ML method that only using word emotions and sentence level labels, and FU and NB methods that using both word emotions and intensities with sentence level labels. The evaluation results at three textual levels are shown in Table 4.

Table 3. Label prediction accuracy (ignoring word emotions and intensities)

	Subset accuracy (RA)	Subset accuracy (FRE-FRS)
Document	0.59385	**0.68279**
Paragraph	0.62934	**0.76349**
Sentence	0.78421	**0.87326**

Hamming Loss is a measure of the value of general accuracy. Hamming Loss is the fewer, the better. So as we can see, in both using and ignoring word emotions with intensities, our method is better than others.

Table 4. Emotion label intensity analysis of Hamming Loss

	Using labeled word emotions and intensities				Ignoring labeled word emotions and intensities	
	ML	FU	NB	FRS	RA	FRE-FRS
Document	0.09845	0.06853	0.21445	**0.03929**	0.28650	**0.16316**
Paragraph	0.10659	0.05117	0.15133	**0.02598**	0.23486	**0.13468**
Sentence	0.11179	0.03927	0.10320	**0.01767**	0.16494	**0.10050**

One-error is a method to measure whether the maximum value we predict is one of the final results, and the results we get are shown in Table 5.

Table 5. Label intensity analysis of one-error

	Using labeled word emotions and intensities			Ignoring labeled word emotions and intensities	
	FU	NB	FRS	RA	FRE-FRS
Document	0.78649	0.76689	**0.04764**	0.79764	**0.39281**
Paragraph	0.58395	0.58807	**0.06903**	0.69162	**0.46293**
Sentence	0.84421	0.37633	**0.07377**	0.66771	**0.52043**

Average precision is widely used in many areas. According to the measure of Average precision [21], the results of the experiments at three levels of text are showed in Table 6.

Table 6. Label intensity analysis of average precision

	Using labeled word emotions and intensities			Ignoring labeled word emotions and intensities	
	FU	NB	FRS	RA	FRE-FRS
Document	0.67249	0.36975	**0.74374**	0.24649	**0.40659**
Paragraph	0.71452	0.44310	**0.77940**	0.33980	**0.46698**
Sentence	0.70985	0.58617	**0.95494**	0.36796	**0.58838**

According to Formulas (17) and (18), one-error is the fewer, the better. In contrast, average precision is the larger, the better. So we can see our algorithm is significantly better than the other methods.

In Fig. 2, we see that all sentiment words have been labeled in the training dataset, and the emotion intensity values of both every sentiment word and every sentence are given. In this case, the evaluation results should be better than the ones without word emotion intensity labels. The results in Tables 4, 5, and 6 have validated our assumption.

However, it is difficult to get such kind of training set. In most cases, labeling emotion intensity for all sentiments words and sentences is unrealistic. Our method can model and predict emotion intensity for a new sentence and get better results without emotion labels and corresponding intensities in training set. Especially, when comparing with the method of regression analysis that ignoring word emotions and intensities, our proposed algorithm shows obvious advantages.

Generally, we argue that the fuzzy logic is suitable for multi-label emotion intensity analysis, which means it is consistent with the logic of human language when expressing emotions. In most related bibliographies with fuzzy mathematics, the introduced examples always depicted intensity analysis of human feeling, such as the oldness degree of 40 years old, or the height degree of a 180 cm man. These questions all got good solutions by using fuzzy logic. As we can see in the Table 2, the emotional logic is not a simple summation. A sentence may have one or more key emotion. In our method, we compare the values of every kind of emotions that may be upper bound or lower bound to the largest value of the word (shown as A) which it belong to. Because the value of sentiment words in different text may not be identical, we do not put them in isolation, but make them interact between each other, which is achieved by the advanced fuzzy rough set. The logic of our proposed algorithm is straightforward. The upper approximation we calculated indicates the most optimistic closeness to the largest emotion intensity, so we take upper bound into this consideration. Similarly, the lower approximation is the most conservative closeness. Finally we combined these two results together to solve the problem. Our model is more suitable for smaller text units, such as sentences and paragraphs. That is because when the text unit is larger, the embedded human emotions are more complex and confused.

5   Related Work

Our work is about multi-label emotion analysis and emotion intensity calculating. For the calculating, we apply fuzzy relation equation and fuzzy-rough set method.

The sentiment analysis researches can be dating back to the early of this century. Pang and Lee [9] showed the effectiveness of classification of emotion by using machine learning methods. Costa et al. [2] verified a method to combine mining algorithms and software agent to build blogs based on sentiment applications. Zhang et al. [23] proved a model to extract emotional characteristics from reviews of products as a weakness finder. Huang et al. [7] proposed a sentiment space model to deal with sentiment classification task by using the semantic information.

Multi-label emotion analysis can be seemed as multi-label learning problem. Fürnkranz et al. [6] proposed a problem transformation method which is called "Calibrated Label Ranking" could draw on the advantages of the pairwise preference learning and the conventional relevance classification technique, in which a separate classifier could be trained for distinguishing whether a label was relevant or not. Boutell et al. [1] built another common problem transformation algorithm called "Binary Relevance". It could transform the multi-label classification problem into the binary classification problem. Some scholars also proposed many other algorithms and methods in the multi-label learning area. This kind of methods makes the traditional supervised machine learning algorithms more suitable to deal with the multi-label problem. Elisseeff and Weston [4] improved the kernel learning algorithm SVM to solve the multi-label data problem.

Although there are a lot of research for multi-label learning and sentiment analysis, little work is done for the emotion intensities in one social media text post. In this paper, our work is regarding the problem as uncertain emotion classification and solving the problem.

Since the fuzzy-rough theory was proposed during the late of 20th century [3], it has been widely used in many areas, especially uncertain classification problems. Wang et al. [16] proposed a new uncertain classification algorithm which combined fuzzy-rough set with decision tree. Shi and Gong [12] built a model for uncertainty characterization called covering-based rough sets by using the advanced fuzzy-rough set. Xiao et al. [19] proposed a method to classify and predict whether the listed companies have financial distress based on the combination of fuzzy-rough set and D-S evidence theory. Sun and Ma [13] gave an approach to decision making problem by combining the soft set with fuzzy-rough theory.

Although fuzzy-rough theory has no existing literature on uncertain emotion classification for social media, some researchers such as Vincenzo and Sabrina [14] has already realized fuzzy logic is associated and consistent with human emotions.

6 Conclusion and Future Work

In this paper, we proposed a new way to solve the multi-label and fine-grained emotion intensity analysis problem. For this particular problem, we used a fuzzy relation equation and an improved fuzzy-rough set theory to model and predict emotion intensity of a sentence, paragraph, and document.

In the future, we would like to build the model of multi-label emotion and intensity analysis on microblog, website reviews, or other social media. The role of adverbs and negative words can be further taken into consideration, which can further improve the performance of multi-label emotion intensity analysis.

Acknowledgments. Project supported by National Natural Science Foundation of China (61370074, 61402091), the Fundamental Research Funds for the Central Universities of China under Grant N140404012.

References

1. Boutell, M., Luo, J., Shen, X., Christopher, M.: Brown: learning multi-label scene classification. Pattern Recogn. (PR) **37**(9), 1757–1771 (2004)
2. Costa, E., Ferreira, R., Brito, P., et al.: A framework for building web mining applications in the world of blogs: a case study in product sentiment analysis. Expert Syst. Appl. (ESWA) **39**(5), 4813–4834 (2012)
3. Dubois, D., Prade, H.: Rough fuzzy sets and fuzzy rough sets. Int. J. Gen. Syst. **17**(2–3), 191–209 (1990)
4. Elisseeff, A., Weston, J.: A kernel method for multi-labelled classification. In: NIPS 2001, pp. 681–687
5. Ernest, C., Józef, D., Witold, P.: Fuzzy relation equations on a finite set. Fuzzy Sets Syst. **7**, 89–101 (1982)
6. Fürnkranz, J., Hüllermeier, E., Mencía, E., Brinker, K.: Multilabel classification via calibrated label ranking. Mach. Learn. (ML) **73**(2), 133–153 (2008)
7. Huang, S., You, J., Zhang, H., Zhou, W.: Sentiment analysis of Chinese micro-blog using semantic sentiment space model. In: Conference: Computer Science and Network Technology (ICCSNT), pp. 1443–1447 (2012)
8. Maji, P., Biswas, R., Roy, A.: Soft set theory. Comput. Math. Appl. **45**(4–5), 555–562 (2003)

9. Pang, B., Lee, L., Vaithyanathan, S.: Thumbs up? Sentiment classification using machine learning techniques. CoRR cs.CL/0205070 (2002)

10. Pawlak, Z.: Rough sets. Int. J. Parallel Prog. (IJPP) **11**(5), 341–356 (1982)

11. Quan, C., Ren, F.: A blog emotion corpus for emotional expression analysis in Chinese. Comput. Speech Lang. (CSL) **24**(4), 726–749 (2010)

12. Shi, Z., Gong, Z.: The further investigation of covering-based rough sets: uncertainty characterization, similarity measure and generalized models. Inf. Sci. (ISCI) **180**(19), 3745–3763 (2010)

13. Sun, B., Ma, W.: Soft fuzzy rough sets and its application in decision making. Artif. Intell. Rev. (AIR) **41**(1), 67–80 (2014)

14. Vincenzo, L., Sabrina, S.: Knowl.-Based Syst. **58**, 75–85 (2014)

15. Wang, C., Feng, S., Wang, D., Zhang, Y.: Fuzzy-rough set based multi-labeled emotion intensity analysis for sentence, paragraph and document. In: Li, J., Ji, H., Zhao, D., Feng, Y. (eds.) NLPCC 2015. LNCS, vol. 9362, pp. 444–452. Springer, Heidelberg (2015). doi:10.1007/978-3-319-25207-0_41

16. Wang, X., Zhai, J., Lu, S.: Induction of multiple fuzzy decision trees based on rough set technique. Inf. Sci. (ISCI) **178**(16), 3188–3202 (2008)

17. Wu, W., Mi, J., Zhang, W.: Generalized fuzzy rough sets. Inf. Sci. (ISCI) **151**, 263–282 (2003)

18. Wu, W., Zhang, W.: Constructive and axiomatic approaches of fuzzy approximation operators. Inf. Sci. (ISCI) **159**(2), 233–254 (2004)

19. Xiao, Z., Yang, X., Pang, Y., Dang, X.: The prediction for listed companies' financial distress by using multiple prediction methods with rough set and Dempster-Shafer evidence theory. Knowl.-Based Syst. (KBS) **26**, 196–206 (2012)

20. Zadeh, L.: Fuzzy sets. Inf. Control (IANDC) **8**(3), 338–353 (1965)

21. Zhang, M., Zhou, Z.: A review on multi-label learning algorithms. IEEE Trans. Knowl. Data Eng. (TKDE) **26**(8), 1819–1837 (2014)

22. Zhang, W., Liang, Y., Xu, P.: Uncertainty Reasoning Based on Inclusion Degree. Tsinghua University Press, Beijing (2007)

23. Zhang, W., Xu, H., Wan, W.: Weakness finder: find product weakness from Chinese reviews by using aspects based sentiment analysis. Expert Syst. Appl. (ESWA) **39**(11), 10283–10291 (2012)

Features Extraction Based on Neural Network
for Cross-Domain Sentiment Classification

Endong Zhu[1], Guoyan Huang[1], Biyun Mo[1], and Qingyuan Wu[2(✉)]

[1] Sun Yat-sen University, Guangzhou, China
{zhuend,huanggy7,moby5}@mail2.sysu.edu.cn
[2] Beijing Normal University, Zhuhai, China
wuqingyuan@bnuz.edu.cn

Abstract. Sentiment analysis is important to develop marketing strategies, enhance sales and optimize supply chain for electronic commerce. Many supervised and unsupervised algorithms have been applied to build the sentiment analysis model, which assume that the distributions of the labeled and unlabeled data are identical. In this paper, we aim to deal with the issue of a classifier trained for use in one domain might not perform as well in a different one, especially when the distribution of the labeled data is different with that of the unlabeled data. To tackle this problem, we incorporate feature extraction methods into the neural network model for cross-domain sentiment classification. These methods are applied to simplify the structure of the neural network and improve the accuracy. Experiments on two real-world datasets validate the effectiveness of our methods for cross-domain sentiment classification.

Keywords: Cross-domain sentiment classification · Neural network · Feature extraction · Transfer learning

1 Introduction

Sentiment analysis aims at classifying sentimental data into polarities (e.g., positive and negative) [1] or multiple emotional labels (e.g., happiness, sadness and surprise) [2] primarily. Traditional classification algorithms have been used to train sentiment classifiers [3]. However, these approaches are domain dependent. A classifier trained from one domain (source domain) might not work well when directly applied to another domain (target domain) [4]. The reason is that different domains focus on different sets of topics, which leads to various distributions of features.

In light of these considerations, we propose effective feature extraction methods based on the neural network model for cross-domain sentiment classification. Neural networks are widely employed in natural language processing, but their training process are time-consuming [5] and the risk of overfitting may be high when the data size is large. One strategy to tackle this problem is to apply

---

The authors contributed equally to this work.

© Springer International Publishing Switzerland 2016
H. Gao et al. (Eds.): DASFAA 2016 Workshops, LNCS 9645, pp. 81–88, 2016.
DOI: 10.1007/978-3-319-32055-7_7

pre-processing on data based on feature extraction methods, which simplify the structure of the neural network, reduce the cost of training process and further improve the accuracy of sentiment classification.

The rest of the paper is organized as follows. The next section reviews some related work. We present our feature extraction methods based on the neural network model for cross-domain sentiment classification in Sect. 3. The datasets and experimental results are discussed in Sect. 4. The last section draws conclusions and points out our future work.

2   Related Work

2.1   Features Extraction

Singular value decomposition (SVD) is a matrix decomposition method in linear algebra, and it has been widely used in the extraction of features. Francesca et al. [6] used SVD to select important features for their probabilistic model. Harb et al. [7] introduced an algorithm *SLRS* based on SVD, which also applied SVD to the document matrix. SVD method can extract important features and reduce the computing complexity by filtering many noise features, i.e., reducing the dimension of the document matrix.

2.2   Cross-Domain Sentiment Classification

Sentiment classification has been extensively studied for reviews [3] and news articles [8–10]. However, these methods focused on detecting emotions specific to one context primarily.

To tackle the issue of different feature distributions in cross-domain datasets, Pan et al. [1] proposed to divide words into two categories for sentiment classification. The first category contains domain-independent (DI) words, and the second one contains domain-specific (DS) words. DI words are those which occur frequently in both the source domain and the target domain, and DS words are those which occur frequently only in one specific domain. Dai et al. [11] introduced a classification model based on co-clustering by the following two processes. Firstly, in-domain documents were used to generate the class structure and propagate the label information. Secondly, co-clustering [12] was applied to out-of-domain data to obtain document and word clusters. Rao [13] developed a contextual sentiment topic model to distinguish context-independent topics from both a background theme and a contextual theme. The limitation of these studies, however, was that they were designed for normal documents which contain sufficient words. In this work, we focus on cross-domain sentiment classification over both normal and short documents.

3   Features Extraction Algorithms

In this section, we describe four effective methods to extract features and apply them to neural networks for cross-domain sentiment classification.

3.1   Naive Methods

The naive methods extract features based on their occurrences in the source domain and the target domain, which are totally unsupervised.

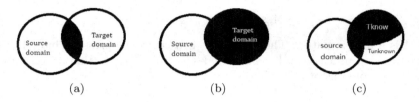

$$\text{(a)} \qquad\qquad \text{(b)} \qquad\qquad \text{(c)}$$

Fig. 1. Algorithm description: (a) words exist in both the source domain and the target domain, (b) words only exist in the target domain, (c) word selection based on limited labeled data in the target domain

Two simple ways are used to extract important features for cross-domain sentiment classification. The first one selects words which exist in both the source domain and the target domain, as shown in Fig. 1(a). The second one selects words which only exist in the target domain, as shown in Fig. 1(b).

3.2   Using Limited Labeled Data in Target Domain

This method extracts features by exploiting a small part of labeled data in the target domain, as shown in Fig. 1(c). The objective is to use as less labeled data in the target domain as possible for training.

We divide the target domain data into two parts with different proportions: the labeled part T_{know} in the target domain is used together with all labeled data in the source domain for training; and the rest unlabeled part T_{unknow} in the target domain is used for testing. Two kinds of features are selected to cross-domain sentiment classification: (1) all features in T_{know}. The reason is that features in T_{know} may be more important than those in the source domain; and (2) features of the source domain which occur in the target domain.

3.3   SVD-Based Method

To filter noise features, this method applies SVD to the $m \times n$ document matrix M, as follows:

$$M_{m\times n} = U_{m\times m}\Sigma_{m\times n}V^T_{n\times n},\qquad(1)$$

where U and V are orthogonal matrices, and Σ is a diagonal matrix where the diagonal elements are eigenvalues of the matrix M. Since the eigenvalues in Σ are sorted in the decreasing order, we can use r large eigenvalues to represent the matrix M approximately. According to the properties of orthogonal matrices, we obtain a new matrix A. The mathematical processes are described as follows:

$$M_{m \times n} \approx \bar{M}_{m \times n} = U'_{m \times r} \Sigma_{r \times r} V^T_{r \times n}, \tag{2}$$

$$M_{m \times n} V'_{n \times r} \approx U'_{m \times r} \Sigma'_{r \times r} \left(V'^T_{r \times n} V'_{n \times r} \right) = U'_{m \times r} \Sigma'_{r \times r}, \tag{3}$$

$$A_{m \times r} = M_{m \times n} V'_{n \times r}, \tag{4}$$

where A is a $m \times r$ matrix. The matrix V' is used to compress M into A, and this compression process can be considered as a feature extraction process if we represent the columns of M as features.

3.4   Feature Alignment Method

Sentiment classification is very domain-specific because users can use domain-specific words to convey emotions in different contexts. For instance, the word "compact" is often used to express a positive attitude in the domain of electronic product reviews. However, the word "interesting" expresses a primarily positive attitude in the books domain. Inspired by the previous work [1], we also evaluate the feature alignment method on neural networks for cross-domain sentiment classification. Algorithm 1 presents the process of selecting domain-independent (DI) and domain-specific (DS) words, where ε is a threshold. Note that all stop words have been deleted beforehand.

> **for** *each word w in the set of all words* **do**
> > Compute the frequency f_s in the source domain ;
> > Compute the frequency f_t in the target domain ;
> > **if** $f_s > \varepsilon$ *and* $f_t > \varepsilon$ **then**
> > > | Add w to the set of DI words;
> > > **end**
> > **else**
> > > | Add w to the set of DS words;
> > > **end**
> **end**

Algorithm 1. Selecting DI/DS words

In our neural network models, we use all DI words as the first part of features directly. To improve the generalization of models, we also apply the clustering algorithm to DS words and use the clusters of DS words as the second part of features. Algorithm 2 presents the process of clustering DS words. Firstly, all DI words are used as the "bridge" to construct an adjacent matrix *wordMat*. Secondly, the dimension-reduction method is conducted on the high-dimensional and sparse adjacent matrix. Finally, the unsupervised method such as spectral clustering and K-means is employed to generate the clusters of DS words.

for *each word w_1 in the set of DS words* **do**

    for *each word w_2 in the set of DI words* **do**

        if *w_1 and w_2 occur in the same document* **then**

           | $wordMat[w_1][w_2] \mathrel{+}= 1$;

        end

    end

end

Conduct dimension-reduction on *wordMat* ;

Conduct clustering on *wordMat* ;

Algorithm 2. Clustering DS words

4   Experiments

To evaluate our feature extraction methods, experiments are designed to cross-domain sentiment classification via neural networks. The neural network using all words as features is implemented for comparison.

4.1   Datasets

SemEval is an English dataset used in the 14th task of the 4th International Workshop on Semantic Evaluations (SemEval-2007) [14]. It contains news headlines extracted from Google news, CNN, and many others. We use the development set (i.e., the source domain) for training and the testing set (i.e., the target domain) for evaluation. The news headlines in the source domain and the target domain are very different in terms of words and topics.

SinaNews is a Chinese dataset which contains news articles extracted from Sina website [13]. We use a total of 2,343 documents published from January to February 2012 for training, and the rest 2,228 documents published from March to April 2012 for testing. Due to that the publication dates vary, the feature distributions in the source and target domains are also quite different.

4.2   Experimental Design

In our experiments, we use the back-propagation neural network (BP neural network) [15] as the classification model to compare the performance of different feature extraction methods. The reasons of employing BP neural network to evaluate the effectiveness of our algorithms are summarized as follows: First, we can use the control rules by non-linear mapping and self-learning of neural networks. Second, feature extraction is quite important for BP neural network to reduce the number of units in the input layer.

The structure of BP neural network is one input layer, one hidden layer and one output layer for simpleness. The number of units in the input layer equals the amount of features, and the number of units in the output layer equals the amount of emotion labels. We use the following formula to determine the number units in the hidden layer [16]:

$$N_{hidden} = \sqrt{N_{input} + N_{output}} + m \quad (-5 \leq m \leq 5), \tag{5}$$

where the value of m is set to 0.

The accuracy at top 1 is employed as the indicator of performance [9], which is essentially the micro-averaged F1 that equally weights precision and recall.

4.3   Results and Analysis

The accuracies of BP neural network using all words as features are 37.0 % and 53.59 % on *SemEval* and *SinaNews*, respectively. This method is used as the baseline model because it outperforms most state-of-the-art models in the previous studies on these two datasets [8, 13].

Table 1. Results of naive methods

Dataset	# of features	Accuracy
SemEval	535	35.5 %
	2380	38.2 %
SinaNews	19774	57.18 %
	21310	56.95 %

The results of naive methods are presented in Table 1. Compared to the 2,749 and 22,946 features in the baseline model on *SemEval* and *SinaNews*, the numbers of features are reduced by the naive methods as presented in Fig. 1(a) and (b). In terms of the accuracy, the performance of models using the naive methods to extract features is worse on *SemEval* when using too few features, and better on *SinaNews* than the baseline model. The results indicate that the effect of the naive methods relies on the scale of the dataset.

The accuracies of using several proportions of labeled data in the target domain for training are shown in Table 2, from which we can observe that the accuracy increases as more labeled data in the target domain are available. However, the gap between the source and target domains may have a negative influence on the performance when using too many features.

Table 2. Accuracies of using limited labeled data in the target domain

Proportion	*SemEval*	*SinaNews*
50 %	42.50 %	57.99 %
25 %	41.60 %	57.49 %
12.5 %	44.29 %	57.48 %
6.25 %	41.47 %	57.01 %
3.125 %	39.93 %	57.29 %

Table 3. Results of SVD-based method

Dataset	# of features	Accuracy
SemEval	2749	37.0 %
	1000	37.9 %
	800	37.8 %
	500	37.1 %
SinaNews	5000	56.86 %
	2000	56.91 %
	1000	56.82 %
	500	57.54 %

Table 4. Results of feature alignment method

Dataset	# of DI words	# of DS word clusters	Accuracy
SemEval	526	50	38.80 %
	526	100	38.40 %
	526	150	39.30 %
	526	200	40.10 %
SinaNews	2168	500	52.69 %
	2168	1000	58.25 %
	2168	1500	58.48 %
	2168	2000	59.57 %

The result of SVD-based method is presented in Table 3, which indicates that the accuracy depends on the degree of compressing the document matrix.

The results of cross-domain sentiment classification based on the feature alignment method are shown in Table 4, from which we can observe that the performance improves as the number of DS word clusters increases.

5   Conclusion

In this paper, we incorporated four effective features extraction methods into BP neural network to reduce the dimension and also improve the accuracy of cross-domain sentiment classification. Extensive evaluations using two real-world datasets validate the effectiveness of the proposed features extraction algorithms. As a trade-off strategy, a good dimension-reduction method should reduce the dimension of features as much as possible with the guarantee of accuracies.

For future work, we plan to develop methods of choosing the optimal parameters for our features extraction algorithms, in addition to combine other methods to tackle the problems in cross-domain sentiment classification.

Acknowledgements. This research has been substantially supported by a grant from the Soft Science Research Project of Guangdong Province (Grant No. 2014A030304013).

References

1. Pan, S.J., Ni, X., Sun, J.-T., Yang, Q., Chen, Z.: Cross-domain sentiment classification via spectral feature alignment. In: Proceedings of the 19th International Conference on World Wide Web, pp. 751–760 (2010)
2. Zhang, Y., Zhang, N., Si, L., Lu, Y., Wang, Q., Yuan, X.: Cross-domain and cross-category emotion tagging for comments of online news. In: Proceedings of the 37th International ACM SIGIR Conference, pp. 627–636 (2014)
3. Pang, B., Lee, L., Vaithyanathan, S.: Thumbs up? sentiment classification using machine learning techniques. In: Proceedings of the Conference on Empirical Methods in Natural Language Processing, pp. 79–86 (2002)
4. Nigam, K., McCallum, A.K., Thrun, S., Mitchell, T.: Text classification from labeled and unlabeled documents using em. Mach. Learn. **39**(2–3), 103–134 (2000)
5. Morchid, M., Dufour, R., Linares, G.: Topic-space based setup of a neural network for theme identification of highly imperfect transcriptions. In: IEEE Automatic Speech Recognition and Understanding Workshopp (2015)
6. Francesca, F., Zanzotto, F.M.: SVD feature selection for probabilistic taxonomy learning. In: Proceedings of the Workshop on Geometrical Models of Natural Language Semantics, pp. 66–73 (2009)
7. Dasgupta, A., Drineas, P., Harb, B., Josifovski, V., Mahoney. M.W.: Feature selection methods for text classification. In: Proceedings of the 13th ACM SIGKDD International Conference, pp. 230–239 (2007)
8. Rao, Y., Lei, J., Liu, W., Li, Q., Chen, M.: Building emotional dictionary for sentiment analysis of online news. World Wide Web J. **17**, 723–742 (2014)
9. Rao, Y., Li, Q., Liu, W., Wu, Q., Quan, X.: Affective topic model for social emotion detection. Neural Netw. **58**, 29–37 (2014)
10. Rao, Y., Li, Q., Mao, X., Liu, W.: Sentiment topic models for social emotion mining. Inf. Sci. **266**, 90–100 (2014)
11. Dai, W., Xue, G.-R., Yang, Q., Yu, Y.: Co-clustering based classification for out-of-domain documents. In: Proceedings of the 13th ACM SIGKDD International Conference, pp. 210–219 (2007)
12. Dhillon, I.S., Mallela, S., Modha, D.S.: Information-theoretic co-clustering. In: Proceedings of the 9th ACM SIGKDD International Conference, pp. 89–98 (2003)
13. Rao, Y.: Contextual sentiment topic model for adaptive social emotion classification. IEEE Intell. Syst. **31**(1), 41–47 (2016)
14. Strapparava, C., Mihalcea, R.: Semeval- task 14: affective text. In: Proceedings of the 4th International Workshop on Semantic Evaluations, pp. 70–74 (2007)
15. Hecht-Nielsen, R.: Theory of the backpropagation neural network. In: International Joint Conference on Neural Networks, pp. 593–605 (1989)
16. Kurkova, V., Kainen, P.C., Kreinovich, V.: Estimates of the number of hidden units and variation with respect to half-spaces. Neural Netw. **10**(6), 1061–1068 (1997)

Personalized Medical Reading Recommendation: Deep Semantic Approach

Tatiana Erekhinskaya[✉], Mithun Balakrishna,
Marta Tatu, and Dan Moldovan

Lymba Corporation, 901 Waterfall Way, Bldg 5, Richardson, TX 75080, USA
{tatiana,mithun,mtatu,moldovan}@lymba.com
http://www.lymba.com

Abstract. Therapists are faced with the overwhelming task of identifying, reading, and incorporating new information from a vast and fast growing volume of publications into their daily clinical decisions. In this paper, we propose a system that will semantically analyze patient records and medical articles, perform medical domain specific inference to extract knowledge profiles, and finally recommend publications that best match with a patient's health profile. We present specific knowledge extraction and matching details, examples, and results from the mental health domain.

Keywords: Deep semantic extraction · Medical information retrieval · Diagnostic inference

1 Introduction

With new scientific findings and studies being reported every day, a therapist is faced with the overwhelming task of identifying, reading, and incorporating new information from a vast volume of publications into their daily clinical decisions. Arming the therapist with the most current literature would help the therapist make the best clinical decisions for their patients throughout the course of diagnosis and treatment.

This paper addresses the task of recommending relevant professional reading for doctors based on their current patient cases. In comparison to standard Information Retrieval task, this task has several complications that make keyword-based search inefficient. First, the query is not a short set of keywords, but a set of relatively large text files, which requires keyword importance evaluation and high performance. Second, the language of patient records is different from the language of papers, which makes keyword matching insufficient. Finally, some publications are more research oriented and do not address therapist needs directly, for example discussing experiments on rats, statistic analysis on population, etc. - the knowledge that does not have immediate clinical implications.

This paper presents a novel NLP-based approach to compute relevance of the candidate papers to the set of cases a therapist has on hand based on deep semantic processing of publications and electronic health records (EHR).

© Springer International Publishing Switzerland 2016
H. Gao et al. (Eds.): DASFAA 2016 Workshops, LNCS 9645, pp. 89–97, 2016.
DOI: 10.1007/978-3-319-32055-7_8

Both EHR and publications are converted into semantic profile. The relevance is computed based on the profiles matching. In addition to relevance, the system computes novelty score to measure how much new knowledge is provided by a candidate publication.

2   Related Work

2.1   Concept Extraction and Expansion

The problem of long queries in medical domain brings the task of extraction important concepts and assigning corresponding importance weight in a ranking formula. MedSearch system [10] was designed to assist ordinary Internet users to search for medical information by accepting queries of extended length. The system rewrites long queries by selectively dropping unimportant terms based on tf-idf scores.

Zheng and Yu [15] also targeted patients as end users. They trained LDA topic models to identify prominent topics. Queries are generated from n-grams, taking the top 5 phrases as queries from the topics that has a combined probability of over 80 %. The authors also employed Conditional Random Fields (CRF) model to identify key concepts, which are most in need of explanation by external education materials. The authors have shown that using full EHR notes is ineffective at retrieving relevant education materials.

Query expansion is a well-known technique in traditional Information Retrieval [13]. Liu and Chu proposed a knowledge-based query expansion technique to support scenario-specific retrieval [9], when query contains general terms like *treatment* that need to be matched to specific terms like *chemotherapy* in the document. The method utilized co-occurrence thesaurus, UMLS and vector space model.

2.2   Usage of Dependencies

The key concepts in the query and in the documents are forming structures that are important for relevance scoring. Choi et al. [7] uses implicit dependencies with the standardized medical concepts to favor the documents that preserve those implicit dependencies to improve ranking performance. The implicit dependence features were harvested from the original query using MetaMap [2]. These semantic concept-based dependence features were incorporated into a semantic concept-enriched dependence model (SCDM).

2.3   Negative Findings

Negative findings in patient records are expressed by means of negation or by using terms which contain negative qualifiers. From IR point of view, negative findings should be recognized and treated in a special way. Namely, EHR and relevant publications should agree on whether the finding is negative, or the negative finding in EHR might be not mentioned in the publication.

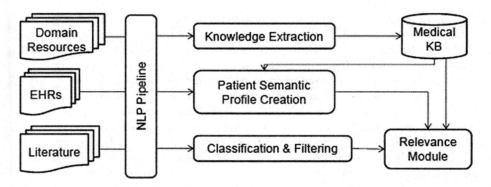

Fig. 1. Dataflow for medical literature recommendation approach.

Ceusters et al. [6] classified these phenomena in terms of the various top-level categories and relations defined in Basic Formal Ontology [8] and taking into account the role of negation in the corresponding descriptions. The authors introduced the *lacks*-relation that allowed them to represent nearly all negative findings that occur in patient charts.

3 Proposed Approach Overview

3.1 Problem Formulation

Given a set of patient cases $\{P_1, P_2, ..., P_k\}$ and past knowledge of the therapist K, the literature recommendation module will return a ranked list $R = [r_1, r_2, ..., r_n]$ of publications with the links between suggested publications and original patient cases $r_i \rightarrow p_j$. Past knowledge K consists of medical profiles of past cases and previously read papers.

The relevance should be computed taking into account the following therapist information needs: (1) diagnosis methods; (2) new, more efficient treatments for known diseases; (3) adverse effects of prescribed treatment; (4) potential risk factors for new health problems.

As the therapist updates a patient's file and adds case notes, the semantic model for the patient will continue to update such that relevant reference articles are presented that may justify the current diagnosis.

The literature recommendation to the clinician can be presented directly at the point of care, as they type in session notes during an ongoing clinical interview as well as in an offline, proactive manner.

3.2 Dataflow Overview

Figure 1 shows the dataflow of the proposed approach. First, the patient records are processed via the NLP Pipeline. The key task is to extract medical concepts: symptoms, diseases, administered treatment, medication, life events, etc.

Then symptoms are normalized, for example, *eating without control* would be matched to *binge eating*. This information about the patient is put into Semantic Patient Profile. Then, the inference module suggests possible diagnosis with some confidence score. This diagnosis can be used as a suggestion for doctors in the beginning of patient care process, as an alternative consideration for doctor-provided diagnosis, and as additional strong keyword for retrieval in case no diagnosis was provided by a therapist. The diagnosis and standardized symptoms are taken from Medical Knowledge Base, that was created based on existing resource like Mesh [1] and SnoMED [14] and extracted from textbooks and manuals. The publications are processed with NLP tools and semantically indexed. In addition, the publications are classified according to the therapist needs. There is a boolean Naive Bayes classifier for each need. The publications that do not match any of the needs are filtered out.

4   NLP Pipeline

The first step of deep semantic processing of medical text is the NLP Processing that spans the lexical, syntactic, and semantic layers of knowledge extraction from text.

Our concept detection methods range from the detection of simple nominal and verbal concepts to more complex named entity and phrasal concepts. This hybrid approach to *concept extraction* makes use of machine learning classifiers, cascade of finite-state automatons, and lexicons to label more than 80 types of concept classes. The concept categories with examples are shown in Table 1. Note, that the categories can be expressed not only with nouns which are easy to extract from ontologies, but with other part of speech words as well, also a concept can have nested concepts in it, as the ones in the bottom of the table.

Table 1. Partial list of recognized medical concept types.

NE type	Examples
Symptom	headache, depressive mood, behavior change, gained weight
Sign	blood pressure, temperature
Diseases & disorders	Aarskog Ose Pande syndrome, obese
Temporal qualities	acute, chronic, episodic, history, x 2 weeks
Body part	leg, mid-calf, cardiac, dental
Life event/condition	toxic exposure, stress, traveling abroad
Addictions	smoker, 1 glass of wine every day
Medical examination	ophthalmoscopy, lung exam, screening
Medical procedure	UPPP, reconstructive surgery, organ transplant
Medication	NSAID, angiotensin-converting enzyme inhibitor
Allergy and sensitivity	hay fever, cryesthesia, anaphylaxis
Medical problem	abnormal lung exam, abdominal pain with palpation

The extracted concepts are normalized using standard formulations in existing knowledge bases via semantic matching. For example, *lost 5 pounds* in EHR is matched to *weight loss* in Medical Subject Headings.

Semantic relations allow the linking of important concepts in a correct way. For example, they help connect temporal information and a medical problem, determine whether a medical problem is related to a patient or belongs to the family history, etc. Co-reference resolution module extracts co-reference chain information to help separate patient specific symptoms and features from other mentions in the patient data.

We define *semantic relations* as abstractions of underlying relations between concepts that occur within a word, between words, between phrases, and between sentences [11]. Semantic relations provide connectivity between concepts, which makes their extraction from text essential for the ultimate goal of machine text understanding. We use a fixed set of 26 relationships, which strike a good balance between too specific and too general [11]. They include the thematic roles proposed by Fillmore and others, and the semantic roles in PropBank, while also incorporating relationships outside of the verb-argument settings, representing semantic connectivity for all content words.

The important module in the pipeline is negation recognition. Negations are used to reverse polarity of a statement. In medical domain it can mean a health issue (e.g. *absent tonsil*) or absence of signs/symptoms (negative findings), which is critically important for providing diagnosis and literature recommendation. The negation module determines the scope and focus of negations and incorporate negations into semantic representation [4,12]. Negations can be expressed with auxilary words like not, without, or with content word, (e.g. denies, stop, cancel, never, absence, absent, etc.)

5   Medical Knowledge Base and Diagnostic Inference

In order to support diagnostic inference, we designed a specific knowledge extraction module that extracts diagnostic requirements such as the diagnostic criteria, diagnostic features, development and course, and the differential diagnosis for each disease described in literature. For example in Reactive Attachment Disorder, eight criteria must be evaluated, a subset is shown in Table 2.

The NLP tools read the detailed descriptions of each disorder and translate them into a graph of concepts and semantic relations. The disorder is represented as a seed node with customized semantic connections to: (1) a list of typical signs and symptoms, (2) any related medical conditions, (3) familial and culture predispositions, (4) typical faith system, (5) IQ, (6) gender, (7) age, (8) any chemical use, (9) psychosocial factors, (10) a detailed representation of the critical criteria and (11) an encoding of the differential diagnosis.

Figure 2 presents a partial view of the semantic representation that we designed to encode the diagnostic requirements such as the diagnostic criteria, diagnostic features, development and course, and the differential diagnosis. We represent the diagnostic information as structured relations with normalized

Table 2. Subset of criteria for Reactive Attachment Disorder.

A	A consistent pattern of emotionally withdrawn behavior toward adult care giver manifested by both of the following: (1) child rarely seeks comfort when distressed, (2) child minimally responds to comfort when distressed
D	The care in Criterion C is presumed to be responsible for the disturbed behavior in Criterion A.
E	The criteria are not met for autism spectrum disorder
F	The disturbance is evident before age 5 years

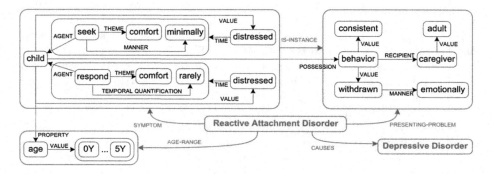

Fig. 2. Semantic representation for a reactive attachment disorder diagnostic criterion.

values for reasoning. Figure 2 also shows the inferred health-specific semantic relations (e.g. AGE-RANGE, SYMPTOM, PRESENTING-PROBLEM, etc.) that were derived using Semantic Calculus [5], a tool for combining the 26 core semantic relations into domain specific relations.

The diagnostic inference module uses this representation to match patient profile and diagnostic criteria. The rest of the section explains the inference on the example of Reactive Attachment Disorder's criteria from Table 2. Criterion A requires that both (1) and (2) be present. For this reason, we encoded inclusion/exclusion, and minimal/maximal semantics for the critical criteria. Criterion D seeks a causation relationship between Criterion A and Criterion C. If any of the factors are true for Criterion C, the diagnostic module checks for a causation relationship with the factors in A. Criterion E introduces the complexity of negation as well as the requirement to assess autism spectrum disorder. To resolve this issue, the system navigates to autism spectrum disorder, evaluates the criteria, and then proceeds with the diagnostic assessment. Finally, Criterion F expects a temporal interval attached to the disturbance event. The system interprets the disturbance as the compilation of the signs and symptoms in order to perform temporal reasoning to decide if they occurred before age 5.

6   Relevance Computation

The relevance module matches publication profiles to semantic patients' profiles and identifies articles that bring new information to the therapist outside the body of knowledge they already have consulted.

Profile comparison algorithm computes the semantic overlap between a patient file and an article by weighed summation of matches for concepts and relations:

$$R = \sum_{i \in concepts(SPS)} w^c_i m^c_i + \sum_{i \in relations(SPS)} w^r_i m^r_i. \qquad (1)$$

In this equation, m denotes match between concept/relation from the semantic patient profile to the publication profile, range from 0 (no match) to 1 (full match) with similarity score in between. Two semantic relations are said to match if their domain and range concepts are the same. Weight w denotes importance. A concept's importance weight is based on its tf-idf score [3] and its linguistic properties. Inferred concepts (e.g. diagnosis) are scored lower than the original ones. Importance weight for relations is based on the domain/range concept importance score and its thematic properties such as its relation type and connection strength.

Figure 3 shows the concept and relation match for the patient file and the article discussing treatment for Reactive Assessment Disorder. The gray concepts show the semantic overlap used to determine relevance.

The system also measures the degree of novelty of the article with respect to past knowledge by identifying the scientific nuggets in the article that provide new information. While article relevance is derived from matching semantic profiles of the patient file and article, the novelty is derived from matching the past knowledge with the article profile. The novelty score is then computed as the semantic difference between the candidate article model and the patient file

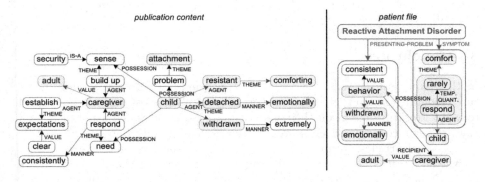

Fig. 3. Semantic representation for a reactive attachment disorder diagnostic criterion.

model augmented with models from previously suggested articles. The information conveyed by an article that could not be mapped to the knowledge stored in the patient's semantic profile is considered to be novel. The system computes the novelty score for an article using the following features: (1) weights new concepts higher than new relations that link known concepts, and (2) prefers explicitly stated knowledge to entailed knowledge from the domain ontology. The overall novelty of a scientific article is computed as the average of the novelty scores associated with each of its meaning constituents (e.g., concepts and semantic relations).

Figure 3 demonstrates the novelty computation operation for an article discussing new treatments for Reactive Attachment Disorder with the patient file from Task 1. The white concepts are the results of the semantic difference operation and indicate the novel information from the article.

7 Evaluation

The evaluation of the approach was done for mental health domain, since this domain has a comprehensive manual - DSM-5 book.

To evaluate the disorder recommendation module, we collected case studies from mental health disorder case study books or online resources. Using this data, we measured the quality of diagnosis recommended at the top-1, top-5, and top-10 levels in terms of accuracy. The disorder recommendation module obtained 62 % (top-1), 82 % (top-5), and 89 % (top-10) accuracy scores.

To evaluate the literature recommendation module, we selected 100 case studies from the test dataset created for the diagnosis module evaluation. Two subject matter experts searched online for articles related to the case studies and tagged two articles for each case study. They then evaluated the articles recommended by our system and scored the relevance and novelty of the articles on a scale of 1–5, with 5 being highly relevant/novel and 1 being not relevant/novel. The literature recommendation module obtained 77 % (top-1), 94 % (top-5), and 95 % (top-10) accuracy scores for relevance, and 21 % (top-1), 44 % (top-5), and 55 % (top-10) accuracy scores for novelty.

8 Conclusion

In this paper, we presented a semantic driven approach to performing literature recommendation that provides therapists with the most current, novel, and relevant literature based on their patient files. We avoided the usual pitfalls of keyword and concept driven search by semantically analyzing patient records and medical articles, performing medical domain specific inference to extract knowledge profiles, and finally recommending publications that best matches a patient's health profile. Deep semantic processing allows expansion, normalization and filtering of the publication content and the patient record. We applied our proposed system to the mental health domain and obtained promising evaluation results for the case studies specified in the DSM-5 book.

References

1. Medical subject headings (mesh). https://www.nlm.nih.gov/pubs/factsheets/mesh.html
2. Aronson, A.R.: Effective mapping of biomedical text to the umls metathesaurus: the metamap program. In: Proceedings of AMIA Symposium, pp. 17–21 (2001). http://view.ncbi.nlm.nih.gov/pubmed/11825149
3. Balakrishna, M., Moldovan, D., Tatu, M., Olteanu, M.: Semi-automatic domain ontology creation from text resources. In: Proceedings of the Seventh International Conference on Language Resources and Evaluation (LREC 2010), Valletta, Malta, May 2010
4. Blanco, E., Moldovan, D.: Semantic representation of negation using focus detection. In: Proceedings of the 49th Annual Meeting of the Association for Computational Linguistics: Human Language Technologies, pp. 581–589. Association for Computational Linguistics, Portland, Oregon, USA, June 2011. http://www.aclweb.org/anthology/P11-1059
5. Blanco, E., Moldovan, D.: Unsupervised learning of semantic relation composition. In: Proceedings of the 49th Annual Meeting of the Association for Computational Linguistics: Human Language Technologies, pp. 1456–1465. Association for Computational Linguistics, Portland, Oregon, USA, June 2011. http://www.aclweb.org/anthology/P11-1146
6. Ceusters, W., Elkin, P., Smith, B.: Negative findings in electronic health records and biomedical ontologies: a realist approach. Int. J. Med. Inform. **76**(Suppl 3), 326–333 (2007). http://www.ncbi.nlm.nih.gov/pmc/articles/PMC2211452/?tool=pubmed
7. Choi, S., Choi, J., Yoo, S., Kim, H., Lee, Y.: Semantic concept-enriched dependence model for medical information retrieval. J. Biomed. Inf. **47**, 18–27 (2014). http://www.sciencedirect.com/science/article/pii/S153204641300141X
8. Grenon, P., Smith, B., Goldberg, L.: Biodynamic ontology: Applying bfo in the biomedical domain. Stud. Health Technol. Inform. **102**, 20–38 (2004)
9. Liu, Z., Chu, W.W.: Knowledge-based query expansion to support scenario-specific retrieval of medical free text. Technical report, Information Retrieval (2005)
10. Luo, G., Tang, C., Yang, H., Wei, X.: Medsearch: A specialized search engine for medical information retrieval. In: Proceedings of the 17th ACM Conference on Information and Knowledge Management, CIKM 2008, pp. 143–152. ACM, NY, USA, New York (2008). http://doi.acm.org/10.1145/1458082.1458104
11. Moldovan, D., Blanco, E.: Polaris: Lymba's semantic parser. In: Proceedings of LREC-2012, pp. 66–72 (2012)
12. Morante, R., Blanco, E.: *SEM 2012 Shared task: resolving the scope and focus of negation. In: Proceedings of the First Joint Conference on Lexical and Computational Semantics (*SEM 2012), pp. 265–274. Montréal, Canada, June 2012
13. Qiu, Y., Frei, H.P.: Concept based query expansion. In: Proceedings of the 16th Annual International ACM SIGIR Conference on Research and Development in Information Retrieval (1993)
14. Spackman, K.A., Campbell, K.E., Côté, R.A.: SNOMED RT: A reference terminology for health care. In: Proceedings of the AMIA Annual Fall Symposium, pp. 640–644 (1997)
15. Zheng, J., Yu, H.: Key concept identification for medical information retrieval. In: Mrquez, L., Callison-Burch, C., Su, J., Pighin, D., Marton, Y. (eds.) EMNLP, pp. 579–584. The Association for Computational Linguistics (2015). http://dblp.uni-trier.de/db/conf/emnlp/emnlp2015.html#ZhengY15

A Highly Effective Hybrid Model for Sentence Categorization

Zhenhong Chen[1], Kai Yang[1], Yi Cai[1(✉)],
Dongping Huang[1], and Ho-fung Leung[2]

[1] School of Software Engineering,
South China University of Technology, Guangzhou, China
ycai@scut.edu.cn
[2] Department of Computer Science and Engineering,
The Chinese University of Hong Kong, Hong Kong, China

Abstract. Sentence categorization is a task to classify sentences by their types, which is very useful for the analysis of many NLP applications. There exist grammar or syntactic rules to determine types of sentences. And keywords like negation word for negative sentences is an important feature. However, no all sentences have rules to classify. Besides, different types of sentences may contain the same keywords whose meaning may be changed by context. We address the first issue by proposing a hybrid model consisting of Decision Trees and Support Vector Machines. In addition, we design a new feature based on N-gram model. The results of the experiments conducted on the sentence categorization dataset in "Good Ideas of China" Competition 2015 show that (1) our model outperforms baseline methods and all online systems in this competition; (2) the effectiveness of our feature is higher than that of features frequently used in NLP.

Keywords: Sentence categorization · Hybrid model · N-grams · Feature

1 Introduction

Sentence categorization is a very important task in text analysis, because the types of sentences contain many useful semantic or syntactic information which can be used in many NLP applications, such as sentiment analysis [13] or question-answering (QA) systems [9], etc. The objective is to classify the type of a sentence as interrogative, negative or imperative and so on. Especially, the identification of negative and interrogative sentences attracts more attention. Because negative sentences express negative sentiment, while an interrogative sentence indicates interrogative attitude to specific parts of a sentence where QA systems need to analyze and give answers. There exist some grammar or syntactic rules to identify these sentences [12,14,18,25]. And keywords like "no", "not", "what", "when" also play an important role in determining sentence types [4,8,27]. However, there are still many sentences have no obvious rules to classify. In addition,

© Springer International Publishing Switzerland 2016
H. Gao et al. (Eds.): DASFAA 2016 Workshops, LNCS 9645, pp. 98–111, 2016.
DOI: 10.1007/978-3-319-32055-7_9

different types of sentences often contain similar keywords. For example, both of the interrogative sentence "Don't you play with us", and the negative sentence "I don't like playing games" have keyword "don't". Besides, many text corpora are from social media like microblog, or e-commerce sites, etc. These user-generated content (UGC) has a variety of informal expressions, which makes this task more challenged.

For this classification problem, there are various classifiers can be used, such as Decision Trees (DTs), Support Vector Machines (SVMs) [25] and so on. These machine learning algorithms are widely used. DTs are good at handling different decision rules and easy to interpret, while SVMs do better in classes that have no intuitive rules to classify. Because some sentences can be determined by rules while others don't have intuitive rules to judge, both of these two algorithms may perform very well on some sentences but not good on the others. In order to prove this assumption and improve classification accuracy, we propose a hybrid model which firstly classify sentences by DTs, and then use SVMs to handle those sentences hard to be judged.

As many machine learning applications, engineering an effective set of features is the main task of sentence cagorization. There are a diversity of features frequent used in natural language analysis, such as lexicons and their frequency, part of speech (POS), phrase position, etc. Among all of these, keywords are one of the most important features. However, different types of sentences may contain same or similar keywords. To enhance the effect of keywords, we design a new kind of feature based on the N-grams. This feature is generated by extracting the combination of keywords (like interrogative words or negative words) and POS of words in their N-grams, and calculating the occurrence probabilities of these combination in different types of sentence. This feature will be used in SVMs.

To validate the efficiency of our hybrid model and the feature, we use the dataset from the "Good Ideas of China" Competition 2015, sentence cagorization task. The dataset mainly contains text from social media. The hybrid model outperforms not only two baseline methods, but also all other systems of teams participated in this task. The quality of the designed feature is also evaluated by comparing it with other frequent-used features. It significantly improves the classification accuracy of SVMs in this task.

The main contributions of this paper are as follows:

- We propose a hybrid model aimed to efficiently determining the types of sentences.
- To the best of our knowledge, it is the first work to design the feature based on N-grams to apply for sentence cagorization.
- We conduct experiments on the dataset from "Good Ideas of China" Competition 2015, sentence cagorization task. As a result, our model outperforms the baseline algorithms and the other competition systems. Besides, the feature shows a high effectiveness compared with other features.

2   Related Works

As sentence categorization is a classification problem, there are several kinds of commonly used classification algorithms: Naive-Bayes (NB), k-Nearest Neighbors (KNN), DTs, and SVMs.

NB classifier is a probabilistic classifier based on Bayes' theorem [19]. It is a highly scalable algorithm which minimizes the probability of misclassification [7]. Different from NB, KNN is a non-parametric classification method [1]. In KNN, an object is assigned to the class most common among its k nearest neighbors.

DTs is one of the widely used approaches to multistage decision making, which uses a tree-like structure of decisions and their possible consequences [20]. For constructing DTs, we can use several commonly algorithm such as ID3 [23], C4.5 [17], ASSISTANT [15] and CART [3]. Among them, C4.5 is a quite popular algorithm which has been ranked 1 in the Top 10 Algorithms in Data Mining [24]. C4.5 is an extension of ID3 algorithm, and it use information entropy to build decision trees in the same way as ID3. Comparing with ID3, C4.5 made a number of improvements. It can handing both continuous and discrete attributes with differing costs [16] and dealing with missing attribute values. Therefore, we adopt C4.5 to generate DTs.

SVMs [6] is a supervised learning model used for classification and regression analysis. SVMs is one of the most robust and accurate methods among all well-known algorithms [24], and it is good at dealing with the problems such as nonlinear, high dimension. Given a set of training data, SVMs will search for a hyperplane by make margin between different classes as large as possible.

However, all the mentioned methods have some specific limitations. In order to balance the advantages and disadvantages of different algorithms, researchers propose different hybrid classification models. Kohavi et al. [11] use a Decision-Tree Hybrid Model to scaling up the accuracy of Naive-Bayes Classifiers. Khashei et al. [10] combine artificial neural networks and multiple linear regression models. Billsus et al. [2] propose a hybrid user model for news story classification. In this paper, we will present a hybrid classification model combing DTs and SVMs.

Negative and interrogative sentence identifications are two main tasks of text orientation identification. For negative sentence identification, Goryachev et al. [8] implement and evaluate four different methods of negation detection. Rowlett et al. [18] discuss the adverbials in negative sentence. Because we will apply our model on the Chinese dataset, we should also consider the characteristic of Chinese sentences. Zhu et al. [27] discuss the different between lexical negation, syntax negation, relative negation and absolute negation. Xu et al. [25] present a method based on semantic comprehension for text orientation identification. It utilizes SVMs to identify the text orientation, and find out those negative sentences. Yao et al. [26] propose a method to compute the sentiment orientation (polarity) of topics. Chen et al. [4] construct a Chinese negation and speculation corpus, then use the corpus to identify the negative sentences.

For interrogative identification, Ultan et al. [22] summarize some general characteristics of interrogative systems. Comorovski et al. [5] discuss the syntax-semantics of interrogative phrases. For Chinese interrogative sentences, Na-na et al. [14] explore the grammar mechanisms and characteristics of interrogatives to determined whether they are used in interrogative sentences or not. Lan et al. [12] discuss the differences between interrogative uses and non-interrogative usages of WH-words. Shi et al. [21] analyze the exclamatory usages of interrogative words in Chinese sentences.

3 Methodology

3.1 Decision Trees

Decision tree (DTs) is a supervised learning model for multistage decision making problems. It is a tree-like structure that predicts the value of a target variable by learning decision rules inferred from the data features. In this structure, leaves represent class labels and branches represent possible decisions that lead to those labels. There are many commonly used learning algorithms, and C4.5 is one of most effective methods. So we use C4.5 to describe the building procedure of DTs. C4.5 uses information entropy ratio to build trees from a training set T with attributes $A=\{A_1,A_2,...,A_m\}$. It selects the attribute A_i with the highest information gain ratio $g_R(T,A_i)$ to build the decision rule of each node from root to leaf:

$$g_R(T, A_i) = g(T, A_i)/H_{A_i}(T) \tag{1}$$

where $g(T, A_i)$ is the information gain and $H_{A_i}(T)$ is the entropy of T on attribute A_i:

$$g(T, A_i) = -\sum_{k=1}^{K} \frac{|C_k|}{|T|} \log \frac{|C_k|}{|T|} + \sum_{n=1}^{N} \frac{|T_n|}{|T|} \sum_{k=1}^{K} \frac{|T_{nk}|}{|T_n|} \log \frac{|T_{nk}|}{|T_n|}, \tag{2}$$

$$H_{A_i}(T) = -\sum_{n=1}^{N} \frac{|T_n|}{|T|} \log \frac{|T_n|}{|T|}, \tag{3}$$

where $|T|$ is the number of T's samples. Supposed T is split into N subsets T_n (n = 1,2,...,N), $|T_n|$ represents the number of the samples belonged to T_n. Supposed a decision tree has K classes C_k(k = 1,2,...,K), $|C_k|$ is the count of the samples that belong to C_k. And T_{nk} is the intersection between T_n and C_k.

DTs is an alternative technique for sentences categorization, because there are many effective attributes of sentences to help determine their types. For instance, interrogative sentences usually contain some fixed usages like "Do you", "Is it", "What are" and so on, while negative sentences have some structures like "don't" followd by verbs. So, if a sentence has one of these structures, it can be classified as an interrogative or negative sentence with a high possibility.

3.2 Support Vector Machines

Support Vector Machines(SVMs) are another kind of supervised learning model used for classification or regression problems, which have been shown to perform high efficiency at traditional text categorization. They aim to find the best decision hyperplane that separates data into different classes. In the two-category case, the basic idea behind training process is to search a decision hyperplane, represented by $\vec{w}$, that not only separates the data of one class from those of the other class, but also maximizes the distance (i.e. margin) between two hyperplanes defined by support vectors; letting $y_i \in \{0,1\}$ be the correct class label of an input sample $\vec{x_i}$, the hyperplane can be written as follows:

$$\vec{w} = \sum_{i=1}^{n} \alpha_i y_i \vec{x_i} \tag{4}$$

where the α_i's value is obtained by solving the dual optimization problem, and n is the number of input samples. The $\vec{x_i}$ is a support vector of the hyperplane $\vec{w}$ if and only if the α_i is greater than zero. And the classification procedure is to decide which side of the hyperplane that input data fall in. Compared with decision trees, SVMs are more capable of handling non-linear classification by implicitly mapping input into high dimensional feature spaces with appropriate kernels like Gaussian kernel, etc.

3.3   N-gram

N-gram Model. N-gram model is one kind of probabilistic language models. It predicts the next word based on the previous (N−1) words in a sequence. This is a Markov model which assumes that the next item w_i depends only on the probability of the last (N−1) sequence $w_{i-(N-1)}^{i-1} = \{w_{i-1},...,w_{i-(N-1)}\}$, and the approximate prediction of a sequence w_1^n is made as following:

$$P(w_1^n) \approx \prod_{k=1}^{n} P(w_k | w_{k-(N-1)}^{k-1}) \tag{5}$$

An N-gram of size two is called bigram, and size three is trigram, and so on. Bigram and trigram models are mostly used.

N-gram Based Feature. As for our sentence cagorization, instead of using N-gram to train language models, we use it to design a new features of sentences. Supposed we have collected many keywords of interrogative or negative sentences like "what", "not" and save them into keyword lists, the new feature is generated as follows: firstly, we extract two kinds of combinations as follows (taking bigrams as an example):

$$(POS_{prev}, keyword) \ and \ (keyword, POS_{next}), \tag{6}$$

where POS_{prev} stands for the POS tag of keyword's previous word while POS_{next} represents the POS tag of the next word. And then, we calculate the occurrence probabilities of these combinations by their frequencies in different types of sentences. In short, the occurrence probabilities are the new kind of feature.

It is worth noting that we construct the N-gram based feature by POS, not by word itself. The reason is that POS may reflect deeper relationships between a keyword and its neighbors in a sentence. Because POS have a good ability to represent various classes of words. The combination of POS and keywords may contain more semantic or syntactic information. We will conduct experiments to explore the effectiveness of this feature. Furthermore, trigram features can also be constructed by the similar ways as the above process.

3.4   A Hybrid Model

In natural language, the types of many sentences can be determined by some grammar or syntactic rules. Because decision trees can efficiently construct decision rules and be easily interpreted, it is a very good model to use. However, there are still a diversity of sentences that can not be correctly judged only by these rules. There are two main cases. The first one is that a lot of sentences have no obvious rules, and their types mainly depend on semantic information. For instance, "Did I tell you?" is a interrogative sentence, while "Didn't I tell you that knocking on the door before entering the office?" is a rhetorical question. There are no direct rules to differentiate the types of these two sentences. Secondly, for corpora, especially the reviews from social media or e-commerce sites, there are many informal expressions lacking of complete syntax constituents. This makes sentence categorization more difficulty.

In order to solve these problems, we propose a hybrid model of DTs and SVMs, shown in Fig. 1 The basic idea is that DTs are firstly used to label sentences that can be determined by grammar or syntactic rules, and then the unlabeled ones are classified by SVMs, which learn decision hyperplanes between different classes by maximizing their margins. In addition, because features play an important role in machine learning algorithms, we design a new kind of features based on N-gram.

Hybrid Classification. As shown in Fig. 1, there are two main steps to predict the types of sentences from test set. First of all, DTs classify sentences by normal grammar or syntactic rules. For instance, a simple rule to determine negative sentences is that whether a sentence contains negative words like "no", "not". If it does not contain, then this sentence maybe classified into the branch of possible non-negative sentences; otherwise, it maybe classified into another branch of various possible negative sentences. And further decision would be made by next rules if existed.

When classification goes to a leave node of DTs, a sentence will either get a label of sentence types, or be unlabeled if it can not be judged by decision rules. For those unlabeled sentences, SVMs will determine their types. A diversity of features will be extracted to train SVMs. They use kernel techniques to search hyperplanes between different types of sentences, and make their margins as larger as possible to get better classification results.

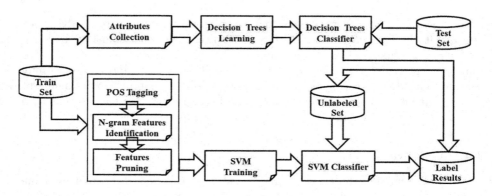

Fig. 1. A hybrid model

N-gram Based Features for SVMs. As most machine learning problems, the main task of sentences categorization is to engineer an effective set of sentences' features. So we design a kind of N-gram based feature for SVMs. As shown in Fig. 1, this feature is generated by three steps: keyword lists collection, POS tagging and N-gram features identification. To collect keyword lists, sentences of the train set are divided into different sets according to their class labels; word frequencies are counted in different sets and topN frequent words are selected into the corresponding keyword lists like negative keyword list or interrogative keyword list. To ensure the effectiveness, keywords of these lists will be manually selected. Besides, every word of sentences will be tagged with part of speech.

As described in Algorithm 1, keyword lists and sentences with POS are used to identify the N-gram based feature. After identifying the feature, its occurrence probabilities OP in every type t of sentences will also be counted (taking bigram based features as an example):

$$OP^t_{bigram_i} = \frac{Count(bigram_i\ occurs\ in\ S_t)}{Count(bigram_i\ occurs\ in\ S_{all})} \tag{7}$$

where $bigram_i$ is the i_{th} combination of the feature. S_t represents sentences of type t, while S_{all} represents all sentences in train set. The probabilities OP are the features used to train SVMs.

Last but no least, there are many other features can also be utilized to train SVMs, such as the length of a sentence, lexicons and their frequencies, punctuation and so on. We will conduct experiments to compare the efficiency between our feature and these frequent-used features, and engineer an effective set of features at last.

4   Experiments

We present a qualitative and quantitative analysis of the proposed hybrid model on sentences categorization task. And we also conduct experiments to evaluate the feature OP by comparing it with other frequent used features.

Algorithm 1. The generation algorithm of N-gram based feature OP

Input: Train set T; Keyword lists L_i; POS of words POS_w;

1   $OP \leftarrow \varnothing$
2   $Set_{N-gram} \leftarrow \varnothing$ //the set of OP's combinations
3   **for** each sentences $s_k \in T$:
4   **do**
5   **for** each keyword $w_j \in L_i$ in s_k:
6   **do**
7                $Set_{N-gram} \leftarrow (POS_{prev}, w_j)$
8                $Set_{N-gram} \leftarrow (w_j, POS_{next})$
9                $OP_{(POS_{prev}, w_j)} \leftarrow$ formula(7)
10            $OP_{(w_j, POS_{next})} \leftarrow$ formula(7)
11   **end**
12 **end**

Output: OP; Set_{N-gram}

4.1 Experiment Setup

Datasets Description. We conduct experiments on the sentences categorization dataset in the latest competition "Good Ideas of China", hosted by China Computer Federation (CCF). This dataset comes from the third task which aims to classify the Chinese sentences into three categories. Interrogative sentences and negative sentences are the first two categories, while sentences of other types are regarded as the third one. The train set and test set are directly provided to all participants.

Preprocessing. When looking into this dataset, we find that they are Chinese microblog text, and several preprocess operations are needed. First of all, text sequences need to be separated into independent sentences by Chinese or English punctuation. Secondly, there are some noisy content like nickname and URL needed to be removed. Thirdly, because Chinese words having no spaces between each other, we will use jieba Chinese module[1] to make segmentation and POS tagging. After having done these preprocess operations, we split the original train set (13,456 sentences) into a off-line train set, a development set and a off-line test set. The final distributions of the dataset are shown in Table 1.

Baseline Methods. We will compare our model with the following classification algorithms:

(1) *Decision Trees*: DTs are not only easy to interpret and explain, but also convenient to handle attributes interactions. The structure is directly based on decision rules, which are suitable for classification that can be relied on certain rules. They are non-parametric, and thus there is no a bunch of

---

[1] https://github.com/fxsjy/jieba.

Table 1. Statistics of "Good Ideas of China" competition 2015, Task3 dataset

	Interrogation	Negation	Others	Total
Train(off-line)	1149	658	7,693	9,500
Dev(off-line)	327	186	1,943	2,456
Test(off-line)	165	96	1,239	1,500
Test(on-line)	—	—	—	2,000

parameters to tune. We build it with C4.5 algorithm and it will give labels for all sentences even if some of them are very hard to classify for having no obvious grammar or syntactic rules.

(2) *Support Vector Machines*: SVMs are widely used baseline methods for natural language applications such as text categorization, sentiment analysis, and so on. They have nice theoretical guarantees and been shown high performance in various natural language tasks. We used Lin's (2011) libSVM[2] package with default RBF (Guassian) kernel for training and testing.

4.2 Model Analysis

For all models, we use the cross-validate approach to tune the parameters on the development set. According to the competition rules, precision P, recall R and F1 score F are defined as follows:

$$P = \frac{tp}{tp + fp}, R = \frac{tp}{tp + fn}, F = \frac{2PR}{P + R} \tag{8}$$

where tp is the number of interrogative and negative sentences which are predicted correctly, while fp is the number of sentences that are predicted as these two types but actually not; fn is the number of sentences which are these two types but not predicted correctly.

The proposed model and the baseline methods are firstly trained and tested on the off-line dataset. Table 2 shows the off-line P, R and F of these three models. From the comparison between two baseline methods, it is shown that DTs have a higher recall, while SVMs has a higher precision. The reason why decision trees have a high recall is that, instead of induct too specific rules that may cause over fitting, the decision rules tends to be suitable for various cases as many as possible, predicting more interrogative or negative sentences. However, there exist some cases hard to determined by rules. Take a sentence from the dataset as an example:

"你要去看电影不"

(Do you want to watch movie)

---

[2] https://www.csie.ntu.edu.tw/~cjlin/libsvm/.

We may directly classify the English sentence as a interrogative sentence based on phrase "Do you". But in the Chinese sentence, there are no interrogative keywords or intuitive rules. Instead it contains a negation keyword "不" (no). So, when determined by general grammar or syntactic rules, the Chinese sentence would be misjudged as a negative sentence. The statistics of Table 3 shows that the misjudgement between interrogative and negative sentences is the main reason for lower precision of DTs.

Table 2. The P, R, F value of three models

Classifiers	Precision	Recall	F1 score
DTs	80.67	94.62	87.09
SVMs	**83.12**	89.88	86.37
Hybrid Model	82.96	**94.69**	**88.44**

Table 3. The P, R, F value of interrogative and negative sentences

Classifiers	Interrogative Sentences			Negative Sentences		
	Precision	Recall	F1 score	Precision	Recall	F1 score
DTs	80.71	95.91	87.66	80.59	92.34	86.07
SVMs	82.65	92.87	87.46	84.05	84.68	84.37
Hybrid Model	83.55	95.31	89.04	81.94	93.62	87.39

Besides, the reason why SVMs have a higher precision is that, instead of using hand-craft rules, this methods focus on directly maximize margin between various classes, and the hyperplanes are more accurate than the rules of DTs. On the other hand, because of the strict classification, SVMs predicted less interrogative or negative sentences and thus have a lower recall.

As shown in Table 2, our proposed hybrid model has the highest F1 score. It seems to combine the advantage of these two baseline methods. To make a deeper looking into this model, Table 4 shows the detail statistics about how these two methods contribute to the whole model's performance.

Table 4. Contributions of decision trees and SVMs

	DT Part	SVMs Part	Total
Precision	91.48	60.87	82.96
Recall	74.04	20.83	94.69
F1 score	81.84	31.04	88.44

DTs is the first part of the hybrid model to predict input, which is the main reason that they have a higher recall that is more than three times as that of SVMs. More importantly, as shown in Table 4, they have a very high precision which is bigger than that in Table 2. This is because, in the hybrid model, DTs only need to predict those sentences that can be judged by rules with high believe, while in the baseline methods, they must predict all sentences even if there don't exist correct rules to classify. Furthermore, for those sentences hard to correctly classified by rules, SVMs makes margin between different classes as large as possible. And because most rule-based sentences have been handled by DTs, SVMs can make a more fine-grained classification.

In short, the proposed hybrid model outperforms the baseline methods. It effectively handle most sentences which can be determined based on rules, and further make a more fine-grained classification for sentences which are hard to judged but have latent differences between various classes.

4.3   Features Analysis

For sentences difficult to judged by rules, we use SVMs to classify. As many machine learning applications, the main task is to engineer an effective set of features. In order to prove the efficiency of our N-gram features, we conduct experiments on a variety of frequent used features, as shown in Table 5.

Table 5. Features for SVMs in the hybrid model

	Sentence length	Lexicon	POS tag	Punctuation	Keywords count	Keywords position	N-gram features OP	P	R	F
1			✓	✓	✓	✓		60.98	19.37	29.40
2			✓	✓	✓	✓	✓	60.68	20.70	30.86
3	✓	✓	✓	✓	✓	✓	✓	**61.19**	20.33	30.52
4					✓	✓	✓	60.87	**20.83**	**31.04**
5					✓	✓		49.38	19.95	28.42

Compared with the first experiment, the second experiment utillizes one more feature, our designed feature *OP*. As a result, the F1 score increases from 29.40 to 30.86, having an improvement of nearly 5 %. There are two possible reasons: (1) adding more features will improve performance, or (2) it is the effectiveness of *OP* that makes a difference. To check the first possible reason, two more features are added in the third experiment, but F1 score decreases from 30.86 to 30.52. Compared with the second experiment, its precision increases but its recall decreases, which indicates that too many features may cause over fitting. To solve this problem, the fourth experiment is conducted with only three parameter. Specially, F1 score increases up to 31.04. Most importantly, the comparison between it and the fifth experiment indicates that the proposed features is the main reason for the improvement of F1 score, which proves its high efficiency.

4.4 Compared with Online Systems

After we have evaluated the effectiveness of our hybrid model and the N-gram based features, we would like to apply our system for sentence categorization task of the competition, and compare it with the online systems of the competition. At the end of the competition, the results are as follows:

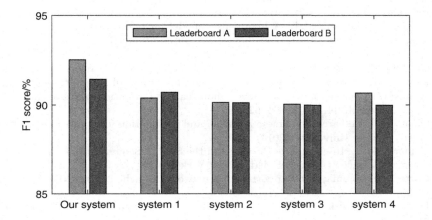

Fig. 2. Comparison with online systems

Figure 2 shows the top5 F1 scores of the sentence categorization task. The competition have two leaderboards: leaderboard A and leaderboard B, where A is shown during the whole process, while B is only shown at the last day. The results of these two leaderboards correspond to two test sets (each set has 1,000 sentences). The motivation is to test the robust of systems. Because leaderboards display name of teams instead name of systems, we use numbers to represent top5 systems. As is shown, our system has the best F1 score on both leaderboards, which indicates that our model is very robust and high effective.

5 Conclusion

We present a hybrid model by sequentially using DTs and SVMs to do sentence categorization. DTs determine types of sentences by rules, which is easily interpreted. SVMs classify sentences, which have not obvious rules to judge, by means of maximizing margin between various classes. On the other hand, one kind of feature is designed based on N-gram to make a better classification. Experimental results on the sentence categorization dataset of "Good Ideas of China" Competition 2015 verify the high effectiveness of our model and the feature. There are also several interesting directions for future research. For instance, N-grams based features may apply for other NLP applications like sentiment analysis. Moreover, sentence categorization can be used for further text mining.

Acknowledgments. This work is supported by National Natural Science Foundation of China (project no. 61300137), and NEMODE Network Pilot Study: A Computational Taxonomy of Business Models of the Digital Economy, P55805.

References

1. Altman, N.S.: An introduction to kernel and nearest-neighbor nonparametric regression. Am. Stat. **46**(3), 175–185 (1992)
2. Billsus, D., Pazzani, M.J.: A hybrid user model for news story classification. In: Kay, J. (ed.) UM99 User Modeling. CISM International Centre for Mechanical Sciences, vol. 407, pp. 99–108. Springer, Heidelberg (1999)
3. Burrows, W.R., Benjamin, M., Beauchamp, S., Lord, E.R., McCollor, D., Thomson, B.: Cart decision-tree statistical analysis and prediction of summer season maximum surface ozone for the vancouver, montreal, and atlantic regions of canada. J. Appl. Meteorol. **34**(8), 1848–1862 (1995)
4. Chen, Z.C.: Research on chinese negation and speculation identification. Master's thesis, Suzhou University (2014)
5. Comorovski, I.: Interrogative Phrases and the Syntax-Semantics Interface, vol. 59. Springer Science & Business Media, New York (2013)
6. Cortes, C., Vapnik, V.: Support-vector networks. Mach. Learn. **20**(3), 273–297 (1995)
7. Devroye, L., Györfi, L., Lugosi, G.: A Probabilistic Theory of Pattern Recognition, vol. 31. Springer Science & Business Media, New York (2013)
8. Goryachev, S., Sordo, M., Zeng, Q.T., Ngo, L.: Implementation and Evaluation of Four Different Methods of Negation Detection. DSG, Boston (2006)
9. Gupta, P., Gupta, V.: A survey of text question answering techniques. Int. J. Comput. Appl. **53**(4), 1–8 (2012)
10. Khashei, M., Hamadani, A.Z., Bijari, M.: A novel hybrid classification model of artificial neural networks and multiple linear regression models. Expert Syst. Appl. **39**(3), 2606–2620 (2012)
11. Kohavi, R.: Scaling up the accuracy of naive-bayes classifiers: a decision-tree hybrid. In: KDD, pp. 202–207. Citeseer (1996)
12. Lan, N.: Basic semantic study on wh-words. In: The Northern Forum, vol. 4, p. 013 (2005)
13. Liu, B., Zhang, L.: A survey of opinion mining and sentiment analysis. In: Aggarwal, C.C., Zhai, C. (eds.) Mining Text Data, pp. 415–463. Springer, Heidelberg (2012)
14. Na-na, T., Han, Y.L.: Mechanisms and characteristics of grammaticalization in interrogative words. Foreign Lang. Res. **5**, 016 (2009)
15. Quinlan, J.R.: Induction of decision trees. Mach. Learn. **1**(1), 81–106 (1986)
16. Quinlan, J.R.: Improved use of continuous attributes in c4.5. J. Artif. Intell. Res. **4**, 77–90 (1996)
17. Quinlan, J.R.: C4.5: programs for machine learning. Mach. Learn. **16**(3), 235–240 (2014)
18. Rowlett, P.: On the syntactic derivation of negative sentence adverbials. J. Fr. Lang. Stud. **3**(01), 39–69 (1993)
19. Russell, S., Norvig, P.: Artificial Intelligence: A Modern Approach. Prentice Hall, Upper Saddle River (1995)
20. Safavian, S.R., Landgrebe, D.: A survey of decision tree classifier methodology. IEEE Trans. Syst. Man Cybern. **21**(3), 660–674 (1991)

21. Shi, L.Z.: Exclamatory usages of question devices in contemporary chinese. Chin. Linguist. **4**, 14–26 (2006)
22. Ultan, R.: Some general characteristics of interrogative systems. Univ. Hum. Lang. **4**, 211–248 (1978)
23. Utgoff, P.E.: Incremental induction of decision trees. Mach. Learn. **4**(2), 161–186 (1989)
24. Wu, X., Kumar, V., Quinlan, J.R., Ghosh, J., Yang, Q., Motoda, H., McLachlan, G.J., Ng, A., Liu, B., Philip, S.Y., et al.: Top 10 algorithms in data mining. Knowl. Inf. Syst. **14**(1), 1–37 (2008)
25. Xu, L., Lin, H., Yang, Z.: Text orientation identification based on semantic comprehension. J. Chin. Inf. Process. **1**, 015 (2007)
26. Yao, T.F., Lou, D.C.: Research on semantic orientation analysis for topics in chinese sentences. J. Chin. Inf. Process. **21**(5), 73–79 (2007)
27. Zhu, Y.J.: Research on chinese language negation. Master's thesis, Tianjin Normal University (2012)

Improved Automatic Keyword Extraction Given More Semantic Knowledge

Kai Yang[1], Zhenhong Chen[1], Yi Cai[1(✉)], DongPing Huang[1],
and Ho-fung Leung[2]

[1] School of Software Engineering, South China University of Technology,
Guangzhou, China
ycai@scut.edu.cn
[2] Department of Computer Science and Engineering,
The Chinese University of Hong Kong, Hong Kong, China

Abstract. Graph-based ranking algorithm such as TextRank shows a remarkable effect on keyword extraction. However, these algorithms build graphs only considering the lexical sequence of the documents. Hence, graphs generated by these algorithm can not reflect the semantic relationships between documents. In this paper, we demonstrate that there exists an information loss in the graph-building process from textual documents to graphs. These loss will lead to the misjudgment of the algorithm. In order to solve this problem, we propose a new approach called Topic-based TextRank. Different from the traditional algorithm, our approach takes the lexical meaning of the text unit (i.e. words and phrase) into account. The result of our experiments shows that our proposed algorithm can outperform the state-of-the-art algorithms.

Keywords: Keyword extraction · Topic model · Graph-based ranking algorithm · Semantic analysis

1 Introduction

Automatic Keyword Extraction is the technology that can generate keywords from documents. This technology is also widely used in text mining and information retrieval and it becomes a popular research field these years. Many graph-based ranking algorithm, such as TextRank [18], have been proposed to extract keywords from documents. These algorithms construct the graph just according to the sequence of words in the documents, which reflects the textual structure inside the documents. However, the sequence of words can not reveal the semantic information embedded in documents. Hence, graphs built by these ranking algorithms take no account of the semantic relationship between words. This motivates us to propose an algorithm taking the semantic information into account.

Several researches have been done to extend the graph-based ranking algorithms. In [11], the importance of words is considered and they add the weight of

H. Gao et al. (Eds.): DASFAA 2016 Workshops, LNCS 9645, pp. 112–125, 2016.
DOI: 10.1007/978-3-319-32055-7_10

vertexes into the graph according to term weighting schemes such as TF and RW [22]. The main point of this approach is adding the statistical information into traditional TextRank. However, the weighting schemes can not really reflect the real semantical relationship inside the document. Liu et al. proposed an extension of TextRank, called Topical PageRank [17], basing on Latent Dirichelt Allocation, which is a popular topic model. They assigned weight to vertexes according to the topic distribution. Latent topic generated by topic model is a set of words that have similarity in semantic meaning. In this paper, we propose a different way to combine topic model and TextRank. We will compare our approach with Topical PageRank in our experiment.

In this work, we attempt to propose a new way to construct graphs for graph-based ranking algorithm. Different from the state-of-the-art algorithm, we build graphs relying not only on the local context, but also the lexical meaning of a text unit(i.e. words or phrase). Our main idea is that the connections between the vertexes having semantical similarity need to be intensified. In order to extract the semantical relationship from textual documents according to the lexical meaning of words or phrase, topic model, e.g. LDA and its variants, is applied in our work. Besides, the approach we proposed is supervised, and we learning the topical knowledge from train corpora. And then these learned knowledge is applied when we build the graphs. In general, the main contribution of us is that we find out a new way to combine semantical information into traditional graph-base keyword extraction algorithm, and we prove that this way can improve the performance of the state-of-the-art approaches.

In this paper, we firstly show the insufficiency of the state-of-the-art graph-based algorithm taking the TextRank for example in Sect. 3. Secondly, we raise our proposed approach and make its mechanism clear in Sect. 4. Finally, we conduct several experiments to verify the effectiveness of our approach and discuss the affection of the parameter setting.

2   Relative Works

2.1   Keyword Extracting Algorithm

Graph-based ranking algorithms are widely used in keyword extraction. These kind of algorithm firstly build the graphs according to the textual documents, and the important vertexes in these graphs are extracted to be the keyword. Many graph algorithms can be applied to find out the important vertexes in the graph. Mihalcea et al. apply PageRank [21], a famous websites ranking algorithm, to find out the important vertexes in [18], and they named these algorithm as TextRank. For the high performance the TextRank reach, there are many extension algorithms based on it. In [11], some term weighting schemes are added into TextRank to weight the vertexes. In [17], Liu et al. proposed Topical PageRank (TPR) where the vertexes are weighted according to the topic distribution generated by topic model. TPR has reached a high performance in keywords extraction, but it doesn't consider the semantical information while the graph-building process. Hence, there are still large space to develop at the state-of-the-art algorithms.

There are some other research working on keyword extraction. In [8], another algorithm based on LDA is proposed. Different from [17], their algorithm do not base on graph. They rank words according to its Coverage, Purity, Phraseness, Completeness, and choose the words which have most ranking score as the keywords. Tomokiyo et al. proposed a statistical way to extract keywords from long text in [23]. They use KL-divergence [16] to calculate the information gain of the phrase. In [24], the relationship between document and words are modeled by a matrix, and then the matrix factorization algorithm is used to find out the latent topic of terms in the document. Finally, they extract keywords according to these latent topic.

2.2   Topic Model

Topic models are a suite of algorithms that uncover the hidden thematic structure in documents. Latent Dirichlet Allocation (LDA) is one of a generative topic model proposed by Blei et al. [2]. As a generative model, LDA assumes that the words in a document are drawn from a set of latent variable called topic which is a distribution over terms.

Collapsed Gibbs Sampling [5,9,10] is a algorithm commonly used to estimate latent parameter in LDA. One of the parameter φ represents the distribution from topic to terms. φ_{kt} represents the probability that term t is assigned to topic k. According to [12], the equation that calculate φ according to Collapsed Gibbs Sampling is as follows:

$$\varphi_{kt} = \frac{n_{k,\neg i}^{(t)} + \beta_t}{\sum_{t=1}^{V}(n_{k,\neg i}^{(t)} + \beta_t)}, \tag{1}$$

where notation $n_{k,\neg i}^{(t)}$ represents the count of word t assigned to topic k excluding the ith word, V is the number of terms appeared in the corpora and β is the hyperparameters of the model. Equation 1 is applied in our approach to find out the most probable terms in a specific topic.

However, standard LDA trend to have a poor performance for topics which mix unrelated or loosely-related concepts. To tackle the problem, some knowledge-based topic models have been proposed [1,6,7]. $DF - LDA$ [1] takes domain knowledge in the form of must-links and cannot-links given by the users. A must-links means that two words should be assigned to the same topic whereas a cannot-links means that two words should not be assigned to the same topic. Besides, there are several models utilizing seed words provided by user [3,14,20]. In some recent research, for example, $GKLDA$ model [7] utilizes the general knowledge such as lexical knowledge to boost the performance. $GKLDA$ can also learn the domain knowledge provided by the user. We choose $GKLDA$ as our model in our approach, and we use the training data as the domain knowledge of $GKLDA$.

3   Existing Problem of Graph-Based Ranking Algorithm

Traditional approaches use slide window algorithm to construct the graph. The slide window moves from the first word to the last word in the document, and words which occur within a window are connected by an edge in the graph. A vertex with more in-edges and out-edges has more probability to be ranked a high score by PageRank algorithm. On the contrary, it also means that a word which have a low frequency is hard to be ranked a high score. However, many keywords in the article have a low frequency. For instance, an author will raise

Table 1. Example of keywords extraction via TextRank and our approach

Document
Machine learning is a subfield of computer science that evolved from the study of pattern recognition and computational learning theory in artificial intelligence. Machine learning explores the study and construction of algorithms that can learn from and make predictions on data. Such algorithms operate by building a model from example inputs in order to make data-driven predictions or decisions, rather than following strictly static program instructions.
Keywords
Keywords assigned by human annotators:
machine learning, computer science, artificial intelligence, data-driven predictions
Keywords assigned by TextRank:
machine learning, data-driven predictions, algorithms, study
Keywords assigned by our approach:
machine learning, artificial intelligence, date-driven, computer science

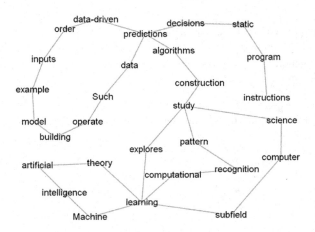

Fig. 1. Word graph of the example

a keyword in the beginning, and then he use several relevant terms to describe this keyword. We give an example in Table 1. Keywords like "computer science" and "artificial intelligence" appear once in the document, and they can not be find out by TextRank. From the graph of this document in Fig. 1, we find that word "computer", "science", "artificial" and "intelligence" have only both one in-edge and out-edge, and this will results the low ranking score of these words. The graph in Fig. 1 can only reflect the context information between words and it has nothing to do with the lexical meaning. Hence, we get a conclusion that the main problem in the state-of-the-art algorithm is the information loss in the transformation from textual document to graph. In the last of our paper, we argue that the semantic meaning must be considered when building the graph, and then we propose an algorithm overcome this problem.

4   Overall Algorithm

This section introduce the proposed overall algorithm. It consists three steps: *learning topical knowledge, extracting keywords using the learned knowledge.* Algorithm 1 shows our proposed algorithm. The input of our algorithm is D_L which consists of document D_i and its respective manual labeled keyword $W_i(D_i, K_i \in D_L)$. Another corpora is testing corpora D_T consisting different documents. w is a variable controlling the window size while creating the word graph. The topic model controls the topic number of the result using variable K. We represent each topic using the first S most relevant words. These inputs will be applied in the following steps.

Algorithm 1. The proposed algorithm.

Input: Topic learning corpora D_L; Test corpora D_T;
Window Size w; Topic number K; The length S of each topic set.

1   //STEP 1: Learning topic knowledge
2   Initialize GKLDA with keywords $K_i \in D_L$
3   $LK \longleftarrow GKLDA(D_L, K)$
4   //STEP 2: generate topic-based word graph
5   **for** each document $D_i \in D_T$:
6   **do** //generate word-graph using slide windows
7           $G \longleftarrow graph(D_i, w)$
8   **for** $(t1, t2)$ in all two-tuples terms of document:
9   **do if** $(t1, t2) \in LK$
10   **do** $G.addEdge(t1, t2)$
11   **End IF**
12   **End For**
13          $RW \longleftarrow PageRank(G)$
14          $TopN \longleftarrow sort(RW, N)$
15 **End For**

Step 1 (learning topical knowledge): We obtain the topical knowledge using GKLDA model, a knowledge-based LDA model using general knowledge. In this step, we use the manual labeled keywords as the prior knowledge initiating GKLDA. Then we obtain topics represented by a set of related words when the model converges after several iterations. Finally, we deal with the obtained topics and change them into the set of two-tuples, which is the topical knowledge we need. We will introduce our knowledge representation in Sect. 5.

Line 2 in Algorithm 1 initialize GKLDA model with prior knowledge K_i, which is a set of keywords in document D_i. Line 3 learn the topical knowledge KL according to the results of GKLDA model whose topic number is set to K.

Step 2 (extracting keywords using the learned knowledge): Our algorithm proposed a new way to generate word graph from documents using the learned knowledge LK. We extract keywords from text corpora D_T according to PageRank algorithm with the learned knowledge. Firstly, we initialize the nodes of word graph and create edges using traditional moving windows algorithm. Secondly, the learned knowledge LK is used and words which appear at the same topic are linked with a edge. Finally, the normal PageRank algorithm is carried out and we can obtain the Top-N words as our keywords according to the ranking. Some post-processing are hidden in Algorithm 1, such as adjacent words combination, we will detail describe these process in Sect. 6.

For Line 5–15, we shows our disposal on each document D_i in test corpora D_T to extract keywords according to learned knowledge LK. Line 7 runs the traditional algorithm to generate word graph according to the order of words. We assigned a windows side w before running our algorithm. Line 8–12 show our proposed way to generate word graph based on the semantic relation between words. We first find out all the word pairs in D_i, and then these pairs are filtered by our learned knowledge LK(Line 9). The remaining words pairs shows a semantic similarity and a new edge is added between the two words in each pairs (Line 10). Line 13 conduct a normal PageRank algorithm to rank the nodes in graph G and obtain ranked words RW. Line 14 sort the ranked words according its ranking and choose Top-N words to become the keywords of D_i.

5   Learning Topical Knowledge

This section details Step 1 in the overall algorithm, which have two steps: GKLDA initialization and topical knowledge learning.

5.1   GKLDA Initialization

We can add domain knowledge into GKLDA model in order to obtain topics with high quality. The domain knowledge of GKLDA is represented by multiple set of words, e.g. price, cheap, expensive, which represents there are semantic relation between 'price', 'cheap' and 'expensive'. In our research, the manual labeled keywords in learning corpus D_L are used to generate domain knowledge. Multi-word keywords which consist of more than two words are separated into single words. We put the single keywords of a specific document into the same set, which means that the keywords labeling the same document can have semantic relations. These generated set is used as domain knowledge to initialize GKLDA.

5.2 Topical Knowledge Learning

In this step, we run the GKLDA model and generate K topics. Each topic is represented by S most probable words according to topic-word distribution φ. From the generated K topic sets, we pair every two words in the set together and form the topical knowledge pairs, i.e. {"hardware","software"}. The topical knowledge is represented by the set of these pairs. Algorithm 2 details the generation of topical knowledge, where w_i^k represents the ith words in topic set k.

Algorithm 2. Generate topical knowledge

---

1 **for** topic k in K topics:
2   **do for** i from 0 to S:
3   **do for** j from $i + 1$ to S:
4   **do** add pair (t_i^k, t_j^k) into knowledge set LK
5   **End For**
6   **End For**
7 **End For**

In general, two words of a pair in learned topical knowledge have some semantic relations because they belong to the same topic. However, research shows that some topics generated by topic model have a inferior quality [4]. In these bad topics, the semantic relation between each words is weak. Adding words in these bad topic into the topical knowledge will reduce the performance of our approach. There are some metric evaluating the quality of topics generated by topic model. Topic Coherent [19] is a metric commonly used to evaluate the performance of topic model, because it shows a well consistence with the judgement of human beings. We set a threshold δ to filter the bad topics. Words in topics whose topic coherence is less than δ will not be learned in topical knowledge. The topic coherence is calculated as

$$C(k, W^{(k)}) = \sum_{s=2}^{S} \sum_{l=1}^{s-1} log \frac{D(w_s^{(k)}, w_l^{(k)}) + 1}{D(w_s^{(k)})}, \qquad (2)$$

Where $W^t = (w_1^t, ..., w_S^t)$ is a list of the first S most probable words in topic t and $D(v, v\prime)$ is the co-document frequency of word v and $v\prime$. Threshold δ is based on empirical value and is set before the algorithm.

6   Extracting Keywords Using the Learned Knowledge

Traditional way to extracting keywords is based on the order of words in a document. However, the order of words can just reflect the structure of the document, but can not show the opinion, emotion and implication behind the text. In our proposed approach, we transform documents into word graphs taking the topical knowledge into account. The topical knowledge learned in last step reflects the semantic information behind the documents.

Some post-processing are carried out in our approach. In most case, the keywords of the document are multi-words rather than single words. We combine the adjacent words which have high ranking in PageRank algorithm and form the multi-keywords.

We use the example showed in Table 1 again to demonstrate the usage of topical knowledge in a keywords extraction task. The traditional TextRank algorithm can not extract keywords such as "artificial intelligence" and "computer science". However, in our approach, we find that these two words are appeared at the same topic in topic models. Hence, the topical knowledge is extracted and showed as follows:

$$\{\{artificial, computer\}, \{artificial, science\},$$
$$\{intelligence, computer\}, \{intelligence, science\}\}$$

We add new edges into the graph showed in Fig. 1 and generate a new graph showed in Fig. 2. Finally, we find out the importance nodes in the graph to be our keywords using PageRank algorithm. The result of our approach is as below: machine learning, artificial intelligence, date-driven, computer science. Our approach can extract keywords precisely in this example.

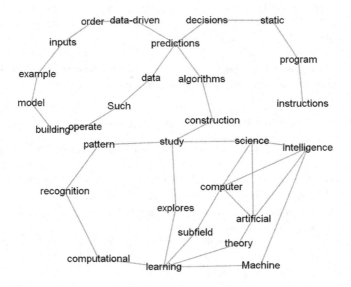

Fig. 2. Word graph of the example

7   Experiments

We prepared two datasets in our experiments. These two datasets are a collection of academic papers which contain the abstracts and the corresponding manually assigned keywords. One is the dataset used in the keyword extraction experiments in [13] and we denote it as $D1$. The academic papers in $D1$ are journal papers from Computer Science and Information Technology. We used 500 aspects and its corresponding keywords for learning topical knowledge, and 500 for extracting keywords. Another dataset is the abstracts of academic paper from ACM, we call it D2. The keywords in D2 assigned by the author himself. For our experiments, 500 papers of D2 is selected to complete the knowledge learning process, and 500 of them is used to the keyword extraction process.

Before testing our proposed algorithm, some preprocessing must be done in the datasets. Firstly, words were converted into lower case, so the words with upper case or lower were treated as the same words in our experiment. Secondly, all the punctuation in the document were eliminated and only the retain the alphabetic and numeric character. Thirdly, we performed text stemming and filter words with useless tagging. In this step, we used WordNet, a large lexical database of English, to perform POS tagging. In this work, we only used nouns, adjective and adverb in modeling, and other words were removed.

The parameters of $GKLDA$ were given default values in our experiments. The total Gibbs sampling iterations was set to 2500 with an initial burn-in of 100 iterations. For the reason that small changes of α and β will not affect the results much [15], we set $\alpha = 1$ and $\beta = 0.1$ as [7] do. Topic number K is a influence factor of our approach, so we individually discuss it in Sect. 7.2 and set $K = 20$ in other experiments.

In our experiments, we firstly compared our approach with some baseline algorithms in Sect. 7.1. Then we discussed the affection of parameters in our models. In Sect. 7.2, we looked into the performance our approach given different topic number K. In Sect. 7.3, the effect of topic-set size S was considered. Furthermore, the size of the slide window can also affect the result of the algorithm, and we discussed it in Sect. 7.4. The evaluation metric we used in all the experiments is F-measure which have also been used in [18].

7.1   Evaluation of Our Proposed Algorithm

In this section, we conducted several experiment to evaluate and compare the proposed algorithm with three baseline models i.e. TextRank, Topical TextRank and N-gram [13]. The first two algorithms are unsupervised and we use the testing part of both two datasets in our experiment. Besides, N-gram is a supervised keyword extraction algorithm. Hence, we carried out N-gram model with both training and testing data in two datasets.

We set the size of slide window w to 2, which showed the highest performance in [13]. The word number of each topic S is set to 10. The threshold σ was set to -1530 and 10 % of the topic was filtered.

Table 2. Topical knowledge of the example

Datasets	Assigned		Correct		Precision	Recall	$F - measure$
	Total	Mean	Total	Mean			
$D1$							
Topical graph TextRank	2,231	4.5	6,761	13.5	33.3	45.9	38.6
TextRank	2,003	4.0	6,803	13.6	29.4	40.9	34.2
Topical PageRank	2,124	4.2	6,728	13.4	31.6	43.3	36.5
N-Gram	1,952	3.9	7,644	15.3	25.5	39.8	31.1
$D2$							
Topical graph TextRank	1,302	2.61	5,044	10.1	25.8	37.7	30.7
TextRank	1,049	2.1	4,911	9.8	21.3	30.4	25.1
Topical PageRank	1,269	2.5	4,832	9.6	26.3	36.8	30.6
N-Gram	1,104	2.2	6,451	12.9	17.1	32.0	22.3

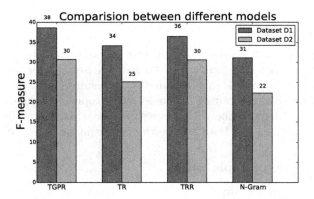

Fig. 3. Comparison of Topical graph PageRank (TGPR) with Topical PageRank (PR), Topical PageRank (TPR) and N-Gram

The detail results of our experiments are showed in Table 2, and we can compare different approach intuitively in Fig. 3. The result shows that the performance of our approach in both datasets $D1$ and $D2$ is better than others. Our approach can promote the traditional TextRank by approximately 15 %. The F-measure results of Topical PageRank are slightly surpassed by our approach. However, our approach is totally different from Topical PageRank, and the combination of these two approach may reach a more better result. We will extend our approach to combine with other approaches in our future works.

7.2 Effect of Topic Number K

For topic models such LDA, the topic number K can be adjusted in order to obtain higher performance. The value of K with which LDA obtain the best performance is various when we deal with other corpora. In this section, we discuss how K affect our algorithm. We set the topic size $K = 10, 20, 40, 60$. The setting of other parameters was the same as Sect. 7.1.

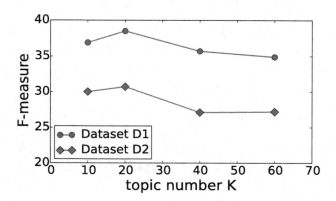

Fig. 4. Effect of different K value.

Figure 4 shows the F-measure of different K setting in two datasets. We observe that when $K = 20$, our approach reach a best performance. Another observation is that when K is larger than 40, the result of our approach become converge. These results are caused by the losing of topical information when the latent topic are divided too finely. Hence, we get a conclusion that lower or larger setting of topic number will both reduce the performance of our approach.

7.3   Effect of Size of Topic Set S

In the proposed approach, we choose the first S most probable words from each topics to represent the corresponding topic. Variable S is also an important factor

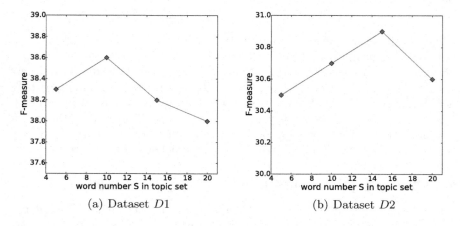

(a) Dataset $D1$ (b) Dataset $D2$

Fig. 5. The effect of different topic set length S

in our approach, because the topic information is incomplete when S is too low, while the topic will mix with irrelevant words when it is set a high value. S was set to 5, 10, 15, 20 in this experiment, and the other parameters were set to the default value as Sect. 7.1.

Figure 5 shows the results of different setting S in two datasets. We find that the result performed in dataset $D1$ and $D2$ is different. Figure 5(a) shows the best setting value is 10, while Fig. 5(b) is 15. We have the conclusion that the setting of topic size S is vary from different datasets.

7.4   Effect of the Window Size w

Window size w control the slide window size when constructing the word graph. In [18], the experiment shows that the model reach the best performance when w is set to 2. The objective of this experiment is to test whether our approach still work in different window size. We give w for $w = 1, 2, 3, 4$ in our experiment. In this experiment, we just use dataset $D1$, and we compare our approach with traditional TextRank.

Figure 6 shows the variation of different window size. We observed that both approach reach its best performance when the window size is set to 2. When the window size was larger than 8, both approach converge into a F-measure value near 32. It shows that our approach can not work will when the window size is too large. In order to find out the reason about that, we looked into the situation when the window size is very large. In that situation, the number of edges will become very large. However, the number of edge added to the graph according to our approach is limited. Hence, our adding edge will not dominate the PageRank algorithm when the window size is too large.

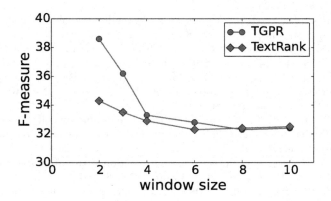

Fig. 6. Effect of window size

8   Conclusions

The objective of our research is to promote the performance of graph-based algorithm in keyword extraction. We solve the problem that the semantic information is ignored during constructing the graphs. We raise a topic based approach to find out the latent semantic relationship between words. Finally, we have conduct a series of experiments to evaluate the effect of our proposed algorithm. It shows that our approach have a outstanding performance comparing to the state-of-the-art algorithm. The results also shows that the semantic information is necessary to be considered while building graphs of documents.

9   Future Works

In future work, we would like to extend our approach to the variants of TextRank or other graph-based algorithms. There are many variants which can be considered such as Topical PageRank. We also want to apply our algorithm for summarization extraction. It needs some further enhancements to fit with this task.

Acknowledgement. This work is supported by National Natural Science Foundation of China (project no. 61300137), and NEMODE Network Pilot Study: A Computational Taxonomy of Business Models of the Digital Economy, P55805.

References

1. Andrzejewski, D., Zhu, X., Craven, M.: Incorporating domain knowledge into topic modeling via Dirichlet forest priors. In: Proceedings of the 26th Annual International Conference on Machine Learning, pp. 25–32. ACM (2009)
2. Blei, D.M., Ng, A.Y., Jordan, M.I.: Latent Dirichlet allocation. J. Mach. Learn. Res. **3**, 993–1022 (2003)
3. Burns, N., Bi, Y., Wang, H., Anderson, T.: Extended twofold-LDA model for two aspects in one sentence. In: Greco, S., Bouchon-Meunier, B., Coletti, G., Fedrizzi, M., Matarazzo, B., Yager, R.R. (eds.) Advances in Computational Intelligence. CCIS, vol. 298, pp. 265–275. Springer, Heidelberg (2012)
4. Chang, J., Gerrish, S., Wang, C., Boyd-Graber, J.L., Blei, D.M.: Reading tea leaves: how humans interpret topic models. In: Advances in Neural Information Processing Systems, pp. 288–296 (2009)
5. Chatterji, S., Pachter, L.: Multiple organism gene finding by collapsed Gibbs sampling. In: Proceedings of the Eighth Annual International Conference on Research in Computational Molecular Biology, pp. 187–193. ACM (2004)
6. Chen, Z., Mukherjee, A., Liu, B.: Aspect extraction with automated prior knowledge learning. In: Proceedings of ACL, pp. 347–358 (2014)
7. Chen, Z., Mukherjee, A., Liu, B., Hsu, M., Castellanos, M., Ghosh, R.: Discovering coherent topics using general knowledge. In: Proceedings of the 22nd ACM International Conference on Information and Knowledge Management, pp. 209–218. ACM (2013)
8. Danilevsky, M., Wang, C., Desai, N., Ren, X., Guo, J., Han, J.: Automatic construction and ranking of topical keyphrases on collections of short documents. In: Proceedings of the SIAM International Conference on Data Mining, 2014 (2014)

9. Geman, S., Geman, D.: Stochastic relaxation, Gibbs distributions, and the Bayesian restoration of images. IEEE Trans. Pattern Anal. Mach. Intell. **6**, 721–741 (1984)
10. Griffiths, T.: Gibbs sampling in the generative model of latent Dirichlet allocation. Technical report, Stanford University (2002)
11. Hassan, S., Mihalcea, R., Banea, C.: Random walk term weighting for improved text classification. Int. J. Semant. Comput. **1**(04), 421–439 (2007)
12. Heinrich, G.: Parameter estimation for text analysis. Technical report (2005)
13. Hulth, A.: Improved automatic keyword extraction given more linguistic knowledge. In: Proceedings of the Conference on Empirical Methods in Natural Language Processing, pp. 216–223. Association for Computational Linguistics (2003)
14. Jagarlamudi, J., Daumé III, H., Udupa, R.: Incorporating lexical priors into topic models. In: Proceedings of the 13th Conference of the European Chapter of the Association for Computational Linguistics, pp. 204–213. Association for Computational Linguistics (2012)
15. Jo, Y., Oh, A.H.: Aspect and sentiment unification model for online review analysis. In: Proceedings of the Fourth ACM International Conference on Web Search and Data Mining, pp. 815–824. ACM (2011)
16. Kullback, S., Leibler, R.A.: On information and sufficiency. Ann. Math. Stat. **22**, 79–86 (1951)
17. Liu, Z., Huang, W., Zheng, Y., Sun, M.: Automatic keyphrase extraction via topic decomposition. In: Proceedings of the Conference on Empirical Methods in Natural Language Processing, pp. 366–376. Association for Computational Linguistics (2010)
18. Mihalcea, R., Tarau, P.: Textrank: Bringing order into text. In: Proceedings of the 2004 Conference on Empirical Methods in Natural Language Processing, EMNLP 2004, A meeting of SIGDAT, a Special Interest Group of the ACL, held in conjunction with ACL 2004, 25-26 July 2004, Barcelona, Spain, pp. 404–411 (2004)
19. Mimno, D., Wallach, H.M., Talley, E., Leenders, M., McCallum, A.: Optimizing semantic coherence in topic models. In: Proceedings of the Conference on Empirical Methods in Natural Language Processing, pp. 262–272. Association for Computational Linguistics (2011)
20. Mukherjee, A., Liu, B.: Aspect extraction through semi-supervised modeling. In: Proceedings of the 50th Annual Meeting of the Association for Computational Linguistics: Long Papers. vol. 1, pp. 339–348. Association for Computational Linguistics (2012)
21. Page, L., Brin, S., Motwani, R., Winograd, T.: The pagerank citation ranking: Bringing order to the web. Technical Report 1999-66, Stanford InfoLab, November 1999. Previous number = SIDL-WP-1999-0120
22. Sparck Jones, K.: A statistical interpretation of term specificity and its application in retrieval. J. Documentation **28**(1), 11–21 (1972)
23. Tomokiyo, T., Hurst, M.: A language model approach to keyphrase extraction. In: Proceedings of the ACL Workshop on Multiword Expressions: Analysis, Acquisition and Treatment, vol. 18, pp. 33–40. Association for Computational Linguistics (2003)
24. Yan, X., Guo, J., Liu, S., Cheng, X., Wang, Y.: Learning topics in short texts by non-negative matrix factorization on term correlation matrix. In: Proceedings of the SIAM International Conference on Data Mining (2013)

Generating Computational Taxonomy
for Business Models of the Digital Economy

Chao Wu[1(✉)], Yi Cai[2], Mei Zhao[2], Songping Huang[2], and Yike Guo[1]

[1] Data Science Institute, Imperial College London, London, UK
{chao.wu,y.guo}@imperial.ac.uk
[2] School of Software Engineering,
South China University of Technology, Guangzhou, China
ycai@scult.edu.cn, {1527757976,798623154}@qq.com

Abstract. We propose to design a semi-automatic ontology building approach to create a new taxonomy of the digital economy based on a big data approach – harvesting data by scraping publicly available Web pages of digitally-focused business. The method is based on a small core ontology which provides the basic level concepts in business model. We try to use computational approaches to extracting Web data towards generating concepts and taxonomy of business models in the digital economy, which can help consequently address the important question while exploring new business models in big data era.

Keywords: Business model taxonomy · Ontology generation · Computational taxonomy

1 Introduction

Big Data and data science promises to change how business is conducted and how innovations emerge. Publicly accessible data, such as Web pages, provide a rich source of structured and unstructured data that we can begin to study and extract knowledge about emerging trends in culture, society and business – and about the digital economy itself.

In the digital economy many new companies emulate their business models, for example Google's exploitation of Web user traffic as a means to generating advertising revenue is commonly mimicked. Many companies also innovate new value streams. The ways in which value is generated in the digital economy are not just unclear, but are also still emerging. Value configurations are very different in digital business, where companies frequently partner with other digital revenue streams. Current business model taxonomies fail to reflect this dynamism.

Taxonomy are data schemas, providing a controlled vocabulary of concepts, each with an explicitly defined and machine processable semantics. By defining shared and common domain theories, taxonomy help both people and machines to communicate concisely, supporting the exchange of semantics and not only syntax. The main problems are how to construct domain-specific taxonomy cheaply and quickly. Still now, the generations of most of the taxonomy for business models depend on human.

© Springer International Publishing Switzerland 2016
H. Gao et al. (Eds.): DASFAA 2016 Workshops, LNCS 9645, pp. 126–133, 2016.
DOI: 10.1007/978-3-319-32055-7_11

The semi-automatic and automatic generation method is far from sophisticated and practical. The manual acquisition of ontologies still remains a tedious, cumbersome task.

What we propose is to address this challenge to designing a semi-automatic ontology building approach which creates a new taxonomy of the digital economy by taking a big data approach – harvesting data by scraping publicly available Web pages of digitally-focused business, and processing this data using text analytics techniques including text feature extraction, natural language processing, and supervised learning. This approach is based on a small core ontology which provides the basic level concepts in business model. Cognitive psychologists find that most human knowledge is represented by basic level concepts which is a family of concepts frequently used by people in daily life. In this work, based on a small core ontology constructed by domain experts, we try to use computational approaches to extracting Web data towards generating concepts and taxonomy of business models in the digital economy, which can help consequently address the research question, "What are the new business models of the digital economy?".

The remainder of this paper is organized as follows. Section 2 overviews of related works. Section 3 presents our approach for taxonomy learning method. Section 4 then presents the experiments and evaluations. Finally, Sect. 5 concludes the paper.

2   Related Work

Taxonomic classification of Web pages by computational means is not novel [2], and text mining has been used extensively in applied sciences, for example in biology to extract linked medical concepts, map biological processes, and even to aid the interpretation of genes for drug development.

"Taxonomy" or broadly "Ontology" in its original sense is a philosophical discipline dealing with the potentialities and conditions of being. Within Computer Science, 'ontologies' have been introduced about a decade ago as a means for formally representing knowledge. Gruber gave out the most popular definition of ontology an "explicit, specification of a conceptualization" [6]. This means that ontologies serve as representation in some pre-defined formalism of those concepts and their relations that are needed to model a certain application domain.

Ontology learning can be defined as the set of methods and techniques used for building ontology from scratch, enriching, or adapting an existing ontology in a semi-automatic fashion using several sources. Several approaches exist for the partial automatization of the knowledge acquisition process. To carry out this automatisation, natural language analysis and machine learning techniques can be used.

Maedche and Staab [7] distinguished different ontology learning approaches focus on the type of input used for learning, such as semi-structured text, structured text, unstructured text. In this sense, they proposed the following classification: ontology learning from text, from dictionary, from knowledge base, from semi-structured schema and from relational schema. Now, most of the domains haven't so much existed semi-structured text, structured text, but there are many unstructured text, such as domain literature, web page. So most of the method is to learn ontology from texts consist of extracting ontologies by applying natural language analysis techniques to texts. The most well-known approaches from this groups are:

1. Ontology pruning is to build a domain ontology based on different heterogeneous sources. It has the following steps. First, a generic core ontology is used as a top level structure for the domain-specific ontology. Second, a dictionary is used to acquire domain concepts. Third, domain-specific and general corpora of texts are used to remove concepts that were not domain specific. This method can quickly construct aim ontology for a specific domain, but for the lack of domain generic core ontology and the efficient method of pruning still now, the effect of exist application is not so good [8].

2. Conceptual clustering, concepts are grouped according to the semantic distance between each other to make up hierarchies. But because of lack the domain context to instruct in the process of distance computation, the conceptual clustering process can't be efficiently controlled. Furthermore, by this method, only taxonomic relations of the concepts in the ontology can be generated [9].

3. Formal concept analysis (FCA), by some technique of NLP, the domain concepts and their attributes can be obtained to form the formal context for the construction of concept lattice. This concept lattice can be viewed as original ontology which just contains classification relations between concepts. After adding non-taxonomic relations, the ontology can be formed. But this difficulty of this method is concept lattice is complicated data structure, when formal context is big, the ontology construction from a set of relevant documents where construction of concept lattice is not easy.

4. Association rules, the association rules have been used to discover non-taxonomic relations between concepts, using a concept hierarchy as background knowledge. Association rules are most used on the data mining process to discover information stored on database. Ontology learning mostly uses unstructured texts but not the structure data in database. So, association rule is just an assistant method to help the ontology generation [7].

5. Pattern-based extraction, a relation is recognized when a sequence of words in the text matches a pattern. But the pattern should be created under the domain expert's instruction. The modification of pattern will bring vibration effect and there is no promise of best pattern [10].

6. Concept learning, a given taxonomy is incrementally updated as new concepts are acquired from real-world texts. Concept learning is a part of the process of ontology learning [11].

The main lacks for all the methods and tools presented in this overview are that there are not integrated methods and tools that combine different learning techniques and heterogeneous knowledge sources with existing ontologies to accelerate the learning process.

Research into taxonomies of e-business has been previously carried out [3, 4], however our approach looks to keep pace with the quickly evolving nature of businesses in today's digital economy by using computational techniques to steer such classification.

Our work is based on a core set of basic concepts. According to the studies of cognitive psychology, there is a family of categories named basic level categories [12, 13]. People most frequently prefer to use basic level concepts constructed from these categories in their daily life, and these concepts are the ones first named and understood by children. For example, when people see a dog, although we also can call it an "animal" or a "terrier", most people would call it a "dog". What is more, most human knowledge is represented by basic level concepts.

3   Method

The approach starts from a core set of concepts built by Human experts, which provides the system with a small number of domain-specific top concepts that represent high-level concepts and are used as seed concepts to discover new concepts and relations [14]. Those concepts and their relations are viewed as the core ontology of the system. Ordinarily, the domain's name can be the top concept of the core ontology. First, the documents in domain corpus need to be preprocessed and converted into plain text format that natural language process tools can conduct. Then, the natural language process tools including stemming, pos tagging and parsing tools used to process the plain texts. Using Gate 3.0 [20] as the tools which is general architecture for text engineering and open source tool, the stemming and pos tagging results of the text will be obtained and stored to the semantic units database.

Then, each sentence of the text will be sent into Stanford Parser [15], which works out the grammatical structure of sentences and the parser tree of each sentence will be analyzed to produce semantic units. The semantic unit is the minimal sentence segment that can be independently viewed as a sentence in a sentence. Then, each of the previous semantic units were POS-tagged and parsed. The process of the corpus processing is shown in Fig. 1.

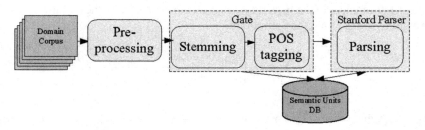

Fig. 1. Domain corpus processing process

As described in [5], the core domain concept or corpus is defined as a tuple $F := (U, T, R, Y)$ where U, T and R are finite sets, whose elements are called users, tags and resources, respectively, and Y is a ternary relation over them, i.e. $Y \subseteq U \times T \times R$.

The target is to learn a hierarchical taxonomy of concepts, which is a tuple $O = (C, P, I, S)$ where C, P and I are finite sets, whose elements are called concepts, properties and instances, respectively, and S is a set of rules, propositions or axioms that specify the relations among concepts, properties and instances. Every concept consists of a category of instances and is described by its properties.

Accordingly, an instance is represented as a vector of tag-value pairs: An instance, r_i, is represented by a vector of tag:value pairs, $r_i = (t_{i,1} : v_{i,1}, t_{i,2} : v_{i,2}, \ldots, t_{i,n} : v_{i,n})$ with $t_{i,k} \in T$, $0 < v_{i,k} \le 1$, $1 \le k \le n$. Where n is the number of the unique tags assigned to resource r_i, $v_{i,k}$ is the weight of tag $t_{i,k}$ in resource r_i. The weight $v_{i,k}$ determines the importance of the tag $t_{i,k}$ to resource r_i. We consider that a tag assigned by more users to a resource is more important because more users think the tag is useful to describe the resource.

A concept is the abstraction of a category of instances and holds the common properties of them: A concept, c_i, is represented by a vector of tag:value pairs, $c_i = (t_i, 1 : v_{i,1}, t_{i,2} : v_{i,2}, \ldots, t_{i,n} : v_{i,n})$ with $t_{i,k} \in T$, $0 < v_{i,k} \le 1$, $1 \le k \le n$. Where n is the number of unique tags, $t_{i,k}$ is a common tag of a category of resources, $v_{i,k}$ is the weight of the tag $t_{i,k}$.

Accordingly, we construct a concept through extracting common tags of a category of instances. These common tags are considered as the properties of the concept. The weights of these tags are their mean values among all instances in a category.

Given a set C of categories and a set F of features, the category utility is defined as follows:

$$cu(C,F) = \frac{1}{m} \sum_{k=1}^{m} p(c_k) \left[\sum_{i=1}^{n} p(f_i|c_k)^2 - \sum_{i=1}^{n} p(f_i)^2 \right]$$

where $p(f_i|c_k)$ is the probability that a member of category c_k has the feature f_i, $p(c_k)$ is the probability that an instance belongs to category c_k, $p(f_i)$ is the probability that an instance has feature f_i, n is the total number of features, m is the total number of categories. Features of instances are represented by tags in folksonomies. Accordingly, in the definition of category utility, the tag set T is used as the feature set F and a tag t_i is used as a feature f_i. As we model, the importance of tags is different in folksonomies. To take the differences of tag importance into account, we modify the definition and add the weight w_i of tag t_i into the definition:

$$cu(C,T) = \frac{1}{m} \sum_{k=1}^{m} p(c_k) \left[\frac{\sum_{i=1}^{n_k} w_i p(t_i|c_k)^2}{n_k} - \frac{\sum_{i=1}^{n} w_i p(t_i)^2}{n} \right]$$

To reflect the mean weight of a tag, w_i is defined as $w_i = \frac{1}{N_{ti}} \sum_{j=1}^{N_{ti}} v_{j,i}$, where N_{ti} is the number of resources annotated as the weighted category utility.

Because basic level categories (and concepts) have the highest category utility, the problem of finding basic level categories (and concepts) becomes an optimization problem using category utility as the objective function. The value of category utility is influenced by the intra-category similarity which reflects the similarity among members of a category. Categories with higher intra-category similarity have higher value of category utility. Accordingly, we put the most similar instances together in every step of our method until the decrease of category utility. To compute the similarity, we use the idf-cosine coefficient [16] which is a commonly used method of computing similarity between two vectors in information retrieval. It is defined as follows:

$$sim(a, b) = \frac{\sum_{k=1}^{n} idf(t_k) \cdot v_{a,k} \cdot v_{b,k}}{\sqrt{\sum_{k=1}^{n} v_{a,k}^2} \cdot \sqrt{\sum_{k=1}^{n} v_{b,k}^2}}$$

where a, b are two concepts, n is the total number of unique tags describing them, and $v_{a,k}$ is the value of tag $t_{a,k}$ in concept a, if a does not have the tag, the value is 0. $idf(t_k)$ is the inverse document frequency of the tag t_k, $idf(t_k) = \log_N(N/N_{tk})$, where N is the total number of resources and N_{tk} is the number of resources annotated by tag t_k, $0 \leq idf$ $(t_k) \leq 1$. When $idf(t_k)$ is 0, the tag t_k is assigned to all resources. In this case, all resources have this tag, the tag is not useful for categorization.

In our algorithm, firstly, we consider every single instance itself as a concept. This type of concept which only includes one instance is considered as the bottom level concepts. Secondly, we compute the similarity between each pair of concepts and build the similarity matrix. Thirdly, the most similar pair in the matrix is identified and merged into a new concept. The new concept contains all instances of the two old concepts and holds their common properties. After that we reconsider the similarity matrix of the remaining concepts. We apply this merging process until only one concept is left or the similarity between the most similar concepts is 0. We then determine the step where the categories have the highest category utility which is the local optimum of category utility. These categories are considered as the basic level categories and the concepts are considered as the basic level concepts. The time complexity is $O(N^2 \log N)$ where N is the number of resources.

To build the taxonomy, we first generate a root concept including all instances. After finding the basic level concepts with Algorithm 1, we add the basic level concepts to the taxonomy as sub-concepts of the root. After several iterations, a cognitively basic ontology is built. The psychological character differentiates the ontology built through our method to the ontology built using methods proposed in previous taxonomy learning research.

4 Result

With the method describe before, we construct business model taxonomy as a pilot study. The construction process starts from core business model ontology with top concepts containing 300 articles from 20 business model related website. Each articles have no more than 5000 words. After once extension by our method, there are respectively over 400 of four types of concepts and relations.

Experiments show that the ontologies generated using our method are more consistent with human thinking. Figure 2 gives an example of the ontology explored through our approach. In our approach, concepts are represented by the common tags of a category of resources. The tags of a concept are inherited by its sub-concepts and a concept has all instances of its descendants. For example, tags "crowdfunding" represents a concept within the category of "crowdsourcing". Such a representation can keep more

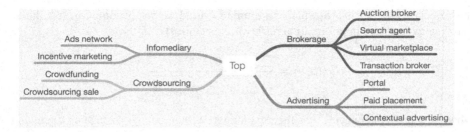

Fig. 2. An ontology generated by our approach.

information and properties of concepts and is consistent with definition of concepts in psychology.

5   Conclusion

In this work, based on a small core ontology constructed by domain experts, we used computational approaches to extracting Web data towards generating concepts and taxonomy of business models in the digital economy. To the best of our knowledge, it is the first work on discovering taxonomy from core concepts for business model. With the result graph-based taxonomy, we can explore the interconnectedness within the digital economy. For example, visual analytics, in the form of interactive graph visualizations, can then be used to explore complex relationships between digital businesses, such as when digital revenue streams cross and are shared within a business model.

Acknowledgment.   The work presented in this paper is supported by NEMODE Network + Pilot Study: A Computational Taxonomy of Business Models of the Digital Economy (P55805).

References

1. Mitchell, G.: How many petabytes is the internet? Focus Science and Technology Magazine (2013). http://sciencefocus.com/qa/how-many-terabytes-data-are-internet
2. Xiaoguang, Q., Davison, B.D.: Web page classification: features and algorithms. ACM Comput. Surv. **41**(2) (2009). Article 12
3. Bartelt, A., Lamersdorf, W.: A multi-criteria taxonomy of business models in electronic commerce. In: Fiege, L., Mühl, G., Wilhelm, U.G. (eds.) WELCOM 2001. LNCS, vol. 2232, pp. 193–205. Springer, Heidelberg (2001)
4. Lambert, S.C.: Do we need a 'real' taxonomy of e-business models? School of commerce research paper series 06-06 (2006)
5. Chen, W.H., Cai, Y., Leung, H.F., Li, Q.: Generating ontologies with basic level concepts from folksonomies. Procedia Comput. Sci. **1**(1), 573–581 (2010)
6. Gruber, T.: A translation approach to portable ontology specifications. Knowl. Acquisition **5**(2), 199–220 (1993)
7. Maedche, A., Staab, S.: Ontology learning for the semantic web. IEEE Intell. Syst. **16**(2), 72–79 (2001). Special Issue on the Semantic Web

8. Kietz, J.U., Maedche, A., Volz, R.: A method for semi-automatic ontology acquisition from a corporate intranet. In: Aussenac-Gilles, N., Biebow, B., Szulman, S. (eds.) CEUR Workshop Proceedings, EKAW 2000 Workshop on Ontologies and Texts, Juan-Les-Pins, France, Amsterdam, The Netherlands, vol. 51, pp. 4.1–4.14 (2000). http://CEUR-WS.org/Vol-51

9. Faure, D., Poibeau, T.: First experiments of using semantic knowledge learned by ASIUM for information extraction task using INTEX. In: Staab, S., Maedche, A., Nedellec, C., Wiemer-Hastings, P. (eds.) Proceedings of the Workshop on Ontology Learning, 14th European Conference on Artificial Intelligence ECAI 2000, Berlin, Germany (2000)

10. Morin, E.: Automatic acquisition of semantic relations between terms from technical corpora. In: Proceedings of the Fifth International Congress on Terminology and Knowledge Engineering (TKE 1999). TermNet-Verlag, Vienna (1999)

11. Hahn, U., Schulz, S.: Towards very large terminological knowledge bases: a case study from medicine. In: Hamilton, H.J. (ed.) Canadian AI 2000. LNCS (LNAI), vol. 1822, pp. 176–186. Springer, Heidelberg (2000)

12. Rosch, E., Mervis, C.B., Gray, W.D., Johnson, D.M., Boyes-Braem, P.: Basic objects in natural categories. Cogn. Psychol. **8**(3), 382–439 (1976)

13. Murphy, G.: The Big Book of Concepts. Bradford Book, Cambridge (2004)

14. Zhou, W., Liu, Z., Zhao, Y., Xu, L., Chen, G., Wu, Q., Qiang, Y.: A semi-automatic ontology learning based on wordnet and event-based natural language processing. In: 2006 International Conference on Information and Automation, ICIA 2006, pp. 240–244. IEEE, December 2006

15. The Stanford Parser. http://nlp.stanford.edu/software/lex-parser.shtml

16. Schultz, J., Liberman, M.: Topic detection and tracking using idf-weighted cosine coefficient. In: Broadcast News Workshop 1999 Proceedings, p. 189 (1999)

How to Use the Social Media Data in Assisting Restaurant Recommendation

Wenjuan Cui[1], Pengfei Wang[1,2], Xin Chen[1], Yi Du[1], Danhuai Guo[1],
Yuanchun Zhou[1(✉)], and Jianhui Li[1]

[1] Computer Network Information Center, Chinese Academy of Sciences,
Beijing 100190, China
zyc@cnic.cn
[2] University of Chinese Academy of Sciences, Beijing, China

Abstract. Online social network applications such as Twitter, Weibo, have played an important role in people's life. There exists tremendous information in the tweets. However, how to mine the tweets and get valuable information is a difficult problem. In this paper, we design the whole process for extracting data from Weibo and develop an algorithm for the foodborne disease events detection. The detected foodborne disease information are then utilized to assist the restaurant recommendation. The experiment results show the effectiveness and efficiency of our method.

Keywords: Recommender system · Event detection · Social media · Foodborne disease

1 Introduction

With the development of information technology, people spend more time surfing on the internet. While people's lives get convenient from the information services, the continuously generated huge volume of data make it difficult to easily get the useful information fulfilling people's requirements. Recommender system was designed aiming to overcome the information overload problem [1]. Except for the traditional recommender systems, the personalized recommendation has been developed and applied in various fields to meet the users' interest [2]. Lots of approaches have been proposed to the recommendation research, in which content based approach, collaborative filtering and hybrid models are most used methods [3–5]. Various recommender systems have been deployed in tremendous applications. However, most of the systems only consider one particular data source. They seldom care about the information from other data sources like the social network.

Twitter, Facebook and other social network applications have been frequently used in recent years. People tend to express their opinions, feeling or just tell friends what they are doing on the social network. Twitter is a popular microblogging service which attracts much attention. People may post tweets at any place and during any time. In general, the length of one tweet has a limit of 140 characters. There may be only little information in a particular tweet, but the

H. Gao et al. (Eds.): DASFAA 2016 Workshops, LNCS 9645, pp. 134–141, 2016.
DOI: 10.1007/978-3-319-32055-7_12

accumulated content can generate a vast amount of information and include important knowledge. The context of the tweet also provides lots of information [6]. Twitter has helped to provide valuable message for various applications. Tumasjan et al. tracked the public political opinions on Twitter and predict the election result [7]. The stock could also be predicted [8]. Sakaki et al. investigated the real-time interaction of earthquakes in Twitter and proposed a method to monitor the tweets and to detect the earthquakes. They can detect the earthquakes with high probability and faster than the government [7].

Tweets can also be used in the area of public health. Users often post the messages like "I got a flu, getting a running nose" or "got a stomachache after eating pizza". Millions of such messages may give a direction for the influenza tracking or other public health problems. Aramaki et al. propose a machine learning method to detect the influenza epidemics from twitter data [9]. Tweets can be used to track the influenza and forecast future influenza rates with high accuracy [10,11]. Twitter data have been used on surveillance of other public health related problems like Dengue and foodborne disease [12].

In this paper, we will consider the relevance of foodborne disease and the restaurants with contaminated food. First, we will crawl the foodborne disease related tweets from Weibo, which is a Chinese social network application similar to Twitter. Then the tweets are classified to filter the part which are the actual foodborne disease related ones. The key-phrases are extracted and a SVM classifier is designed to detect the foodborne disease events. Meanwhile, the locations are also determined and the restaurants with contaminated food are identified. Finally the foodborne disease factor is used in the restaurant recommendation.

2   Related Work

Foodborne disease (or foodborne illness) refers to any disease resulting from the consumption of contaminated food. Foodborne Disease Outbreak (FBDO) is defined as the two or more cases of a similar illness resulting from the ingestion of a common food. According to CDC 2011 Estimates, each year roughly 1 in 6 Americans (or 48 million people) gets sick, 128,000 are hospitalized, and 3,000 die of foodborne diseases [13]. In the past years, the tracking and detection of foodborne disease were mainly carried on by surveillance systems. But the traditional surveillance systems have the limit of time lag in the detection of FBDO. Recently, the social media data have been introduced to the surveillance of foodborne disease [14–16].

The users for Twitter or Weibo get more and more and the functions of these social network applications get complex. People tend to express their feelings and describe what they are doing on the social network applications. The event detection from the short text data such as tweets has been of significant value. The influenza and foodborne disease could also be detected by the Twitter or Weibo data. The response time is proved to be faster than the traditional surveillance systems.

As the foodborne diseases are mostly caused by contaminate food, the twitter data can also help with the identification of restaurants related with foodborne diseases. Sadilek et al. combine the Twitter data with foodborne disease and restaurants [17]. They collect the tweets posted in mobile devices and get the geo-location of the tweets, and then find the tweets which are near to some restaurants. After that, they design a human guided machine learning method to classify the foodborne disease events. Finally, the detected tweets are associated with the restaurants. Their application is deployed in Las Vegas [18]. Their method performs well comparing with the health department. However, they only collect the mobile data with specified geo-locations. For the data which do not contain specific geo-locations, their method does not work. In this paper, we will design a method to solve this problem.

The paper is organized as follows. Section 1 introduces the background and Sect. 2 shows some related works. In Sect. 3, the process for the foodborne disease event detection will be illustrated step by step. Section 4 will show the experiment results and we will conclude in Sect. 5.

3 The Method

3.1 Data Filtering

To get the relevance of foodborne diseases and the restaurants with contaminated food, we should first get the tweets with the information of foodborne diseases. There is a government surveillance system for the foodborne diseases which collects foodborne disease cases in 13 provinces in China from 2013. Each case is reported by the doctors from the sentinel hospital. We analyze the symptoms of these cases and get a list of keywords for them. The frequency of appearance for each keyword is calculated and the keywords with high frequency are selected.

With the help of Weibo API, we then crawl the tweets in Weibo which contain the keywords of the symptoms for the foodborne diseases. While the tweets contain huge amount of information, there are also lots of noisy data. The extracted tweets may be from the public accounts which offer health advices. To reduce the effect of the useless tweets, we build a support vector machine (SVM) classifier to filter the tweets, which is based on some properties of the tweets and the users. After the Weibo data are filtered, we should identify the foodborne disease event.

3.2 Foodborne Disease Event Detection

As the contents of tweets are texts, we should first convert the natural language into mathematical format in order to process the tweets using machine learning methods. Constructing the vector representation for the words is a good way to capture the syntactic and semantic word relationships. The toolkit word2vec is an open source software developed by Google, which is used to convert the words into vectors. It is efficient while billions of words could be trained in a day, which makes it possible for us to train our data from Weibo. We construct the vector representations for the words in tweets using word2vec. Then the sematic similarities between the tweets could be calculated by the similarities in the vector space [19, 20].

There are huge amount of data continuously generated in Weibo and the topics change fast. Except for the tweet containing the keywords for foodborne diseases, the tweets in its context may contain other important information for the foodborne diseases, such as the location where the foodborne diseases are caused. Assuming the tweets for a user is a tweet list $S = \{T_1, T_2, ..., T_k, ..., T_n\}$, where T_k is the tweet which contains the keywords for the foodborne disease symptoms and T_i is posted earlier than T_j if $i < j$.

We design an algorithm to dynamically choose more relevant tweets in the context to get a larger candidate corpus. We carry on a tokenization for the tweets and construct the vector representation for each tweet. For two tweets T_i and T_j, the vectors for them are v_i and v_j respectively. We define the semantic similarity between two tweets as the cosine similarity of their word vectors. The equation is illustrated in formulas (1).

$$Sim(T_i, T_j) = \cos(v_i, v_j) \tag{1}$$

The similarity between two tweets gets higher while the value of cosine is larger. We compute a dynamic window in the context for a particular tweet T_k which contains the keywords for the foodborne disease symptoms. The algorithm is shown as Algorithm 1.

For a tweet T_i, we compute the similarity between T_k and T_i. If the similarity is greater than the threshold U, the tweet T_i will be added into the candidate tweet set C. We find the similar tweets with T_k before and after it. Our method takes account of the semantic similarity between the tweets and makes sure that the tweets in the candidate set are relevant to the foodborne disease. At the same time, less noisy data are introduced.

After the candidate tweet set is built, we try to extract the key-phrases for the tweets. The easiest method is based on TF/IDF. But it only considers the statistical properties of the phrases. The relationships between the phrases are not considered and the phrases with low frequency will be ignored. In this paper, we use TextRank, which is a graph based key-phrase extraction algorithm [21]. It divides the text into several segments and builds the graph model for them. The voting schema is used to rank the phrases of the text and the key-phrases are extracted according to the rank.

We manually label some tweets which are indeed related with foodborne disease. And then we use the extracted key-phrases for the tweets together with the keywords for the symptoms of the foodborne disease to train a SVM classifier to detect the foodborne disease event.

3.3 Restaurant Recommendation

When the foodborne disease events are detected, we want to associate the results with the restaurant recommendations. To find the restaurants which are related

Algorithm 1. Dynamic context window calculation

Input: A tweet list for a user $S = \{T_1, T_2, ..., T_k, ..., T_n\}$; The tweet T_k which contains
the keywords for the foodborne disease symptoms; The threshold U for the similarity
between tweets; The decreasing rate for the similarity measure η; The upper bound
P for the dynamic window and the lower bound Q for the dynamic window.

Output: The candidate tweet set C

    Initialize $T = T_k, C = \emptyset$

    Push T_k into C

    for i=1 to P **do**

        if $Sim(T, T_{k-i}) > U$ **then**

            Push T_{k-i} into C

            Update $T = T + T_{k-i}$, Update $U = U * \eta$

        else

            break;

        end if

    end for

    Update $T = T_k$

    for j=1 to Q **do**

        if $Sim(T, T_{k+j}) > U$ **then**

            Push T_{k+j} into C

            Update $T = T + T_{k+j}$, Update $U = U * \eta$

        else

            break;

        end if

    end for

    return C

with the foodborne diseases, the geo-location for the foodborne disease event
should be determined. There may be location description in the registration
information for the Weibo users. But this location is the zone where the user
lives, not exactly the location of the foodborne disease event occurs. There are
also GPS data for some mobile users, but the data are also sparse. On the other
hand, we observe that lots of the tweets related with foodborne disease contain
the information of the restaurant or the name of the food. And some users also
refer to the location when they post a tweet. We utilize this information with
the aid of other data to find the restaurants related to the foodborne disease.

We define the restaurant and food information in the tweet as information
A. If the tweet contains the restaurant, we will get the location of the restaurant
from the website of dianping(https://www.dianping.com/), which has the information for the restaurants and their detailed locations. If the tweet contains the
name of the food, we could find all the restaurants with this food and their location from Baidu API. The tweets may also contain the geo-location when they
are posted. We define this kind of information as information B. We will use the
administrative location data on tcmap(http://www.tcmap.com.cn) to get the
detailed location for information B. The location in the registration information
of the Weibo users are defined as information C. Note that information A and

B may be missed, but information C is usually available. We then design an algorithm which utilizes the information A, B and C to get the location L of the foodborne disease events. The location L is determined as in formulas (2).

$$\begin{cases} L = \{A_i | minDist(A_i, B_j), A_i \in A, B_j \in B\}, & A \neq \emptyset, B \neq \emptyset \\ L = \{A_i | A_i \in C, A_i \in A\}, & A \neq \emptyset, C \neq \emptyset, B = \emptyset \\ L = \{B_j | B_j \in C, B_j \in B\}, & B \neq \emptyset, C \neq \emptyset, A = \emptyset \end{cases} \quad (2)$$

When A and B are both available, we compute the nearest two points $A_i \in A$ and $B_j \in B$ and A_i is the desired location. When A and C are available and B is empty, the locations in A which are also in C are the desired locations. When B and C are available and A is empty, the locations in B which are also in C are the desired locations.

After the locations for the foodborne disease events are determined, we find all the restaurants which are related with the foodborne disease events. We give a score for each restaurant and the score is lower for a restaurant related with the foodborne diseases. Then we insert the foodborne disease related information as a factor in the restaurant recommendation algorithm.

4   Experiments

In the experiments, we extract the tweets from 933,313 users in Beijing, China which contain the 31 keywords for the symptoms of foodborne diseases. The tweets are posted between August 2014 and October 2014. We construct the word vectors for the tweets using word2vec, and then use Algorithm 1 to get the candidate tweet set. There are totally about 80 million tweets in the set.

We use the ten features(#followings, #followers, length of personal description, #all tweets, #average retweeting, #average recommendation, #average comments, average length of the tweets, time of posting, # average links in tweets) to construct a SVM classifier to filter the tweets. After data filtering, we get about 31 % of the tweets in the candidate tweet set.

Except for Algorithm 1, there is also a method which selects a fixed number of tweets before and after the tweet T_k which is foodborne disease related. We select the 200 tweets before and after T_k into the candidate tweet set and compare it with our proposed Algortithm 1.

Table 1. Comparison of fixed context window calculation and Algorithm 1

Method	Accurate rate
Fixed context window calculation	27.4 %
Dynamic context window calculation	39.2 %

From Table 1 we can see that our proposed algorithm can get the candidate tweet set semantically and get higher accurate rate for the foodborne disease event detection.

We make statistics to the information A, B, C mentioned in Sect. 3.3. There are 13 % of all the tweets which contain both A and B, 19 % containing A and C but not B, and 16 % containing B and C but not A. For the case where A, B and C are all available, we calculate the accurate rate for the location determination. The result is shown in Table 2.

Table 2. The accurate rate for location determination

Total number	Correct number	Accurate rate
500	332	66.4 %
1000	647	64.7 %
1500	1009	67.3 %
2000	1280	64.0 %

From all the above experiment results, we see that the accurate rate for location determination and event detection are not high. That is partly caused by the characteristics of the tweets.

5   Conclusion

In this paper, we show how to use the social media data to extract valuable information about foodborne diseases. Algorithms are designed to get the foodborne disease related tweets and detect the foodborne disease events. The experiments show that our methods are effective. However, the accurate rate for the event detection is not high due to the sparsity and continuously changed topics of Weibo. In the future work, we will try to improve the detection algorithm and use the algorithm to assist the restaurant recommendation.

Acknowledgments. This work was supported by the National Natural Science Foundation of China under Grant No. 61402435,41371386,91224006, and the Knowledge Innovation Program of Chinese Academy of Sciences under Grant No. CNIC_QN_1507.

References

1. Adomavicius, G., Tuzhilin, A.: Toward the next generation of recommender systems: A survey of the state-of-the-art and possible extensions. IEEE Trans. Knowl. Data Eng. **17**(6), 734–749 (2005)
2. Sharma, L., Gera, A.: A survey of recommendation system: research challenges. Int. J. Eng. Trends Technol. (IJETT) **4**(5), 1989–1992 (2013)
3. Koren, Y., Bell, R., Volinsky, C.: Matrix factorization techniques for recommender systems. Computer **8**, 30–37 (2009)
4. Su, X., Khoshgoftaar, T.M.: A survey of collaborative filtering techniques. Adv. Artif. Intell. **2009**, 4 (2009)

5. Shi, Y., Larson, M., Hanjalic, A.: Collaborative filtering beyond the user-item matrix: A survey of the state of the art and future challenges. ACM Comput. Surv. (CSUR) **47**(1), 3 (2014)

6. Xie, H., Li, Q., Mao, X.: Context-aware personalized search based on user and resource profiles in folksonomies. In: Sheng, Q.Z., Wang, G., Jensen, C.S., Xu, G. (eds.) APWeb 2012. LNCS, vol. 7235, pp. 97–108. Springer, Heidelberg (2012)

7. Tumasjan, A., Sprenger, T.O., Sandner, P.G., Welpe, I.M.: Predicting elections with twitter: What 140 characters reveal about political sentiment. In: ICWSM 2010, pp. 178–185 (2010)

8. Li, X., Xie, H., Song, Y., Li, Q., Zhu, S., Wang, F.: Does summarization help stock prediction? news impact analysis via summarization. IEEE Intell. Syst. **30**(3), 26–34 (2015)

9. Aramaki, E., Maskawa, S., Morita, M.: Twitter catches the flu: detecting influenza epidemics using twitter. In: Proceedings of the Conference on Empirical Methods in Natural Language Processing, pp. 1568–1576. Association for Computational Linguistics (2011)

10. Signorini, A., Segre, A.M., Polgreen, P.M.: The use of twitter to track levels of disease activity and public concern in the US during the influenza A H1N1 pandemic. PloS ONE **6**(5), e19467 (2011)

11. Culotta, A.: Detecting influenza outbreaks by analyzing twitter messages (2010). arXiv preprint arXiv:1007.4748

12. Gomide, J., Veloso, A., Meira Jr., W., Almeida, V., Benevenuto, F., Ferraz, F., Teix-eira, M.: Dengue surveillance based on a computational model of spatio-temporallocality of twitter. In: Proceedings of the 3rd International Web Science Conference, p. 3. ACM (2011)

13. Center for Disease Control Prevention (CDC): CDC estimates of foodborne illness in the United States. Retrieved 23 March 2011

14. Newkirk, R.W., Bender, J.B., Hedberg, C.W.: The potential capability of social media as a component of food safety and food terrorism surveillance systems. Foodborne Pathog. Dis. **9**(2), 120–124 (2012)

15. Harris, J.K., Mansour, R., Choucair, B., Olson, J., Nissen, C., Bhatt, J., Brown, S.: Health department use of social media to identify foodborne illness-chicago, illinois, 2013–2014. MMWR Morb. Mortal Wkly. Rep. **63**(32), 681–685 (2014)

16. Xie, H., Yu, L., Li, Q.: A hybrid semantic item model for recipe search by example. In: 2010 IEEE International Symposium on Multimedia (ISM), pp. 254–259. IEEE (2010)

17. Sadilek, A., Brennan, S., Kautz, H., Silenzio, V.: nEmesis: Which restaurants should you avoid today? In: First AAAI Conference on Human Computation and Crowd- Sourcing (2013)

18. Sadilek, A., Kautz, H., DiPrete, L., Labus, B., Portman, E., Teitel, J., Silenzio, V.: Deploying nemesis: Preventing foodborne illness by data mining social media (2016)

19. Mikolov, T., Chen, K., Corrado, G., Dean, J.: Efficient estimation of word representations in vector space (2013). arXiv preprint arXiv:1301.3781

20. Mikolov, T., Sutskever, I., Chen, K., Corrado, G.S., Dean, J.: Distributed representations of words and phrases and their compositionality. In: Advances in Neural Information Processing Systems, pp. 3111–3119 (2013)

21. Mihalcea, R., Tarau, P.: Textrank: Bringing order into texts. Association for Computational Linguistics (2004)

A Combined Collaborative Filtering Model for Social Influence Prediction in Event-Based Social Networks

Xiao Li, Xiang Cheng, Sen Su$^{(\boxtimes)}$, Shuchen Li, and Jianyu Yang

State Key Laboratory of Networking and Switching Technology,
Beijing University of Posts and Telecommunications, Beijing, China
{lixiao,chengxiang,susen,lsc,jyyang}@bupt.edu.cn

Abstract. Event-based social networks (EBSNs) provide convenient online platforms for users to organize, attend and share social events. Understanding users' social influences in social networks can benefit many applications, such as social recommendation and social marketing. In this paper, we focus on the problem of predicting users' social influences on upcoming events in EBSNs. We formulate this prediction problem as the estimation of unobserved entries of the constructed user-event social influence matrix, where each entry represents the influence value of a user on an event. In particular, we define a user's social influence on a given event as the proportion of the user's friends who are influenced by him/her to attend the event. To solve this problem, we present a combined collaborative filtering model, namely, Matrix Factorization with Event Neighborhood (MF-EN) model, by incorporating event-based neighborhood method into matrix factorization. Due to the fact that the constructed social influence matrix is very sparse and the overlap values in the matrix are few, it is challenging to find reliable similar event neighbors using the widely adopted similarity measures (e.g., Pearson correlation and Cosine similarity). To address this challenge, we propose an additional information based neighborhood discovery (AID) method by considering three event-specific features in EBSNs. The parameters of our MF-EN model are determined by minimizing the associated regularized squared error function through stochastic gradient descent. We conduct a comprehensive performance evaluation on real-world datasets collected from DoubanEvent. Experimental results demonstrate the superiority of the proposed model compared to several alternatives.

1 Introduction

In the past few years, event-based social networks (EBSNs), such as Plancast[1] and DoubanEvent[2], have proliferated to be the online platforms for users to organize, attend and share social events to be held in offline physical venues [18]. EBSNs link online and offline social worlds, providing not only typical

---

[1] http://www.plancast.com.

[2] http://www.douban.com.

© Springer International Publishing Switzerland 2016
H. Gao et al. (Eds.): DASFAA 2016 Workshops, LNCS 9645, pp. 142–159, 2016.
DOI: 10.1007/978-3-319-32055-7_13

online social networking services, but also face to face offline communication by attending events. Previous studies [6,26,29] have shown that users could influence others to attend events in EBSNs, especially for close social ties. For instance, when an organizer publishes an event, users can express their willingness to join the event by RSVP ("yes" or "maybe")[3] and broadcast posts about their participating information to their friends (i.e., followers), who hesitate in making decisions. When a user's friends see his/her participating post, they might want to attend the event together with the user.

Social influence aims to study the behavioral change of a person because of the perceived relationships with other people in social networks. Since it has a wide range of applications, such as social recommendation [27] and social marketing [12], considerable works have been conducted to the influence analysis or prediction in social networks (e.g., Twitter and Facebook) [2,24,25]. However, as a newly emerged online social service, EBSN has its unique characteristics, such as event location and event organizer, which leads to the social influence analysis or prediction approaches used in conventional social networks might be ineffective in EBSNs. Nevertheless, social influences of EBSN users can provide valuable insights. For an upcoming event (i.e., the event which has not been held but has been published in the EBSNs), the event organizer hopes to maximize the attendees. This goal makes him/her desire to target the influencers on this event. These influencers are able to let many friends to attend this event by sharing the event. In this case, the event organizer needs to know users' influences to their friends. Therefore, understanding users' influences is a key issue in EBSNs.

In this paper, we focus on the social influence prediction problem in EBSNs. We formulate this prediction problem as the estimation of unobserved entries of the constructed social influence matrix S, where each entry (u, e) represents the social influence of user u on event e. Notice that, we focus on predicting users' influences on upcoming events which could provide valuable information for event organizers. In particular, similar to the definition of item-level social influence in conventional social networks [7], we define user u's influence on event e as the proportion of u's friends who are influenced by u to attend e. Different from the structure-level influence [21] and the topic-level influence [17,23], the predicted event-level influence can be used in two angles. On one hand, given an upcoming social event, we could find out the influencers to attract more friends for attending the event. On the other hand, given a user, we could recommend events for him/her to share, which can improve the interactions between the user and their friends. Matrix factorization is a straightforward approach to solve this prediction problem. By using users' observed influences, we could predict their influences on the upcoming events which have already some RSVPs ("yes" or "maybe"). However, as matrix factorization does not detect associations among the closely related items (i.e., users or events), the prediction performance of this approach might be poor. To improve the prediction accuracy, a potential

---

[3] The RSVP ("yes" or "maybe") indicates that a user wants to attend or is interested in an event. We assume that a user will attend the events which he/she has expressed RSVP ("yes") to.

approach is to integrate neighborhood method with matrix factorization [13]. However, since the social influence matrix S is very sparse and the overlap values in S are few, it is hard to find reliable similar neighbors using the widely adopted similarity measures (e.g., Pearson correlation and Cosine similarity). Therefore, how to discover reliable event neighbors in S is a challenging problem.

In EBSNs, event content, event location and event organizer are the major components of an event which affect users' decisions in attending the event. Therefore, if two events are similar on these three aspects, we can consider these two events as similar events. To this end, we propose an additional information based neighborhood discovery (AID) method to identify event neighborhood. To find the neighborhood of a targeted event e, we first capture three event-specific features (i.e., event content, event location and event organizer) and compute the similarities between e and other events on each feature. Then, we take the most similar events on each feature as a neighborhood of e. Such that, we obtain three neighborhood sets corresponding to the three features. Finally, a neighborhood aggregation strategy is proposed to derive the final neighborhood. In particular, in such strategy, we pick up the events contained in at least two neighborhood sets to make up the final neighborhood of e.

Based on AID, we present a combined collaborative filtering model, namely, Matrix Factorization with Event Neighborhood (MF-EN) model, to predict users' social influences on upcoming events in EBSNs. The model incorporates event-based neighborhood method into matrix factorization and thus can take advantages of both matrix factorization and neighborhood method. Model parameters are determined by minimizing the associated regularized squared error function through stochastic gradient descent. In summary, the major contributions of our work are listed as follows:

- We present a novel combined collaborative filtering model, namely, Matrix Factorization with Event-User Neighborhood (MF-EN) model, which incorporates event-based neighborhood method into matrix factorization, for social influence prediction in EBSNs. To the best of our knowledge, this is the first attempt to define and solve the event-level social influence prediction problem in EBSNs.
- To find reliable similar neighbors, we propose an additional information based neighborhood discovery (AID) method by considering event-specific features (i.e., event content, event location and event organizer) in EBSNs.
- We evaluate the performance of our prediction model on real-world datasets collected from DoubanEvent, which is the biggest event-based social network in China. Experimental results demonstrate the superiority of our MF-EN model compared to several alternatives.

The remainder of this paper is organized as follows. In Sect. 2, we give the definition of the social influence prediction problem, and show our model framework. In Sect. 3, the AID method is discussed in detail. We present our MF-EN model for social influence prediction in Sect. 4, followed by experimental evaluation in Sect. 5. We review the related work in Sect. 6. Finally, Sect. 7 concludes the paper.

2   Preliminaries

2.1   Event-Based Social Network

Users can establish, join and share events held offline in physical venues in event-based social networks (EBSNs). Users, events and organizers are three essential types of entities in EBSNs. As shown in Fig. 1, users in an EBSN denoted as U_1, U_2, U_3 and U_4 are interconnected via social links to form an online network. They are the participants of social events. Events denoted as E_1, E_2 and E_3 contain textual content information and locations (i.e., L_1 and L_2) where they are held offline. Organizers denoted as O_1 and O_2 are a special kind of users who establish as well as attend social events. They are the owners of social events and an organizer may hold more than one event.

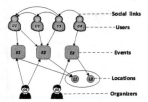

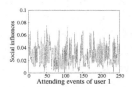 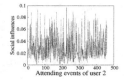

Fig. 1. The description of EBSNs

Fig. 2. Influence varies with users and events in EBSNs

2.2   Social Influence in EBSNs

Users may influence others to attend events in EBSNs. Let us consider the following scenario. Bob discovers a drama in DoubanEvent. Since he is not quite sure whether it is worth watching, Bob hesitates in making a decision to watch it until his friend Alice broadcasts her participation information on this drama. Given that Alice has broadcasted several wonderful social events before, Bob is quite confident about her taste in dramas and finally attends the event together with Alice.

We randomly select two active users in DoubanEvent, which is the largest event-based social network in China. For each user, we calculate the proportion of their influenced friends on each of his/her attending events and plot them in Fig. 2. From Fig. 2, we can observe that: different users have different social influences to their friends; users' social influences are different on different events.

2.3   Problem Definition

In EBSNs, there is a set of users $U = \{u_1, u_2, ..., u_M\}$, a set of events $E = \{e_1, e_2, ..., e_N\}$ and a set of organizers $O = \{o_1, o_2, ..., o_C\}$. For each user $u \in U$, it has a user profile (e.g., user id and username), friends (i.e., followers) collection

$F(u)$ and a set of past and upcoming events $HE_u \subseteq E$ that user u has expressed RSVP ("yes"). For each event $e \in E$, it has an event id, content information, an event organizer $o(e)$, a set of users who have expressed RSVP ("yes") (denoted as $HU_e \subseteq U$) and a physical location $l_e = \{lon_e, lat_e\}$ in terms of longitude and latitude where event e is held. For each organizer $o \in O$, it has a set of events E_o organized by him/her.

Xu et al. [26] have validated that mutual influences have effects on the event participation between friends in EBSNs by using statistics analysis. Similar to the definition of item-level social influence in conventional social networks [7], we define user u's social influence on event e as s_{ue}, which equals to $p_{ue}\big/|F(u)|$ (i.e., the proportion of u's influenced friends on event e), where p_{ue} denotes the number of u's friends who are influenced by u to attend event e. The social influence prediction problem can be formally defined as estimating the unknown s_{ue} according to the observed influence values. Notice that, we focus on predicting the social influence of a user on an upcoming event, on which some users have expressed RSVP ("yes"), and this user has already expressed RSVP ("yes") on some past or upcoming events before.

2.4   Model Framework

In this paper, focusing on the social influence prediction problem in EBSNs, we present a combined collaborative filtering predicting model, namely, Matrix Factorization with Event Neighborhood (MF-EN) model, which incorporates neighborhood method into matrix factorization to improve the prediction accuracy. As shown in Fig. 3, the model is composed of three major components: social influence matrix construction, additional information based neighborhood discovery and MF-EN predicting model.

Social Influence Matrix Construction. Using each user's attending events, we construct the user-event social influence matrix S. Obviously, there are $M \times N$ entries in S, and each entry is denoted as s_{ue}. Recall that s_{ue} is user u's social influence on event e. In practice, only some elements of S can be observed and

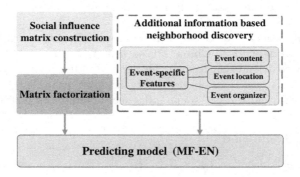

Fig. 3. The framework of the MF-EN model

the unobserved elements (of upcoming events) are represented by $\hat{s}_{ue}$. The social influence prediction problem is to predict $\hat{s}_{ue}$ by the observed s_{ue}.

Additional Information Based Neighborhood Discovery. We propose an additional information based neighborhood discovery (AID) method which takes event-specific features (i.e., event content, event location and event organizer) into consideration for neighborhood discovery.

MF-EN Model. Our proposed MF-EN model incorporates event-based neighborhood method into matrix factorization. Model parameters are learned by solving the associated regularized squared error function through stochastic gradient descent.

3 Additional Information Based Neighborhood Discovery

In EBSNs, since lots of new events are published every day and each user can only attend a small proportion of events, the constructed social influence matrix S is very sparse, and the overlap values in S are few. Therefore, it is challenging to find reliable similar neighbors by using the widely adopted similarity measures such as Pearson correlation and Cosine similarity. To address this challenge, by leveraging three event-specific features in EBSNs, we propose an additional information based neighborhood discovery (AID) method for event neighborhood discovery.

Event content, event location and event organizer are the major components of an event which affect users' decisions in attending the event in EBSNs. Therefore, we consider two events are similar events if they are similar on these three aspects. To this end, our AID method captures three event-specific features (i.e., event content, event location and event organizer) to perform event neighborhood discovery. Given a targeted $\hat{s}_{ue}$ in social influence matrix S, for each event e' in HE_u, which is a set of past and upcoming events user u has expressed RSVP ("yes"), we first compute the similarities of event content, event location and event organizer between e and e', denoted as $ES_c(e, e')$, $ES_l(e, e')$ and $ES_o(e, e')$, respectively. Then, we take the most similar events on each feature as a neighborhood of e. Such that, we derive three neighborhood sets corresponding to the three features. Finally, we utilize a neighborhood aggregation strategy to find the final neighborhood. In particular, in this strategy, we take events contained in at least two neighborhood sets as the final neighborhood of e. We discuss how to determine the size of the neighborhood on each feature in Sect. 5. In the following, we will introduce how to compute the similarities between events on each event-specific feature in detail.

3.1 Event Content

Event content is the key component of an event, which plays a major role in determining users' decisions in attending the event. In order to derive the content similarity between events, we put the event content including event title, description and category as a document and obtain the event distribution on latent topics by employing the Latent Dirichlet Allocation (LDA) model [3], which is an unsupervised machine learning technique to identify latent topics from a large document.

Based on the assumption that documents are mixtures of topics, LDA models document d as a probability topic distribution, denoted as θ_d, and each topic z is represented as a probability distribution over terms in the vocabulary, denoted as ϕ_z. Like previous study [16], we first format content text of event e to document d_e, by removing the stop words from each corpus. Then each event has a corresponding document, which is taken as the input of the LDA model. Finally, we can obtain all events' document-topic distributions Θ and topic-word distributions Φ, where the topic distribution represents the varieties of an event. We suppose there are K latent topics. The generative process of LDA is as follows:

1. For each topic $z \in \{1, 2, ...K\}$, draw $\phi_z \sim Dirichlet\,(\beta)$, which is a multinomial distribution over terms.
2. For each event document d_e
   (a) Draw a topic distribution $\theta_{d_e} \sim Dirichlet(\alpha)$.
   (b) For each word $w_{d_e,n}$ in document d_e,
       (i) Draw a topic $z_{d_e,n} \sim Mult(\theta_{d_e})$,
       (ii) Draw a word $w_{d_e,n} \sim Mult(\phi_{z_{d_e,n}})$.

Given the hyperparameters α and β, the joint distribution of an event document is specified as

$$p\left(w_{d_e}, z_{d_e}, \theta_{d_e}, \Phi | \alpha, \beta\right) = \prod_{n=1}^{N_e} p\left(w_{d_e,n} | \phi_{z_{d_e,n}}\right) p\left(z_{d_e,n} | \theta_{d_e}\right) p\left(\theta_{d_e} | \alpha\right) p\left(\Phi | \beta\right), \quad (1)$$

where N_e is the number of words of event e.

The complete likelihood of N event documents is derived based on the assumption that all the documents are independent of each other:

$$p(W, Z, \Theta, \Phi | \alpha, \beta) = \prod_{e=1}^{N} p(w_{d_e}, z_{d_e}, \theta_{d_e}, \Phi | \alpha, \beta). \quad (2)$$

The model has two unknown parameters to be inferred: the document-topic distributions Θ, and the topic-word distributions Φ. We utilize Gibbs sampling [11] to estimate these parameters, which is a special sort of Markov-chain Monte Carlo (MCMC) simulation.

Given the topic distributions θ_{d_e} and $\theta_{d_{e'}}$ of events e and e', we can use the Jensen-Shannon divergence [11] to compute the content similarity between them. The Jensen-Shannon divergence is defined as follows:

$$D_{JS}(\theta_{d_e}, \theta_{d_{e'}}) = \frac{1}{2}[D_{KL}(\theta_{d_e}, \frac{\theta_{d_e} + \theta_{d_{e'}}}{2}) + D_{KL}(\theta_{d_{e'}}, \frac{\theta_{d_e} + \theta_{d_{e'}}}{2})], \quad (3)$$

where $D_{KL}(\cdot)$ is the Kullback-Leibler divergence. In particular, the value Jenson-Shannon divergence ranges from 0 to 1 and increases when the distinction between θ_{d_e} and $\theta_{d_{e'}}$ becomes larger.

Finally, given two events e and e', we define the content similarity $ES_c(e, e')$ $(0 \leq ES_c(e, e') \leq 1)$ between them as:

$$ES_c(e, e') = 1 - D_{JS}(\theta_{d_e}, \theta_{d_{e'}}).$$
(4)

3.2   Event Location

In EBSNs, event location specifies the physical place where the event is held. Compared with conventional social networks, offline social interaction is a unique characteristic of EBSNs. Therefore, event location is an important factor affecting whether a user attends an event. To better understand the impact of event location on users' behaviours, we perform a data analysis on a real-world dataset crawled from DoubanEvent. We calculate the distances between all pairs of events with different locations which each user has attended and plot the probability density function of the distance in Fig. 4 in log-log scale. As shown, we can observe that the distance probability distribution approximately follows a power law, which means that most of the event pairs which a user has attended are within a short distance. To this end, we take event location as an important factor, and consider two events are similar if their locations are close.

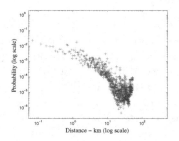

Fig. 4. Distance probability distribution

Let $dis(l_e, l_{e'})$ denote the Euclidean distance between event e and e'. We use the Gauss formula which describes the exponential decay with the distance to measure the location similarity $ES_l(e, e')$ $(0 < ES_l(e, e') \leq 1)$ between event e and e', which is defined as:

$$ES_l(e, e') = \exp\{-\frac{dis(l_e, l_{e'})^2}{2}\}.$$
(5)

3.3   Event Organizer

In EBSNs, organizers are the owners of events. Whether a user attends an event is also affected by the event organizer [8]. For example, if the organizer is a user's favorite singer, the user would be interested in attending the singer's concerts. Therefore, we consider, two events are similar if they are held by the same organizer. In particular, we define the organizer similarity $ES_o(e, e')$ between event e and e' as a binary value function,

$$ES_o(e, e') = \begin{cases} 1 & o(e) = o(e') \\ 0 & \text{others} \end{cases}, \qquad (6)$$

where $o(e)$ denotes event e's organizer and $o(e) = o(e')$ means that event e and e' are held by the same organizer.

4   MF-EN Predicting Model

To improve the accuracy of social influence prediction in EBSNs, we present a combined collaborative filtering model, namely, Matrix Factorization with Event Neighborhood (MF-EN) model. In this section, we first introduce the basis of our model (i.e., matrix factorization), then show our MF-EN model.

4.1   Matrix Factorization

Matrix factorization is the basis of our combined model. It can be seen as the baseline approach to solve our social influence prediction problem. For the user-event social influence matrix S, matrix factorization maps both users and events into a joint latent factor space of dimensionality d, such that the influence matrix is modeled as inner products in that space [14]. User u is associated with a vector $p_u \in \mathbb{R}^d$ and event e is associated with a vector $q_e \in \mathbb{R}^d$. Both vectors p_u and q_e are referred to as d-dimensional latent factors. The social influence $\hat{s}_{ue}$ of user u on the upcoming event e can be predicted according to the following equation,

$$\hat{s}_{ue} = p_u^T q_e. \qquad (7)$$

Parameters p_u and q_e are generally learned by solving the following regularized least squares problem:

$$\min_{p_*, q_*} \sum_{(u,e) \in \kappa} \left(s_{ue} - p_u^T q_e \right)^2 + \lambda \left(\| p_u \|^2 + \| q_e \|^2 \right), \qquad (8)$$

where κ is the collection of the whole observed values of the social influence matrix and the constant λ is a parameter determining the extent of regularization.

4.2   MF-EN: Matrix Factorization with Event Neighborhood Model

Similar to matrix factorization, neighborhood method [4,5] can also be used to predict the unobserved values of the social influence matrix. These two types of methods have their own advantages and disadvantages [13,22]. Neighborhood method is effective at detecting localized relationships, which focus on computing the relationships between similar neighbors. They always ignore the vast majority of values and predict values only dependent on a few significant neighborhood. While matrix factorization is effective at estimating global information and poor at detecting strong associations on closely related neighbors.

In this paper, we incorporate both event-based neighborhood method into matrix factorization by a combined collaborative filtering model, namely, Matrix Factorization with Event Neighborhood (MF-EN) model, which takes advantages of both matrix factorization and neighborhood method to enrich each other. Let $\bar{s}_u$ denote the average social influence values of user u. The model is shown as follows:

$$\hat{s}_{ue} = p_u^T q_e + |N(u,e)|^{-\frac{1}{2}} \sum_{f \in N(u,e)} w_{ef}(s_{uf} - \bar{s}_u), \tag{9}$$

where $N(u,e)$ is the neighbor set of event e among HE_u, which is found by the AID method. Moreover, w_{ef} is the parameter which denotes the influence weights of event f to e.

Equation 9 provides a 2-tier model for social influence prediction: The first tier $p_u^T q_e$ considers the global interaction between users and events; The second tier contributes the fine grained adjustments that the event neighborhood plays roles in. All parameters in Eq. 9 can be determined by minimizing the associated regularized squared error function through stochastic gradient descent:

$$\min_{p_*,q_*,w_*} \sum_{(u,e)\in\kappa} \left(s_{ue} - p_u^T q_e - |N(u,e)|^{-\frac{1}{2}} \sum_{f \in N(u,e)} w_{ef}(s_{uf} - \bar{s}_u) \right)^2 + \lambda_1 \left(\|p_u\|^2 + \|q_e\|^2 + \sum_{f \in N(u,e)} w_{ef}^2 \right), \tag{10}$$

where λ_1 determines the extent of regularization.

5   Experiments

In this section, we evaluate the proposed model based on real-world EBSN datasets. We first describe the experimental setup including the datasets, evaluation metrics and comparison methods. Then, we evaluate the prediction accuracy of our proposed model.

5.1   Experimental Setup

Datasets. The datasets used in our experiments are collected from the website of DoubanEvent. We get the following data: (1) event information, including event id, content information (category, title and textual description), organizer id, physical location (longitude, latitude) and the set of attendees who are recorded in order of time when they express RSVP ("yes"); (2) user information, including user id, username, city and followers IDs. To make data sufficient for evaluation, we remove users who have attended fewer than 5 events (about 5 % of the total users) and events whose participants are fewer than 8 (about 3 % of the total events). After preprocessing, we get 11123 users, 29342 events, 153408 friend links and 356052 user-event pairs. The sparsity of the resulting dataset is 99.9 %.

To capture the social influence in EBSNs, similar to previous social influence studies [10,24], we consider that a user is influenced by a friend when he/she attends an event after that friend's attending. In particular, we assume that all social influences are independent from each other, thus when a user gets multiple broadcastings of an event information before he expresses RSVP ("yes") on it, we simplify the case as that the user is influenced by the latest one.

In our experiments, we randomly sample different number of users and select events attended by these users to form datasets with different sizes, including 1000 users dataset, 5000 users dataset and 11123 users dataset. In particular, we use the 5000 users dataset for parameters setting. We randomly select 50 %, 70 % and 90 % of the observed entries in social influence matrix S of different sizes of datasets (i.e., 1000 users dataset, 5000 users dataset and 11123 users dataset) for training, and the rest for testing.

Evaluation Metrics. To evaluate the accuracy of our proposed method, we adopt two popular evaluation metrics, namely, *Root Mean Square Error* (*RMSE*) and *Mean Absolute Error* (*MAE*), which are defined as:

$$RMSE = \sqrt{\frac{1}{|S'|} \sum_{(u,e) \in S'} (s_{ue} - \hat{s}_{ue})^2}, \qquad MAE = \frac{1}{|S'|} \sum_{(u,e) \in S'} |s_{ue} - \hat{s}_{ue}|, \quad (11)$$

where $|S'|$ denotes the size of the testing set S'. The smaller *RMSE* or *MAE* value indicates better accuracy.

Comparison Methods. We compare our MF-EN model with the following 7 methods.

- **Logistic Regression (LR):** If we regard the event-specific features as variables, and the values of social influences as the response, then the social influence prediction on upcoming events can be formulated as the regression problem. Thus, we use the LR model to combine the event-specific features linearly and learn the regression coefficients of these features from the training data.
- **Event Influence Mean (EM):** This method uses the mean influence value of the corresponding event to predict the unobserved values:

$$\hat{s}_{.,e} = \sum_{u \in HU_e} s_{ue} \Big/ |HU_e|. \quad (12)$$

- **User Influence Mean (UM):** Similar to EM, this method uses the mean influence value of the corresponding user to predict the unobserved values:

$$\hat{s}_{u,.} = \sum_{e \in HE_u} s_{ue} \Big/ |HE_u|. \tag{13}$$

- **Classical Event-Based Neighborhood Method (P-EN):** The unobserved values are predicted based on the values of events' neighbors discovered by using Pearson correlation:

$$\hat{s}_{ue} = |N(u,e)|^{-\frac{1}{2}} \sum_{f \in N(u,e)} w_{ef}(s_{uf} - \bar{s}_u). \tag{14}$$

Model parameters are determined by Eq. 15. Actually, we have also evaluated Cosine similarity as the similarity measure which gives poorer results compared with Pearson correlation.

$$\min_{w_*} \sum_{(u,e) \in \kappa} (s_{ue} - A_{ue})^2 + \lambda_2 \cdot \sum_{f \in N(u,e)} w_{ef}^2. \tag{15}$$

- **Event-Based Neighborhood Method Using Additional Information (AI-EN):** Different from P-EN, events' neighbors are discovered by using our AID method.
- **Matrix Factorization (MF):** In this method, we predict the unobserved values using matrix factorization (i.e., Eq. 7) and parameters are determined by Eq. 8.
- **HF-NMF:** If we consider the event content as the content of the web post, and the proportion of friends who are influenced by the user to attend the event as the user's social influence on the post, then the hybrid factor non-negative matrix factorization (HF-NMF) approach proposed in [7] can be used to predict the user's social influence on an upcoming event.

5.2 Experimental Results

5.2.1 Parameters Setting

Parameters of LDA. In order to achieve the content similarity between events, we select the optimal LDA parameters: α, β and the number of topics K. According to the setting in [15], we set $\alpha = 50 \big/ K$ and $\beta = 0.01$. We evaluate $RMSE$ and MAE under different values of K from 10 to 100. The results plotted in Fig. 5 show that the performance of LDA increases with the growth of K and there is little performance improvement after $K = 70$. However, the time consumption increases sharply when K is larger than 70. To balance the accuracy and computation complexity, we fix the value of K in LDA to 70 in our experiments.

Parameters of MF. We set $\lambda = 0.01$ for matrix factorization. How to set the dimension number d is important for the prediction performance. If d is too small, we cannot discriminate users and events in the latent space. If d is too large, the computation complexity will be greatly increased. Thus, we evaluate $RMSE$ and MAE of MF by varying the number of latent dimensions d from

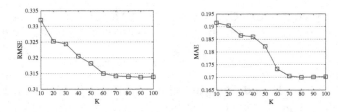

Fig. 5. Prediction performance of LDA with different topic numbers

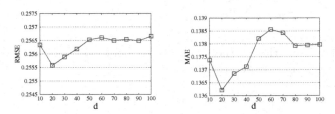

Fig. 6. Prediction performance of MF with different latent dimensions

10 to 100. The performance results are plotted in Fig. 6. According to Fig. 6, we set the dimension number as 20, where *RMSE* and *MAE* achieve the best performance.

Parameters of Neighborhood Size. In our AID method, the size of the event neighborhood set on each event-specific feature determines the final event neighborhood size. For simplicity, the size of the event neighborhood on each event-specific feature is set to be equal, which is denoted as k. Recall that, for a given event e, its neighborhood on the feature of event organizer is consisted of the events which have the same organizer with e. Therefore, the size of the neighborhood on the event organizer feature is not restricted by k (i.e., it might be smaller or larger than k). The prediction performance of our proposed methods under different neighborhood size k is shown in Fig. 7. We can observe that the size affects the prediction performance and the optimal size of these methods is different. To obtain the best performance, we set k of AI-EN and MF-EN to 100 and 70, respectively.

5.2.2   Advantages of the AID Method

In this section, we study the performance of the AID method. We compare the performance of neighborhood methods whose neighbors are found by our proposed AID approach (i.e., AI-EN) with methods whose neighbors are directly obtained by using the Pearson correlation (i.e., P-EN).

Our proposed AID method considers three event-specific features for event neighborhood discovery. In order to demonstrate the advantages of the combination of these features in the event neighborhood discovery, we also compare the performance of AI-EN with methods whose neighbors are discovered only

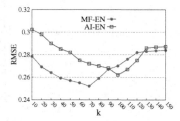

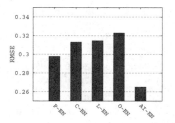

Fig. 7. Performance with different k

Fig. 8. Performance of neighborhood methods

by a single feature (e.g., event content). The event-based neighborhood methods using the single feature of event content (C), event location (L) and event organizer (O) for event neighborhood discovery are denoted as C-EN, L-EN and O-EN, respectively.

In Fig. 8, we plot the performance of these event-based neighborhood method. We can observe that AI-EN which discovers the event neighborhood by aggregating the neighborhood sets on the event-specific features can obtain the best performance. The reason lies in that: (1) The social influence matrix S is sparse, and similar neighbors found by the Pearson correlation are unreliable; (2) Only using the single feature of EBSNs (e.g. event content) cannot find reliable event neighborhood.

5.2.3 Performance Comparison

The prediction errors measured by RMSE and MAE of the comparison methods on different datasets are shown in Table 1 with best results highlighted in boldface. We make 4 observations from the results.

First, the proposed MF-EN model, which incorporates event-based neighborhood method into matrix factorization by considering multiple event-specific features in EBSNs, achieves the best performance measured by both RMSE and MAE in our experiments. Notice that, similar to the work [7], the values of users' event-level social influences are very small in our datasets (usually less than 0.2). Therefore, a small decrease of the RMSE (or MAE) can give significant performance improvements.

Second, the comparisons between MF-EN and MF reveal that the performance improves when incorporating event-based neighborhood method into matrix factorization. For example, MF-EN is better than MF and EN. This is because the memory-based and model-based collaborative filtering approaches have their own advantages and can enrich each other.

Third, in most of the experiments, HF-NMF performs better than MF, while MF-EN perform better than HF-NMF. This is because HF-NMF incorporates the event content information into matrix factorization. However, since MF-EN integrates event-based neighborhood method with matrix factorization by

Table 1. RMSE and MAE of all methods

Training %	Metrics	LR	EM	UM	AI-EN	MF	HF-NMF	MF-EN
1000 users								
50 %	RMSE	0.352436	0.338952	0.325979	0.267345	0.261651	0.258213	**0.255415**
	MAE	0.215672	0.19437	0.18908	0.135672	0.135824	0.135162	**0.133524**
70 %	RMSE	0.348563	0.336231	0.316227	0.255742	0.2495	0.249485	**0.246755**
	MAE	0.210435	0.190721	0.170242	0.132275	0.129875	0.129716	**0.127657**
90 %	RMSE	0.334436	0.326821	0.309297	0.251824	0.245768	0.244702	**0.241334**
	MAE	0.201534	0.189877	0.169782	0.131835	0.127337	0.126837	**0.124007**
5000 users								
50 %	RMSE	0.360245	0.341215	0.332983	0.276524	0.272651	0.266536	**0.259040**
	MAE	0.222875	0.198547	0.197218	0.137282	0.137211	0.135921	**0.133142**
70 %	RMSE	0.351895	0.324997	0.320721	0.262412	0.255582	0.255272	**0.252169**
	MAE	0.218854	0.189451	0.189882	0.137410	0.136214	0.135415	**0.132086**
90 %	RMSE	0.346537	0.310115	0.318723	0.258927	0.253286	0.252438	**0.249027**
	MAE	0.206981	0.172589	0.170071	0.135446	0.134275	0.133727	**0.129224**
11123 users								
50 %	RMSE	0.369857	0.349987	0.349927	0.278562	0.275315	0.274423	**0.270140**
	MAE	0.232471	0.205817	0.203802	0.138741	0.138634	0.137912	**0.135934**
70 %	RMSE	0.357635	0.329752	0.343552	0.264921	0.260089	0.259355	**0.253261**
	MAE	0.226934	0.190168	0.198021	0.137486	0.136987	0.136581	**0.134632**
90 %	RMSE	0.351537	0.318853	0.338571	0.261374	0.254758	0.254527	**0.251426**
	MAE	0.214325	0.178954	0.191872	0.135961	0.135364	0.135036	**0.131236**

considering some unique characteristics of EBSNs, such as event location and event organizer, they obtain better results than HF-NMF.

Last, the percentage of the training data has a significant impact on the prediction performance. In particular, the more training data, the lower prediction errors (i.e., measured by RMSE and MAE) the method can achieve. The reason lies in that the performance of matrix factorization is poor in the case where there is very little training data. Moreover, the prediction performance of EM is worse than all the collaborative filtering approaches.

6   Related Work

Event-Based Social Network. Liu et al. [18] firstly introduce and define the EBSN, a new type of social network which connects the online and offline social worlds. Some works study the event recommendation problem [19,31] in EBSNs. Qiao et al. [19] present a Bayesian latent factor model for event recommendation by incorporating heterogenous social relationships, geographical features and implicit ratings. Yu et al. [28] study the problem of identifying the most influential and preferable set of invitees by extending the credit distribution model. To predict the event attendance, Du et al. [8] propose a singular value decomposition with multi-factor event-based neighborhood algorithm. Formulating the group-oriented event participation problem as a novel discriminant framework, Xu et al. [26] exploit the impact of dynamic mutual influence for the social event participation prediction. Different from these works, we attempt to quantify the social influences of users in EBSNs.

Social Influence Analysis. Influence is a potential factor which affects users' behaviors. Considerable works have been conducted to qualitatively validate the existence of influence [2, 20]. Anagnostopoulos et al. [2] apply the statistical test (i.e., shuffle test) to identify whether influence is a source of social correlation using the time factor in a social network system. Chin et al. [6] investigate the user behaviour on attending offline events and find that social influences exist in EBSNs and users choose to attend an event partly because their friends will attend this event. Recently, some works have been proposed to quantify the social influence in different social networks [1, 10, 25]. Zhang et al. [30] argue that the influence is continuously dynamic and infer the continuous dynamic social influence for temporal behavior prediction. Goyal et al. [10] propose both static and time-dependent model to capture influence probabilities from a log of past propagations. They consider user u influences user v on an action if they are friends and u performs this action before v. Zhang et al. [29] propose a unified metric to quantify the mutual user influence between social relation and geographical distance in location-based social networks (LBSNs). They evaluate the social influence of each user-pair in the participant set from a random walk perspective. Different from this work, we attempt to estimate each user's social influence to their friends on an event whose potential participants are unknown. There are also some studies focusing on the social influence in a more fine-grained level. Tang et al. [17] analyze the topic-level social influence using the probabilistic model, in which they state, individuals' influences to others could vary greatly across different topics. Weng et al. [25] measure the topic-sensitive influences of users in Twitter by taking both the topical similarity between users and the link structure into account. Embar et al. [9] present online, multi-dimensional approach for topic-specific social influence analysis. Cui et al. [7] consider a user's social influences are different on different posts, thus they define the item-level social influence and propose a hybrid factor non-negative matrix factorization approach (HF-NMF) to solve the influence prediction problem in conventional social network (e.g., Facebook and Twitter). Inspired by their work, we focus on predicting the event-level social influence in EBSNs. Since the method proposed in [7] does not consider the unique characteristics of EBSNs (e.g., event location), it cannot satisfactorily solve our event-level social influence prediction problem. To the best of our knowledge, this is the first attempt to quantify the event-level social influence in EBSNs.

7   Conclusion

In this paper, we study the problem of predicting users' social influences on upcoming events in event-based social networks (EBSNs). In particular, we define a user's social influence on a given event as the proportion of his/her friends who are influenced by the user to attend the event. To solve this problem, we present a combined collaborative filtering model, namely, Matrix Factorization with Event Neighborhood (MF-EN) model, which takes advantages of both matrix factorization and neighborhood method. In the MF-EN model, to find reliable similar neighbors, we propose an additional information based neighborhood discovery (AID) method, which takes three event-specific features (i.e., event content, event location and event organizer) into consideration. We conduct extensive experiments on real-world datasets collected from DoubanEvent. The experimental results demonstrate that our proposed model outperforms several alternatives. In our future work, we plan to incorporate the user-specific features into our model.

Acknowledgement. The work was supported by National Natural Science Foundation of China under Grant 61502047.

References

1. Agarwal, N., Liu, H., Tang, L., Yu, P.S.: Identifying the influential bloggers in a community. In: WSDM, pp. 207–218 (2008)
2. Anagnostopoulos, A., Kumar, R., Mahdian, M.: Influence and correlation in social networks. In: SIGKDD, pp. 7–15 (2008)
3. Blei, D.M., Ng, A.Y., Jordan, M.I.: Latent dirichlet allocation. J. Mach. Learn. Res. **3**, 993–1022 (2003)
4. Cai, Y., Lau, R.Y., Liao, S.S., Li, C., Leung, H.F., Ma, L.C.: Object typicality for effective web of things recommendations. Decis. Support Syst. **63**, 52–63 (2014)
5. Cai, Y., Leung, H.F., Li, Q., Min, H., Tang, J., Li, J.: Typicality-based collaborative filtering recommendation. IEEE Trans. Knowl. Data Eng. **26**(3), 766–779 (2014)
6. Chin, A., Tian, J., Han, J., Niu, J.: A study of offline events and its influence on online social connections in douban. In: GreenCom and iThings/CPSCom, pp. 1021–1028 (2013)
7. Cui, P., Wang, F., Liu, S., Ou, M., Yang, S., Sun, L.: Who should share what?: item-level social influence prediction for users and posts ranking. In: SIGIR, pp. 185–194 (2011)
8. Du, R., Yu, Z., Mei, T., Wang, Z., Wang, Z., Guo, B.: Predicting activity attendance in event-based social networks: content, context and social influence. In: Ubicomp, pp. 425–434 (2014)
9. Embar, V.R., Bhattacharya, I., Pandit, V., Vaculín, R.: Online topic-based social influence analysis for the wimbledon championships. In: KDD, pp. 1759–1768 (2015)
10. Goyal, A., Bonchi, F., Lakshmanan, L.V.: Learning influence probabilities in social networks. In: WSDM, pp. 241–250 (2010)
11. Heinrich, G.: Parameter estimation for text analysis. Technical report (2005)
12. Kempe, D., Kleinberg, J., Tardos, É.: Maximizing the spread of influence through a social network. In: SIGKDD, pp. 137–146 (2003)
13. Koren, Y.: Factorization meets the neighborhood: a multifaceted collaborative filtering model. In: KDD, pp. 426–434 (2008)
14. Koren, Y., Bell, R., Volinsky, C.: Matrix factorization techniques for recommender systems. Computer **8**, 30–37 (2009)

15. Lin, J.: Divergence measures based on the shannon entropy. IEEE Trans. Inf. Theory **37**(1), 145–151 (1991)
16. Liu, B., Xiong, H.: Point-of-interest recommendation in location based social networks with topic and location awareness. In: SDM, vol. 13, pp. 396–404 (2013)
17. Liu, L., Tang, J., Han, J., Jiang, M., Yang, S.: Mining topic-level influence in heterogeneous networks. In: CIKM, pp. 199–208 (2010)
18. Liu, X., He, Q., Tian, Y., Lee, W.C., McPherson, J., Han, J.: Event-based social networks: linking the online and offline social worlds. In: KDD, pp. 1032–1040 (2012)
19. Qiao, Z., Zhang, P., Cao, Y., Zhou, C., Guo, L., Fang, B.: Combining heterogenous social and geographical information for event recommendation. In: AAAI, pp. 145–151 (2014)
20. Singla, P., Richardson, M.: Yes, there is a correlation: -from social networks to personal behavior on the web. In: WWW, pp. 655–664 (2008)
21. Strogatz, S.H.: Exploring complex networks. Nature **410**(6825), 268–276 (2001)
22. Su, X., Khoshgoftaar, T.M.: A survey of collaborative filtering techniques. Adv. Artif. Intell. **2009**, 4 (2009)
23. Tang, J., Sun, J., Wang, C., Yang, Z.: Social influence analysis in large-scale networks. In: SIGKDD, pp. 807–816 (2009)
24. Wen, Y.T., Lei, P.R., Peng, W.C., Zhou, X.F.: Exploring social influence on location-based social networks. In: ICDM, pp. 1043–1048 (2014)
25. Weng, J., Lim, E.P., Jiang, J., He, Q.: Twitterrank: finding topic-sensitive influential twitterers. In: WSDM, pp. 261–270 (2010)
26. Xu, T., Zhong, H., Zhu, H., Xiong, H., Chen, E., Liu, G.: Exploring the impact of dynamic mutual influence on social event participation. In: SDM, pp. 262–270 (2015)
27. Ye, M., Liu, X., Lee, W.C.: Exploring social influence for recommendation: a generative model approach. In: SIGIR, pp. 671–680 (2012)
28. Yu, Z., Du, R., Guo, B., Xu, H., Gu, T., Wang, Z., Zhang, D.: Who should i invite for my party?: combining user preference and influence maximization for social events. In: Ubicomp, pp. 879–883(2015)
29. Zhang, C., Shou, L., Chen, K., Chen, G., Bei, Y.: Evaluating geo-social influence in location-based social networks. In: CIKM, pp. 1442–1451 (2012)
30. Zhang, J., Wang, C., Wang, J., Yu, J.X.: Inferring continuous dynamic social influence and personal preference for temporal behavior prediction. PVLDB **8**(3), 269–280 (2014)
31. Zhang, W., Wang, J.: A collective bayesian poisson factorization model for cold-start local event recommendation. In: KDD, pp. 1455–1464 (2015)

Learning Manifold Representation from Multimodal Data for Event Detection in Flickr-Like Social Media

Zhenguo Yang[1]([⊠]), Qing Li[1,2], Wenyin Liu[2], and Yun Ma[1]

[1] Department of Computer Science, City University of Hong Kong,
Hong Kong, China
yzgcityu@gmail.com, itqli@cityu.edu.hk, mayun371@gmail.com
[2] Multimedia-software Engineering Research Center,
City University of Hong Kong, Hong Kong, China
liuwenyin@gmail.com

Abstract. In this work, a three-stage social event detection model is devised to discover events in Flickr data. As the features possessed by the data are typically heterogeneous, a multimodal fusion model (M^2F) exploits a soft-voting strategy and a reinforcing model is devised to learn fused features in the first stage. Furthermore, a Laplacian non-negative matrix factorization (LNMF) model is exploited to extract compact manifold representation. Particularly, a Laplacian regularization term constructed on the multimodal features is introduced to keep the geometry structure of the data. Finally, clustering algorithms can be applied seamlessly in order to detect event clusters. Extensive experiments conducted on the real-world dataset reveal the M^2F-LNMF-based approaches outperform the baselines.

Keywords: Social media analytics · Multimedia content analysis · Multimodal fusion · Manifold learning · Event detection

1 Introduction

The popularity of Flickr-like photo-sharing social media services has resulted in huge amounts of user-contributed images available online, attracting researchers to link the data to numerous real-world concepts. Social event detection (SED) from Flickr data aims to discover the real-world events that are attended by people and can be represented by user-contributed multimedia data. Instances of such events could include concerts, public celebrations, annual conventions, local gatherings, sports events, etc.

SED from Flickr data is deemed to be more challenging than the event detection tasks in textural data [4,11] as the data is heterogeneous. For instance, the Flickr images possess context features including time-taken, user identity, location, tags and visual content, etc. Such features will be helpful for capturing the similarities among the social media documents and, in turn, for identifying

© Springer International Publishing Switzerland 2016
H. Gao et al. (Eds.): DASFAA 2016 Workshops, LNCS 9645, pp. 160–167, 2016.
DOI: 10.1007/978-3-319-32055-7_14

event clusters and their associated documents. However, the heterogeneous features are hard to be exploited by traditional clustering or classification models seamlessly. To address the problem, early fusion and late fusion are widely-used strategies. Late fusion is expensive in terms of learning effort and the result may not be good since each modality might be poor. As a result, early fusion strategy is more popular. However, early fusion models usually construct multiple affinity graphs [1, 2] with intensive computations, making them not adaptive in dealing with social media data due to its large quantity and high updating rate.

In this work, a social event detection model is designed to discover events from photo-sharing social media sites, which consists of three stages. In the first stage, we propose a multimodal fusion (M^2F) model to learn fused features from the heterogeneous feature modalities. Particularly, M^2F exploits a unimodal soft-voting strategy to learn comparable and robust vote features, which expresses the data samples by their neighborhood information. Furthermore, M^2F exploits a reinforcing model to learn the vote propagations among the multimodal features and achieve fused features. In the second stage, we exploit a Laplacian non-negative matrix factorization model, denoted as LNMF, to extract compact manifold representation from the fused features. In particular, the Laplacian regularization term is constructed based on the multimodal features, which tends to learn similar manifold representation for the samples that are close in the fused feature space. In the third stage, clustering algorithms can be applied on the manifold representation learned by M^2F-LNMF to discover event clusters. Particularly, incorporating density knowledge or label information in the initialization of the center-based clustering algorithm will give significant improvement on the performance.

The rest of the paper is organized as follows. In Sect. 2, the related work is reviewed. In Sect. 3, the proposed three-stage social event detection model is presented. In Sect. 4, extensive experiments are conducted and analyzed. Finally, we offer conclusions in Sect. 5.

2 Related Work

2.1 Social Event Detection in Multimedia Social Media Streams

To detect events from Flickr data, researchers usually employed classification models. Liu et al. [7] trained various models like KNN, SVM, decision tree and random forest to classify the images into the events they depict. Chen et al. [5] developed a system to discover semantic concept in videos by exploiting Web images and their associated tags, and trained a SVM model for predictions. However, most of them are designed for domain-specific events that are well-defined in advance, making them not adaptive in dealing with the variety of events in social media. Nitta et al. [8] constructed similarity graphs and applied community detection to identify subgraphs that could be landmarks or events. However, these approaches require pair-wise similarity calculations, which are computation-intensive and memory-consuming.

2.2 Multimodal Feature Fusion

A number of early fusion models have been proposed. For instance, Cai et al. [2] computed a multimodal Laplacian matrix by integrating the individual affinity matrix on each modality, and further learned a low-dimensional feature space by introducing a penalty for each modality. Julien et al. [1] proposed a unifying graph-based framework, which combines both visual and textual information. Petkos et al. [9] trained a classifier to acquire an indicator matrix showing whether two images could be in the same event. Furthermore, spectral clustering was performed on the indicator matrix to discover event clusters. However, these approaches have high computational complexity in constructing multiple affinity matrices, which makes them hardly to be applied in dealing with large-scale of social media data.

3   Multimodal Fusion and Manifold Learning for SED

In this section, the proposed three-stage SED model is presented. Particularly, we preprocess the data by introducing the multimodal similarity metrics before the three-stage event detection process, and present each stage in each subsection.

3.1   Similarity Metrics for the Feature Modalities

Considering Flickr images possess multiple context features, such as *Time*, *Location*, *Tags*, *User identity*, etc., we define the similarity metrics for them as a preprocessing step. Specifically, *Time* similarity is measured by using an inverse function [13] on the time interval. *Tags* similarity is calculated on the associated tags by using *Jaccard index*. *User* similarity is defined as a binary indicator to show whether the images were taken by the same user. *Location* similarity can be calculated by using *Haversine* formula on the *latitude* and *longitude* attributes. Particularly, we adopt perceptual hashes (*pHash*) to evaluate the visual similarity as it is advantageous in terms of efficiency and memory cost.

3.2   Multimodal Fusion Model (M^2F)

In order to exploit the rich context features associated with social media data, we propose a multimodal fusion (M^2F) model, which consists of unimodal feature voting step and vote feature reinforcing step. In the first step, M^2F collects the votes from neighboring dictionary images for a given image, which can be obtained by computing their similarities on each feature modality based on the defined metrics. The image dictionary can be obtained from the images with labels. For a number of M feature modalities, d patterns in the image dictionary, and n samples in the image collection, vote matrices for the modalities, denoted as $F^1 \in \mathbb{R}^{d \times n}, ..., F^M \in \mathbb{R}^{d \times n}$ respectively, can be obtained based on the unimodal feature voting processes.

Furthermore, M^2F exploits a reinforcing model to learn the vote propagations among the feature modalities and achieve fused features in the second step, as specified in Eq. (1),

$$\begin{cases} (F^1)^{(n)} = p_1(F^1)^{(0)} + (1-p_1)(F^2)^{(n-1)} \\ (F^2)^{(n)} = p_2(F^2)^{(0)} + (1-p_2)(F^3)^{(n-1)} \\ \vdots \\ (F^m)^{(n)} = p_m(F^m)^{(0)} + (1-p_m)(F^1)^{(n-1)} \end{cases} \tag{1}$$

where $(F^1)^{(n)}$, $(F^2)^{(n)}$,...,$(F^m)^{(n)}$ indicate the results at n-th iteration, $n = 0$ denotes the original affinity matrices, and p_m is the parameter for the m-th modality. The iterative process will be terminated until $(F^1)^{(n)}$ is convergent to $(F^1)^{(n-1)}$. As a result, the fused features $X \in \mathbb{R}^{M \times N}$ can be assigned as the converged result of $(F^1)^{(n)}$.

3.3 LNMF-Based Manifold Learning

Based on the fused features achieved by M^2F, manifold learning models such as NMF, PCA, ICA, etc., can be applicable to deal with multimodal tasks. Considering the non-negativity of the fused features achieve by M^2F, we introduce graph regularized non-negative matrix factorization model [2] by defining a Laplacian term based on the fused features, which is denoted by LNMF, to extract compact representation. Formally, given the fused features learned by M^2F in matrix form, denoted as $X \in \mathbb{R}^{d \times n}$, manifold learning aims to learn k-dimensional ($k < d$) hidden data representation, denoted as $H \in \mathbb{R}^{k \times n}$. by approximating the original data matrix with two non-negative matrices $W \in \mathbb{R}^{d \times k}$ and $H \in \mathbb{R}^{k \times n}$. Particularly, a Laplacian term constructed from the fused features is introduced to keep the geometry structure of the data in the manifold learning process.

(1) Objective Function. The objective function of LNMF is defined as follows,

$$\begin{aligned} \underset{W,H}{\text{minimize}} \quad & \|X - WH\|^2 + \frac{\lambda}{2}\sum_{i,j}^n A_{i,j}\|h_i - h_j\|^2 \\ \text{subject to} \quad & W \geq 0, H \geq 0. \end{aligned} \tag{2}$$

where the first term aims to minimize the reconstruction errors and the second one is the graph regularization term. $A_{i,j}$ is the affinity between image i and j, which can be calculated as follows,

$$A_{i,j} = \frac{\sum_{f=1}^d F_{f,i}F_{f,j}}{\sqrt{\sum_{f=1}^d (F_{f,i})^2}\sqrt{\sum_{f=1}^d (F_{f,j})^2}} \tag{3}$$

where $F \in \mathbb{R}^{d \times n}$ is the element in the fused features. Particularly, the graph regularization term can be represented by a Laplacian term as follows,

$$\begin{aligned} \frac{1}{2}\sum_{i,j}^n A_{i,j}\|H_i - H_j\|^2 &= \sum_{i=1}^n D_{i,i}H_iH_i^T - \sum_{i,j}^n A_{i,j}H_iH_j^T \\ &= Tr(HDH^T) - Tr(HAH^T) = Tr(HLH^T) \end{aligned} \tag{4}$$

where $Tr(\cdot)$ denotes the trace of matrix. Furthermore, we arrive at the following formulation,

$$
\begin{aligned}
&\underset{W,H}{\text{minimize}} && \|X - WH\|^2 + \lambda Tr(HLH^T) \\
&\text{subject to} && W \geq 0, H \geq 0.
\end{aligned}
\tag{5}
$$

(2) Optimizations. We denote the objective function of LNMF as $J(W, H)$, which can be minimized in a gradient descent manner by adopting additive updating rules as follows.

$$
\begin{cases}
(W^{(n+1)})_{i,j} = (W^{(n)})_{i,j} - \gamma(\nabla_{W^{(n)}} J(W, H))_{i,j} \\
(H^{(n+1)})_{i,j} = (H^{(n)})_{i,j} - \gamma(\nabla_{H^{(n)}} J(W, H))_{i,j}
\end{cases}
\tag{6}
$$

where the indicator (n) denotes the n-th iteration, γ is the step size parameter controlling the learning rate, $\nabla_{W^{(n)}} J(W, H)$ and $\nabla_{H^{(n)}} J(W, H)$ denotes the partial derivatives of $J(W, H)$ with respect to $W^{(n)}$ and $H^{(n)}$ respectively. Updating W and H will be terminated until converged.

$$
\begin{cases}
\nabla_{W^{(n)}} J(W, H) = -2XH^T + 2W^{(n)}HH^T \\
\nabla_{H^{(n)}} J(W, H) = -2W^T X + 2W^T W H^{(n)} + 2\lambda H^{(n)} L
\end{cases}
\tag{7}
$$

3.4 Event Clustering

To detect event clusters in the image collections, clustering algorithms such as Density-Based Spatial Clustering of Applications with Noise (DBSCAN) and K-Means, can be applied seamlessly on the manifold feature representation learned by M^2F-LNMF. Considering the randomness of the initial centers of K-Means, we initialize the cluster centers by exploiting unsupervised density knowledge or the semi-supervised label information, denoted by DKM and SKM respectively. For simplicity, we denote the social event detection algorithms applying DBSCAN, K-Means, DKM and SKM in the third stage as M^2F-LNMF-DBSCAN (MLD), M^2F-LNMF-K-Means (MLK), M^2F-LNMF-DKM (MLDK) and M^2F-LNMF-SKM (MLS), respectively.

4   Experiments

In this section, we conduct a series of experiments on MediaEval Social Event Detection 2014 [10], which contains 110,541 Flickr images that are related to 6,635 events in total. The images in the dataset have been associated with some context features, such as image identifier, geo-tags, time-stamp, user identifier and tags, etc. However, 80 % of the geo-tags are not available.

4.1 Baseline Algorithms

The baselines include a number of recently proposed methods on SED and multimodal fusion tasks, such as graph-based Multimodal Spectral Clustering

(MMSC) [2], SVD-based Multimodal Clustering (SVD-MC) [13], Semi-supervised Multimodal Clustering (SMC) [14], Constrained Incremental Clustering via Ranking (CICR) [12]. In addition to LNMF, GRBM [6], PCA, ICA are applicable in the manifold learning stage, and we implement them for comparisons and denote them as M^2F-LNMF (ML), M^2F-GRBM (MG), M^2F-PCA (MP), M^2F-ICA (MI), respectively. All the methods are tested on MediaEval SED 2014 tasks via the Normalized Mutual Information (NMI), which is a standard technique for evaluating the quality of clusters. The value of NMI is between 0 and 1, and a larger value is preferred.

4.2 Experimental Results

The experimental results for the SED tasks are shown in Table 1, from which some interesting observations can be concluded. Firstly, the proposed M^2F-LNMF-based approaches, i.e., MLK, MLD, MLDK and MLS, outperform the baselines, giving significant improvement on the performance. Secondly, compared to SMC, which has no manifold learning process, the performance is improved by 7 % at most, indicating the effectiveness of the LNMF model. Thirdly, MLD and MLS exploit density knowledge and labeled data to initialize the centers, outperforming the approaches using either of them. Note that the superscript "*" denotes the result outperforms the best one from the baselines.

Table 1. NMI achieved by the algorithms

	MMSC	SVD-MC	CICR	SMC	SMR	JMSR	MLD	MLK	MLDK	MLS
NMI	0.5982	0.8940	0.9024	0.9113	0.9413	0.9417	0.9475*	0.9426*	0.9536*	**0.9751***

4.3 Evaluation on Manifold Learning and Event Clustering Models

In addition to LNMF, we implement PCA, ICA, GRBM [6] in the second stage of the SED model for comparisons. The experimental results are shown in Table 2, which indicates LNMF outperforms the other models on the SED tasks. On the other hand, DKM and SKM that have incorporated density knowledge or label information achieve better performance than using either DBSCAN or K-Means merely.

Table 2. NMI achieved by the combinations of the manifold learning and clustering models

	DBSCAN	K-Means	DKM	SKM
M^2F-GRBM (MG)	0.8127	0.8198	0.8345	0.8366
M^2F-ICA (MI)	0.9275	0.9164	0.9487*	0.9589*
M^2F-PCA (MP)	0.9365	0.9407	0.9512*	0.9678*
M^2F-LNMF (ML)	**0.9475***	**0.9426***	**0.9536***	**0.9751***

4.4   Evaluation on the Dictionary Scale

In the process of M^{2}F, an image dictionary is used for the unimodal feature voting, and we evaluate the impact of the dictionary scale as shown in Fig. 1, where MPS denotes MP-SKM, and so on. From the figure some interesting observations can be revealed. Firstly, the NMI values achieved by MLS and MPS increase and tend to be sable with the scale of dictionary increases. The reason can be explained by that a larger dictionary will capture more patterns of the data. However, a too large dictionary could not capture more than the actual number of patterns in the data collection, i.e., the dictionary must contain replicated patterns. As a result, As a result, LNMF and PCA can deal with the replicated pattern problem well, while ICA is negatively impacted as shown in the figure. On the other hand, a dictionary that is too small may not be complete in expressing the data collections. Secondly, the proposed MLS is more effective for the current SED tasks outperforming the baselines.

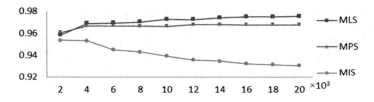

Fig. 1. Impact of the dictionary scale

5   Conclusion

In this paper, we propose a three-stage SED model, i.e., M^{2}F-based multimodal fusion, LNMF-based manifold learning and event clustering. Firstly, fused features integrating the multimodal features are achieved by M^{2}F. Furthermore, compact manifold representation is learned by LNMF, keeping the geometry structure of the data in the learning process. Finally, clustering algorithms can be applied on the manifold learned by M^{2}F-LNMF seamlessly to discover event clusters. Particularly, the hybrid clustering algorithm gives significant improvement on the performance. The experiments conducted on the real-world dataset manifest the effectiveness of the M^{2}F-LNMF based event detection approaches.

Acknowledgments. We would like to thank Dr. Zheng Lu, Mr. Min Cheng and Mr. Yangbin Chen for the discussions.

References

1. Ah-Pine, J., Csurka, G., Clinchant, S.: Semi-supervised visual and textual information fusion in CBMIR using graph-based methods. ACM Trans. Inf. Syst. (TOIS) **33**(2), 9 (2015)

2. Cai, X., Nie, F., Huang, H., Kamangar, F.: Heterogeneous image feature integration via multi-modal spectral clustering. In: IEEE Conference on Computer Vision and Pattern Recognition, pp. 1977–1984. IEEE Press (2011)
3. Cai, D., He, X., Han, J., Huang, T.S.: Graph regularized nonnegative matrix factorization for data representation. IEEE Trans. Pattern Anal. Mach. Intell. **33**(8), 1548–1560 (2011)
4. Cai, Y., Li, Q., Xie, H., Wang, T., Min, H.: Event relationship analysis for temporal event search. In: Meng, W., Feng, L., Bressan, S., Winiwarter, W., Song, W. (eds.) DASFAA 2013, Part II. LNCS, vol. 7826, pp. 179–193. Springer, Heidelberg (2013)
5. Chen, J., Cui, Y., Ye, G., Liu, D., Chang, S.F.: Event-driven semantic concept discovery by exploiting weakly tagged internet images. In: International Conference on Multimedia Retrieval, pp. 1–8. ACM (2014)
6. Hinton, G.E., Salakhutdinov, R.R.: Reducing the dimensionality of data with neural networks. Science **313**(5786), 504–507 (2006)
7. Liu, X., Huet, B.: Heterogeneous features and model selection for event-based media classification. In: 3rd ACM International Conference on Multimedia Retrieval, pp. 151–158. ACM (2013)
8. Nitta, N., Kumihashi, Y., Kato, T., Babaguchi, N.: Real-world event detection using flickr images. In: Gurrin, C., Hopfgartner, F., Hurst, W., Johansen, H., Lee, H., O'Connor, N. (eds.) MMM 2014, Part II. LNCS, vol. 8326, pp. 307–314. Springer, Heidelberg (2014)
9. Petkos, G., Papadopoulos, S., Kompatsiaris, Y.: Social event detection using multimodal clustering and integrating supervisory signals. In: 2nd ACM International Conference on Multimedia Retrieval, p. 23. ACM (2012)
10. Petkos, G., Papadopoulos, S., Mezaris, V., Kompatsiaris, Y.: Social event detection at MediaEval 2014: Challenges, datasets, and evaluation. In: MediaEval 2014 Workshop (2014)
11. Rao, Y., Li, Q.: Term weighting schemes for emerging event detection. In: 2012 IEEE/WIC/ACM International Conferences on Web Intelligence and Intelligent Agent Technology (WI-IAT), vol. 1, pp. 105–112 (2012)
12. Sutanto, T., Nayak, R.: Ranking based clustering for social event detection. In: MediaEval 2014 Workshop, 1263, pp. 1–2 (2014)
13. Yang, Z., Li, Q., Lu, Z., Ma, Y., Gong, Z., Pan, H.: Semi-supervised multimodal clustering algorithm integrating label signals for social event detection. In: IEEE International Conference on Multimedia Big Data, pp. 32–39. IEEE (2015)
14. Yang, Z., Li, Q., Lu, Z., Ma, Y., Gong, Z., Pan, H., Chen, Y.: Semi-supervised multimodal fusion model for social event detection on web image collections. Int. J. Multimed. Data Eng. Manage. **6**(4), 1–22 (2015)

Deep Neural Network for Short-Text Sentiment Classification

Xiangsheng Li[1], Jianhui Pang[1], Biyun Mo[1],
Yanghui Rao[1(✉)], and Fu Lee Wang[2]

[1] Sun Yat-sen University, Guangzhou, China
{lixsh6,pangjh3,moby5}@mail2.sysu.edu.cn,
raoyangh@mail.sysu.edu.cn
[2] Caritas Institute of Higher Education, New Territories, Hong Kong
pwang@cihe.edu.hk

Abstract. As a concise medium to describe events, short text plays an important role to convey the opinions of users. The classification of user emotions based on short text has been a significant topic in social network analysis. Neural Network can obtain good classification performance with high generalization ability. However, conventional neural networks only use a simple back-propagation algorithm to estimate the parameters, which may introduce large instabilities when training deep neural networks by random initializations. In this paper, we apply a pre-training method to deep neural networks based on restricted Boltzmann machines, which aims to gain competitive and stable classification performance of user emotions over short text. Experimental evaluations using real-world datasets validate the effectiveness of our model on the short-text sentiment classification task.

Keywords: Neural network · Restricted Boltzmann machine · Pre-training · Short-text sentiment classification

1 Introduction

With an explosive growth of social media services, many online users can conveniently express their feelings through various channels. Facing such large-scaled sentimental documents, it is important for us to detect the sentiments from them automatically. Sentiment classification aims to identify and extract the user attitude towards an object. Unlike classifying sentiments over normal documents, short-text sentiment classification tends to be ambiguous without enough contextual information. Thus, conventional machine learning algorithms are difficult to be applied to short text directly.

Neural networks, although were applied to various natural language processing tasks, their performance were highly dependent on the adaptiveness of initializations if trained by raw features with random initialized weights [1]. Recently, Hinton and Salakhutdinov [2] indicated that deep neural networks have a better

© Springer International Publishing Switzerland 2016
H. Gao et al. (Eds.): DASFAA 2016 Workshops, LNCS 9645, pp. 168–175, 2016.
DOI: 10.1007/978-3-319-32055-7_15

ability of learning features, and the learned features have a better description of the original data, which are more appropriate for classification. However, deep neural network also has its disadvantages such as instability, gradient vanish and overfitting. Thus, Bengio et al. [3] proposed a method of greedy layer-wise training for deep networks which could effectively alleviate the problems of neural networks. In this paper, we employ the greedy layer-wise training as a pre-training method of our deep neural network model, in addition to validate its effectiveness on sentiment classification of short text. The general process of our model is as follows: First, we convert the high dimensional data into low dimensional expressions, which has the ability to reconstruct the original input vectors. Second, we use the logistic regression classifier at the last layer to predict emotional labels. Experimental evaluations validate that our model based on the pre-training method outperforms the baselines, as well as obtains much more stable performance than the conventional neural network on sentiment classification over short text.

The remainder of this paper is organized as follows. In Sect. 2, we summarize the related work on sentiment classification, short text modeling and deep neural network models. In Sect. 3, we describe our restricted Boltzmann machine and its training algorithms. Experimental evaluation is conducted on two datasets in Sect. 4. In Sect. 5, we present our conclusions.

2 Related Work

2.1 Sentiment Classification

Given unlabeled documents, the objective of sentiment classification is to estimate a score for each emotional label automatically. Pang et al. [4] evaluated three state-of-the-art machine learning algorithms on sentiment classification for movie reviews. Katz et al. [5] proposed a supervised unigram model (i.e., SWAT) to exploit user emotions with individual words. However, the performance of these methods are quite limited. Recently, Bao et al. [6] developed an emotion term (ET) method and an emotion topic model (ETM) to associate emotions with words and topics, respectively. Rao et al. [7,8] also focused on detecting user emotions by topics. The limitation of these models is that they need abundant features to gather enough statistics.

2.2 Short Text Modeling

To tackle the issue of lacking contextual information and the sparsity of content in short text, Sahami and Heilman [9] utilized the external documents collected from the web to expand the features. Banerjee et al. [10] applied the existing knowledge bases such as WordNet or Wikipedia to mine the semantic association between words. However, these methods are dependent on the quality of the external documents or knowledge bases, but we need a model which can automatically extract features implied from the sentence.

2.3   Deep Neural Network

As one typical stream of classification models, neural networks take the frequency of words in the short sentence or normal documents as the input and transfer it to a sequence of layers. The neural network can automatically extract features from the word-level up to the sentence and document levels. To compactly represent highly non-linear and highly-varying functions, deep multi-layer neural networks with more levels of non-linearities have been developed [3]. However, deep neural networks were found to be more difficult to train and the performance was even worse than neural networks with one or two hidden layers [11].

3   Restricted Boltzmann Machine

Restricted Boltzman machine (RBM) [2] is a generative stochastic neural network that is widely employed in deep learning and many other areas. As a variant of Boltzmann machines, RBM restrains that the neurons must form a bipartite graph. Figure 1 presents the framework of RBM models, which is a symmetric structure with a visible layer, a hidden layer and a bias unit.

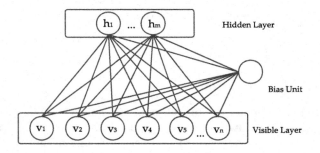

Fig. 1. Framework of RBM models

As shown in Fig. 1, $\mathbf{v}$ and $\mathbf{h}$ denote the visible layer and the hidden layer, which represent the input data and low dimensional features, respectively. $\mathbf{w}$ is the connective weight between the two layers. A joint configuration $(\mathbf{v}, \mathbf{h})$ of the whole RBM has an energy [12] as follows:

$$E(\mathbf{v}, \mathbf{h}|\boldsymbol{\theta}) = - \sum_{i \in visible} a_i v_i - \sum_{j \in hidden} b_j h_j - \sum_{i,j} v_i h_j w_{ij}, \tag{1}$$

where v_i and h_j are binary states of visible unit i and hidden unit j, with biases a_i and b_j, and the weight w_{ij}. $\boldsymbol{\theta} = \{w_{ij}, a_i, b_j\}$ represents the parameters of RBM, which are all real numbers. Given these parameters, the joint distribution between the hidden layer and the visible layer is

$$P(\mathbf{v}, \mathbf{h}|\boldsymbol{\theta}) = \frac{e^{-E(\mathbf{v}, \mathbf{h}|\boldsymbol{\theta})}}{Z(\boldsymbol{\theta})}, \tag{2}$$

where Z is a partition function by summing all possible pairs of visible and hidden vectors, i.e., $Z(\boldsymbol{\theta}) = \sum_{\mathbf{v},\mathbf{h}} e^{-E(\mathbf{v},\mathbf{h}|\boldsymbol{\theta})}$.

Note that there is no direct connections between hidden units in RBM, the states of all hidden units are conditionally independent given the visible states. Due to the symmetrical structure of RBM, the states of all visible units are also conditionally independent given the hidden states. Thus, the activation probabilities of the hidden unit and the visible unit are represented as:

$$P(h_j = 1|\mathbf{v}, \boldsymbol{\theta}) = \sigma(b_j + \sum_{i \in visible} v_i w_{ij}), \tag{3}$$

$$P(v_i = 1|\mathbf{h}, \boldsymbol{\theta}) = \sigma(v_i + \sum_{j \in hidden} w_{ij} h_j). \tag{4}$$

where $\sigma(x)$ is the logistic sigmoid function, i.e., $1/(1 + exp(-x))$.

3.1   Objective Function

Our objective is to make the visible vectors close to the original input vectors as much as possible. The likelihood function of the input vectors given $\boldsymbol{\theta}$, i.e., the probability that RBM assigns to a visible vector is estimated by

$$P(\mathbf{v}|\boldsymbol{\theta}) = \frac{1}{Z(\boldsymbol{\theta})} \sum_h e^{-E(\mathbf{v},\mathbf{h}|\boldsymbol{\theta})}. \tag{5}$$

Thus, the optimization problem is to minimize the negative log-likelihood function on the train data, as follows:

$$\mathcal{L}(\boldsymbol{\theta}) = \min_{\boldsymbol{\theta}} \sum_{t=1}^{T} -log P(\boldsymbol{v}^{(t)}|\boldsymbol{\theta}), \tag{6}$$

where T is the size of batch training.

We then apply the stochastic gradient descent to solve the above function. The gradient of the log probability of the training vector is given by

$$\frac{\partial \mathcal{L}}{\partial \boldsymbol{\theta}} = -\sum_{t=1}^{T} \left(\langle \frac{\partial}{\partial \boldsymbol{\theta}} E(\mathbf{v}^{(t)}, \mathbf{h}|\boldsymbol{\theta}) \rangle_{data} + \langle \frac{\partial}{\partial \boldsymbol{\theta}} E(\mathbf{v}, \mathbf{h}|\boldsymbol{\theta}) \rangle_{model} \right), \tag{7}$$

where the angle brackets are used to denote the expectation under the distribution specified by the subscript. Finally, we can estimate the gradient in terms of w_{ij}, a_i and b_j by

$$\frac{\partial P(\boldsymbol{v}|\boldsymbol{\theta})}{\partial w_{ij}} = \langle v_i h_j \rangle_{data} - \langle v_i h_j \rangle_{model}, \tag{8}$$

$$\frac{\partial P(\boldsymbol{v}|\boldsymbol{\theta})}{\partial a_i} = \langle v_i \rangle_{data} - \langle v_i \rangle_{model}, \tag{9}$$

$$\frac{\partial P(\boldsymbol{v}|\boldsymbol{\theta})}{\partial b_j} = \langle h_j \rangle_{data} - \langle h_j \rangle_{model}. \tag{10}$$

Next, we employ the contrastive divergence gradient algorithm [13] to conduct pre-training for deep neural networks.

3.2   Pre-training Process

The pre-training process embedded in our model aims to make the reconstructed data close to the original data as much as possible, in which, all layers except the last classifier layer could be considered as encoder machines. Given a closer vector expression of the original data, the last layer for classification will become more robust. Algorithm 1 presents the method of pre-training via the contrastive divergence gradient [13]. The whole process of our model can be summarized as pre-training using the contrastive divergence gradient algorithm, and fine-tuning by the traditional back-propagation algorithm.

Algorithm 1. Pre-training using contrastive divergence gradient

---

Input:
  1. x_0: training samples
  2. T: the size of batch training
  3. m, n: the amounts of hidden and visible units
Output:
  Gradient approximation: $\theta = \{w_{ij}, a_i, b_j\}$

Initialize $v_1 = x_0$; w, a and b are random values
for $t = 1$ to T do
    for $j = 1, \cdots, m$ do sample $h_{1j}^{(t)} \in \{0, 1\} \sim P(h_{1j}^{(t)} = 1|v_1^{(t)})$
    for $i = 1, \cdots, n$ do $v_{2i}^{(t)} = P(v_{2i}^{(t)} = 1|h_1^{(t)})$

Update parameters with a learning rate ϵ:
$\Delta w \leftarrow \Delta w + \epsilon(P(h_1 = 1|v_1)v_1 - P(h_2 = 1|v_2)v_2)$
$\Delta a \leftarrow \Delta a + \epsilon(v_1 - v_2)$
$\Delta b \leftarrow \Delta b + \epsilon(P(h_1 = 1|v_1) - P(h_2 = 1|v_2))$

Some strategies are also adopted in pre-training [14]. First, the hidden states are represented as the binary value rather than probabilities, because the hidden neuron is suitable to convey only one bit information as in biology. Second, the visible states are encoded by probabilities which may reduce the reconstruction error of the input data. After the process of pre-training, we use the traditional back-propagation algorithm to fine-tune the whole network, in which, the simple logistic regression is used in the last layer for sentiment classification.

4   Experiments

4.1   Datasets

SemEval is an English dataset used in the 14th task of the 4th International Workshop on Semantic Evaluations [15]. The dataset include news headlines and user scores over emotions of *anger, disgust, fear, joy, sadness* and *surprise*.

Among 1,246 news headlines with the total score larger than 0, the first 1,000 and the rest 246 of them are used for training and testing, respectively.

ISEAR is a collection of 7,666 sentences annotated by 1,096 participants with different cultural backgrounds [16]. The emotional labels of this dataset are *anger, disgust, fear, joy, sadness, shame* and *guilt*. We randomly select 60 percent of sentences as the training set, 20 percent as the validation set, and the rest as the testing set.

4.2   Experiment Setting

The experiment is conducted to validate the effectiveness of the deep neural network with pre-training (DNN) on sentiment classification of short text. We also implement the conventional neural network with only one hidden layer (NN_one) and a deeper structure with two hidden layers (NN_two) without pre-training for comparison. Gaussian random values with a mean value of 0 and different variances are used to initialize these models, so as to evaluate the stability of performance. To make an appropriate comparison with other baseline models, the accuracy at top 1 (*Accu@*1) is employed as the performance indicator [6].

4.3   Results and Analysis

The performance of DNN, NN_one and NN_two with different variances of initializations on *SemEval* and *ISEAR* is presented in Tables 1 and 2, respectively. The results indicate that DNN is more stable than others and NN_one outperforms NN_two for both datasets. In terms of *Accu@*1 values, NN_one performs better on the small-scaled *SemEval* dataset than *ISEAR*. Compared to NN_two, we also observe that the process of pre-training embedded in DNN with two hidden layers not only improves *Accu@*1 values but also alleviates the problem of instability. The reason is that pre-training could improve the effectiveness of parameter estimation for the deep neural network model, where the features approach an approximate low dimensional expression.

Table 1. Mean *Accu@*1(%) on *SemEval*

Variance	0.05	0.1	0.2	0.3	0.4	0.5
NN_one	36.2	38.5	37.9	37.3	35.2	35.2
NN_two	36.2	36.2	22.7	20.2	19.5	24.2
DNN	36.2	36.2	36.2	36.2	36.2	36.2

We also compare the performance of these methods with other existing models in Table 3, in which the best variance of initialization is determined by cross-validation for DNN, NN_one and NN_two. Compared to the baseline models of SWAT [5], ET and ETM [6], the *Accu@*1 of DNN improves 8%, 16.8%, 41.4% on *SemEval* and 100%, 2%, 8% on *ISEAR*, respectively.

Table 2. Mean $Accu@1(\%)$ on $ISEAR$

Variance	0.05	0.1	0.2	0.3	0.4	0.5
NN_one	42.4	50.1	49.0	47.2	27.9	35.4
NN_two	14.0	14.3	17.6	17.6	20.4	15.0
DNN	49.3	52.7	51.2	50.5	47.6	47.1

Table 3. Performance of different models

(a) *SemEval*

Model	$Accu@1(\%)$
NN_one	38.5
NN_two	36.2
DNN	36.2
SWAT	33.5
ET	31.0
ETM	25.6

(b) *ISEAR*

Model	$Accu@1(\%)$
NN_one	50.1
NN_two	20.4
DNN	52.7
SWAT	26.3
ET	51.7
ETM	48.8

5 Conclusion

In this paper, we incorporated a pre-training method into the deep neural network model for sentiment classification of short text. To evaluate the effectiveness of our model, we compared it with conventional neural network models and other existing methods using two different datasets. The results indicated that the deep neural network with pre-training performed competitively in terms of accuracy and robustness. This is because pre-training makes the output of each layer become a low dimensional code of the original data, which is more significant than conventional outputs.

For future work, we plan to improve the structure of our deep neural network by refining the sparse code of layers. With the rapid development of online communications, the method of extracting significant information from a non-standard short document also deserves further research.

Acknowledgements. This research was supported by the National Natural Science Foundation of China (61502545, 61472453, U1401256, U1501252), the Fundamental Research Funds for the Central Universities, and a grant from the Research Grants Council of the Hong Kong Special Administrative Region, China (UGC/FDS11/E06/14).

References

1. Bengio, Y.: Learning deep architectures for ai. Found. Trends® Mach. Learn., vol. 2(1), pp. 1–127 (2009)
2. Hinton, G.E., Salakhutdinov, R.R.: Reducing the dimensionality of data with neural networks. Sci. **313**(5786), 504–507 (2006)

3. Bengio, Y., Lamblin, P., Popovici, D., Larochelle, H.: Greedy layer-wise training of deep networks. Adv. Neural Inf. Process. Syst. (NIPS) **19**, 153 (2007)
4. Pang, B., Lee, L., Vaithyanathan, S.: Thumbs up? sentiment classification using machine learning techniques. In: Proceedings of the Conference on Empirical Methods in Natural Language Processing, pp. 79–86 (2002)
5. Katz, P., Singleton, M., Wicentowski, R.: Swat-mp: the semeval- systems for task 5 and task 14. In: Proceedings of the 4th International Workshop on Semantic Evaluations (SemEval), pp. 308–313(2007)
6. Bao, S., Xu, S., Zhang, L., Yan, R., Su, Z., Han, D., Yu, Y.: Mining social emotions from affective text. IEEE Trans. Knowl. Data Eng. **24**(9), 1658–1670 (2012)
7. Rao, Y., Li, Q., Mao, X., Wenyin, L.: Sentiment topic models for social emotion mining. Inf. Sci. **266**, 90–100 (2014)
8. Rao, Y., Li, Q., Wenyin, L., Wu, Q., Quan, X.: Affective topic model for social emotion detection. Neural Netw. **58**, 29–37 (2014)
9. Sahami, M., Heilman, T.D.: A web-based kernel function for measuring the similarity of short text snippets. In: Proceedings of the 15th International Conference on World Wide Web (WWW), pp. 377–386 (2006)
10. Banerjee, S., Ramanathan, K., Gupta, A.: Clustering short texts using wikipedia. In: Proceedings of the 30th Annual International ACM SIGIR Conference on Research and Development in Information Retrieval (SIGIR), pp. 787–788 (2007)
11. Tesauro, G.: Practical issues in temporal difference learning. Mach. Learn. **8**(3–4), 33–53 (1992)
12. Hopfield, J.J.: Neural networks and physical systems with emergent collective computational abilities. Proc. Natl. Acad. Sci. **79**(8), 2554–2558 (1982)
13. Hinton, G.E.: Training products of experts by minimizing contrastive divergence. Neural Comput. **14**(8), 1771–1800 (2002)
14. Hinton, G.: A practical guide to training restricted boltzmann machines. Momentum **9**(1), 926 (2010)
15. Strapparava, C., Mihalcea, R.: Semeval- task 14: Affective text. In: Proceedings of the 4th International Workshop on Semantic Evaluations (SemEval 2007), pp. 70–74 (2007)
16. Scherer, K.R., Wallbott, H.G.: Evidence for universality and cultural variation of differential emotion response patterning. J. Pers. Soc. Psychol. **66**(2), 310 (1994)

BDMS 2016

VMPSP: Efficient Skyline Computation Using VMP-Based Space Partitioning

Kaiqi Zhang[1]([✉]), Donghua Yang[2], Hong Gao[1],
Jianzhong Li[1], Hongzhi Wang[1], and Zhipeng Cai[3]

[1] School of Computer Science and Technology,
Harbin Institute of Technology, Harbin, China
{zhangkaiqi,honggao,lijzh,wangzh}@hit.edu.cn
[2] Academy of Fundamental and Interdisciplinary Sciences,
Harbin Institute of Technology, Harbin, China
yang.dh@hit.edu.cn
[3] Department of Computer Science, Georgia State University, Atlanta, USA
zcai@gsu.edu

Abstract. The skyline query returns a set of interesting points that are not dominated by any other points in the multi-dimensional data sets. This query has already been considerably studied over last several years in preference analysis and multi-criteria decision making applications fields. Space partitioning, the best non-index framework, has been proposed and existing methods based on it do not consider the balance of partitioned subspaces. To overcome this limitation, we first develop a cost evaluation model of space partitioning in skyline computation, propose an efficient approach to compute the skyline set using balanced partitioning. We illustrate the importance of the balance in partitioning. Based on this, we propose a method to construct a balanced partitioning point VMP whose ith attribute value is the median value of all points in ith dimension. We also design a structure RST to reduce dominance tests among those subspaces which are comparable. The experimental evaluation indicates that our algorithm is faster at least several times than existing state-of-the-art algorithms.

1 Introduction

The skyline query [1] returns a set of interesting points that are not dominated by any other points in the multi-dimensional data sets. A point q is dominated by a point p if and only if p is not worse than q in all dimensions and strictly better in at least one dimension. Without loss of generality, here we assume that lower value is better to users in all dimensions.

The skyline query is important in database community and has already been considerably studied over last several years in preference analysis, multi-criteria decision making applications, and so on. It aims to reduce search space when there is not existing a scoring function.

H. Gao et al. (Eds.): DASFAA 2016 Workshops, LNCS 9645, pp. 179–193, 2016.
DOI: 10.1007/978-3-319-32055-7_16

The existing algorithms can be summarized into two categories, index-based algorithms and index-independent algorithms, respectively. Index-based algorithms [8–11] strive primarily to avoid scanning the entire data sets by utilizing pre-construct index structure. Some points are usually checked to be a skyline point or a non-skyline point by index structure early on. The main issues are that it needs a great deal of time and space to pre-construct the data points index structures and store them respectively and it can't deal with dynamic or stream data sets.

The other category is index-independent algorithms, which compute skyline results without any pre-construct structures. The early algorithms [2,6,7] are mainly based on sorting. They first sort the data sets by a monotone function to guarantee that previous points are never dominated by their following points. These approaches strive to find out early the skyline points which can prune out most of the non-skyline points. [3] first proposed the schema of space partitioning to compute skyline set, and implemented two algorithms which both outperform significantly than prior methods. This schema divides recursively the entire space into several disjoint subspaces and compute local skyline results in their subspace according to the relationship of these subspaces. We collect all the local skyline points as the results of skyline computation. It reduces dramatically the dominance tests. While it randomly selects a skyline point as partitioning point, this method cannot guarantee the stability. Based on the schema, BSkyTree [4] makes some progress by selecting the skyline point with the closest value to the main diagonal as the partitioning point. They both do not consider the importance of the balance in partitioning.

This paper focuses mainly on developing a cost evaluation model of space partitioning in skyline computation and designing an efficient approximated approach based on it. We illustrate the importance of the balance in partitioning and provide a low-cost balanced partitioning point VMP whose ith attribute value is the median value of all points in ith dimension. Finally, we utilize a structure named RST to reduce the dominance tests among comparable subspaces. We also implement an algorithm VMPSP using our VMP partitioning point and RST structure. Our experiments indicate that our algorithm VMPSP is faster at least several times than existing state-of-the-art algorithms.

In this paper, the key contributions are summarized as follows:

- We analyze the defections of existing sorting-based and space-partitioning-based algorithms.
- We develop a cost evaluation model of space partitioning in skyline computation, and illustrate the importance of balance of partitioned subspaces.
- We propose a new more balanced partitioning point Virtual Median Point (VMP) whose ith attribute value is the median value of all points in ith dimension. It can reduce the cost of partitioning. We also design a structure named Recursive Search Tree(RST) to reduce the dominance tests among comparable subspaces.
- We implement an algorithm VMPSP by constructing VMP as the partitioning point and utilizing RST structure.

– We evaluate our proposed algorithm VMPSP by comparing it with state-of-the-art algorithms in dimensionality and cardinality over real and synthetic data sets.

The rest of this paper is organized as follows. Section 2 introduces some key proposed algorithms based on sorting or space partitioning. Section 3 presents existing definitions and properties about skyline and space partitioning schema. Section 4 develops a cost evaluation model of space partitioning in skyline computation and illustrates the importance of balance in partitioning. A construction of balanced partitioning point VMP and a structure of reducing dominance tests among comparable subspaces RST are presented in this section. Section 5 illustrates our algorithm VMPSP using our constructed partitioning point VMP and the structure RST. Section 6 evaluates our proposed algorithm with existing approach in dimensionality and cardinality scalability for real and synthetic data sets. Finally, our conclusion is summarized in Sect. 7.

2   Related Work

The skyline query mainly resolves the problem that how to reduce the search space by discarding the non-interesting objects from multi-dimensional databases when the scoring function is not existing. It is important in multi-criteria decision making and analysis in database community. There are a great many of existing literatures in this field, while we just summarize sketchily most proposed key main-memory algorithms into two categories as follows.

2.1   Index-Based Algorithms

Index-based skyline computation techniques utilize the pre-construct indexes on all data sets to accelerate the skyline query by avoiding scanning the entire data sets. [1] first roughly imagines the algorithms utilizing B-tree or R-tree index structures which can prune out non-skyline points immediately without accessing the entire data sets. Since then a set of techniques [8–11] by using indexes are proposed to improve the efficiency of the skyline computation. [9] exploits bitmap to represent data sets and perform a series simple of bitwise operations to get skyline results. By the observation which the nearest neighbor point to the origin must be a skyline point, [8] develops NN algorithm. NN first computes the nearest neighbor point to the origin which must be a skyline point from the entire data sets by utilizing R-tree. Then divides remaining points except those dominated by the nearest neighbor point into several overlapping subsets. Next, keeps on computing NNP from these subsets recursively. Finally it is necessary to eliminate duplicate skylines from these NNP. Based on NN, an improvement version named BBS is presented in [11]. It adopts branch-and-bound strategy and avoids retrieving duplicates to improve performance. The index-based state-of-the-art algorithm is ZSearch [10]. It improves significantly the overall performance according to ZBtree in which data points is organized.

Obviously, the disadvantage of index-based algorithms is need a great deal of time and space to pre-construct the data points index structures and store them respectively. When the dimension is high, it is unfeasible since the time cost of preconstruction is dramatically more than the skyline query response time using indexes. Also, it is unfeasible to construct indexes for stream or dynamic data sets. So it is urgent to find out a method that efficiently computes the skyline results without preconstruction.

2.2 Index-Independent Algorithms

The other type of skyline computation is to retrieve the skyline results straightly without any pre-prepare structures. The first algorithm is proposed along with the skyline problem named block-nested-loops(BNL) in [1]. It uses a memory buffer *window* to contain the *candidate* skyline points which have been yet not dominated by others. The point accessed by the stored order in the data sets must compare with *candidates* one by one until it is dominated. If it is dominated by any one of the *candidates*, discard the point and validate the next point in the data sets. Otherwise insert the point into the *window* and remove those *candidates* which are dominated by it. The *candidates* in the *window* are definitely all skyline points until the data sets in store have been all accessed. BNL is much inefficient since the *candidates* in the buffer may be not the skyline points, conducts a great extremely many of unnecessary dominance tests among non-skyline points in the *window*.

Based on this, Sort-First Skyline(SFS) algorithm [2] is presented. It guarantees that the points added into the *window* are certainly skyline points by sorting the data sets using a monotone function(such as *entropy function* and *sum function*). In the sorted data sets, a point must certainly not dominate its previous points. So, we are sure that the *candidates* in the *window* must be the skyline points of the data sets. In the light of this, SFS algorithm accelerates significantly the query response time to BNL. Later, Some important approaches, such as LESS [7] and SaLSa [6], have been developed to improve the performance of SFS based on the framework of sorting the data sets first. LESS conducts the elimination by maintaining a small memory buffer when the points are sorting. SaLSa selects monotone functions carefully and adopts a stop point to terminate the algorithm early such that not all points in the sorted list need to be accessed.

Although these methods, especially SaLSa, have a better performance, the greatest disadvantage of above sorting-based algorithms is sensitive to the size of skyline set since every accessed point must compare with all *candidates* until it is dominated. The skyline size becomes larger as the cardinality or dimensionality higher such that the performance of above algorithms deteriorates severely. The total numbers of dominance tests are unimaginable huge, as well as the cost of sorting is not negligible when data sets are large.

In order to overcome the aforementioned problems, a crucial algorithm OSPS is devised in [3] which first proposes a new framework named *space partitioning* to organize already found skyline points as a skyline tree. Any accessed point need not compare with all the found skylines instead do the dominance tests with only a small part of skylines in ROSP even if the point belongs to skyline. This framework is also effective to other studies [5,12] which are related with skyline. [3] proposes a method that selects randomly a skyline point as the *partitioning point*. Although avoid the occurrence of the worst case like SFS, it is failed to guarantee the stability. Also the random partitioning point cannot partition the space into balanced subspaces. BSkyTree [4] both consider dominance and incomparability relationships which reduces substantially the numbers of dominance tests and have become state-of-the-art algorithm of the skyline computation in main-memory environment. It selects the skyline point with the closest value to the main diagonal and makes the subpartitions more balanced than the random selection method. While only the first partitioned subspaces are relatively balanced, the subsequent partitioning conduct severely unbalanced situation. Besides, BSkyTree cannot organize yet found skylines as a recursive search tree to accelerate dominance tests.

3   Preliminaries and Observation

In this section, we introduce some definitions and properties which are all given in prior works [1,3,4] as well as present our observations that can be used to improve the overall performance of skyline computation. Given a data set S in a d-dimensional positive real space R^d, for any point p in S, we denote the value of point p in ith dimension by p_i.

3.1   Skyline

Definition 1. (Dominate) *Given two distinct points p and q in the data set S, p dominates q, denoted by $p \succ q$ iff $\forall\, i \in [1, d]$, $p_i \leq q_i$ and $\exists i \in [1, d]$, $p_i < q_i$. Otherwise, p doesn't dominate q, denoted by $p \nsucc q$. We call point p as the first check point and q as the second check point from here on.*

Definition 2. (Skyline) *For any point p in S, p is a skyline point iff $\nexists\, q \in S$ and $q \neq p$, such that $q \succ p$. The skyline set consists of all skyline points.*

Definition 3. (Incomparable) *Given any two points p, $q \in S$, if $p \nsucc q$ and $q \nsucc p$, then p and q are incomparable each other, denoted by $p \sim q$.*

3.2   The Framework of Space Partitioning

Before the framework of space partitioning is proposed, most prior algorithms focus heavily on *point-wise* dominance tests as well as how to fast check out whether a point is dominated or not. We can know nothing but whether the *second check point* is dominated by the *first check point* from one *point-wise* dominance test. If not, we must repeat the same operation for the next point. While the schema of space partitioning can provide more information even if the *second check point* is not dominated by the *first check point*. And it promotes *point-wise* dominance tests to *space-wise* dominance tests which reduces significantly dominance tests among incomparable point pairs.

Definition 4. (Partitioning Point *[3]) Given a point $\widehat{p}$ in R^d, $\forall\ i \in [1,d]$, $\widehat{p}_i$ partitions R^d into 2 disjoint complementary subspaces $S_i^+(\widehat{p})$ and $S_i^-(\widehat{p})$ respectively. $\forall\ p \in S$, if $p_i < \widehat{p_i}$, $p \in S_i^+(\widehat{p})$, otherwise, $p \in S_i^-(\widehat{p})$. Therefore, the space R^d is partitioned into 2^d non-overlapping subspaces by $\widehat{p}$ which is formally named as partitioning point. We define a subspace set $C_{\widehat{p}}^{R^d}$ consists all of aforementioned 2^d subspaces partitioned by $\widehat{p}$ in R^d. Any point in S except $\widehat{p}$ is mapped into unique one space of $C_{\widehat{p}}^{R^d}$. We name a d-bit vector as the address of any subspace V, denoted by $A_{\widehat{p}}^V$, $\forall\ i \in [1,d]$, if $V \subset S_i^+(\widehat{p})$, set the ith bit value of $A_{\widehat{p}}^V$ be 0, i.e., $A_{\widehat{p}}^V[i]{=}0$; Otherwise $A_{\widehat{p}}^V[i]{=}1$.*

Based on Definition 4, any *partitioning point* in R^d can divide S into 2^d separate subsets corresponding to 2^d disjoint subspaces of R^d of which the *addresses* range from 0 to 2^d-1. We can also represent a subspace by its *address*.

Definition 5. (Space-wise Dominate *[3]) For any two subspaces V, $W \in C_{\widehat{p}}^{R^d}$, if $\forall\ i \in [1,d]$, $A_{\widehat{p}}^V[i] \leqslant A_{\widehat{p}}^W[i]$ then V space-wise dominates W, for brevity of representation, replace it with dominate, denoted by $V \succ W$. Otherwise, V cannot dominate W, denoted by $V \nsucc W$.*

Definition 6. (Incomparable *[3] and Comparable)For any two subspaces V, $W \in C_{\widehat{p}}^{R^d}$, if V and W cannot dominate each other, we call that they are incomparable, denoted by $V \sim W$. Otherwise, W and V are comparable. Specifically, subspace V is comparable with itself. The space set of all spaces dominating V is named as the dominating space set, denoted by $D_{\widehat{p}}^V = \{W \in C_{\widehat{p}}^{R^d} \mid W \succ V\}$, the space set of all spaces dominated by V is named as the dominated space set, denoted by $U_{\widehat{p}}^V = \{W \in C_{\widehat{p}}^{R^d} \mid V \succ W\}$, and the space set of all spaces being incomparable with V is named as the incomparable space set, denoted by $I_{\widehat{p}}^V = \{W \in C_{\widehat{p}}^{R^d} \mid V \sim W\}$.*

For the definitions of space-wise dominate and incomparable, there are a few differences with prior works. We describe some key properties about space partitioning in the following.

Property 1. For any two different subspace V, $W \in C_{\widehat{p}}^{R^d}$, if V and W are incomparable, then $\forall\ v \in V$, $\forall\ w \in W$, v and w must be incomparable, here v and w are both points.

Property 2. For any two different subspace V, $W \in C_{\widehat{p}}^{R^d}$. $V \succ W$ iff $(A_{\widehat{p}}^V \mid A_{\widehat{p}}^W) = A_{\widehat{p}}^W$.

Property 3. For any two different subspace V, $W \in C_{\widehat{p}}^{R^d}$. $V \sim W$ iff $(A_{\widehat{p}}^V \mid A_{\widehat{p}}^W) \neq A_{\widehat{p}}^V$ and $(A_{\widehat{p}}^V \mid A_{\widehat{p}}^W) \neq A_{\widehat{p}}^W$.

We ignore detailed process of proof about above properties, which have already been proofed in [3]. Property 1 can promote *point-wise* dominance tests to *space-wise* dominance tests. We need first check out whether space V and W are incomparable or not, if it is, all dominance tests between the points in V and the points in W can be safely skipped through only one space-wise dominance tests. If not, *i.e.*, they are comparable with each other, then must do point-wise dominance tests between the points in the two spaces. Based on this, space partitioning can significantly reduce dominance tests among incomparable point pairs. Specifically, if the space $V_{00\ldots0}$ is not empty, then all the points in space $V_{11\ldots1}$ can be discarded immediately since they must be dominated by any point in space $V_{00\ldots0}$. Given a space V belonging to $C_{\widehat{p}}^{R^d}$, and any other space except V in $C_{\widehat{p}}^{R^d}$ can be summarized into 3 types, $D_{\widehat{p}}^V$, $U_{\widehat{p}}^V$ and $I_{\widehat{p}}^V$, respectively. For any point p in V need to be checked to judge whether it is a skyline point, obviously, those points in $U_{\widehat{p}}^V$ can be immediately ignored since they can certainly not dominate p according to partitioning definitions. Also, above properties imply that it is also correct to skip the dominance tests with all the points in $I_{\widehat{p}}^V$. We combine $U_{\widehat{p}}^V$ and $I_{\widehat{p}}^V$ as the space set which can be ignored dominance tests of space V, denoted by $IT_V = \{W \in C_{\widehat{p}}^{R^d} \mid W \in U_{\widehat{p}}^V \text{ or } W \in I_{\widehat{p}}^V\}$. While it is necessary to compare with the points in $D_{\widehat{p}}^V$ and the other points in itself space V.

We combine $D_{\widehat{p}}^V$ and space V as the space set which cannot be ignored dominance tests of space V, denoted by $NIT_V = \{W \in C_{\widehat{p}}^{R^d} \mid W \in D_{\widehat{p}}^V \text{ or } W = V\}$. It is easy to imply that if space V and W are comparable with each other, then $V \in NIT_W$ or $W \in NIT_V$. Here, we can claim that any accessing point p in V, the points in IT_V is safely skipped based on above analysis. While it is necessary to check out with the points in NIT_V.

4   Constructing Virtual Median Point as Partitioning Point

This section introduces the importance of partitioning points and illustrates the schema of space partitioning is a kind of typical divide and conquer algorithm that sub-problems are non-independent. To simplify the complexity of selecting optimized partitioning points, we first illuminate the importance of balance in partitioning, in addition provide a method to construct balanced partitioning points. Then, we adopt a structure name recursive search tree to reduce dominance tests among those subspaces which are comparable.

4.1   The Cost Evaluation Model of Space Partitioning in Skyline Computation

The schema of space partitioning needs to divide recursively current space into several disjoint subspaces until their sizes are small enough, and compute the local skyline points in all subspaces. Finally, all local skyline points are combined as the global skyline set, *i.e.*, the results of skyline computation. The schema is a typical kind of divide and conquer algorithm that sub-problems are non-independent. Now we design a cost evaluation model of space partitioning in skyline computation as follows:

$$Cost(\mathcal{A}) = C_p(\mathcal{A}) + C_c(\mathcal{A})$$

$\mathcal{A}$ is any algorithm of utilizing the schema of space partitioning, $C_p(\mathcal{A})$ and $C_c(\mathcal{A})$ represent the cost in partitioning the spaces and in dealing with comparable subspaces, respectively. And $C_c(\mathcal{A})$ relies primarily on the method of partitioning.

Theorem 1. *Using divide and conquer method to solve a problem, if it is divided into several independent disjoint sub-problems and the cost of dividing and conquering is proportional to the size of data set. Then the optimized dividing method is that all the sub-problems have the same size.*

Proof. For any partitioning method, suppose that the problem can be divided into k sub-problems, the time complexity is $T(kn) = T(n + \alpha_1) + T(n + \alpha_2) + \cdots + T(n+\alpha_k) + kn$ and $\forall\, i \in [1, k]$, $\alpha_i \in [-n, n]$ as well as $\Sigma_{i=1}^{k}\alpha_i = 0$. The cost of every iteration is $T(kn)$, the optimized partitioning has the least iteration times. Thus it is easy to conclude that the times of partitioning is minimum when $\alpha_1 = \alpha_2 = \cdots = \alpha_i = 0$.   □

In general, the lowest cost of partitioning the points is proportional to the size of data set because all points must compare with partitioning point at least once even though ignore the cost of determining partitioning point. Based on aforementioned analysis, the optimized space partitioning approach makes that all subspaces have the same size and can compute their local skylines independently. While this partitioning is certainly not existing because there must be comparable partitioned subspaces.

Considering the complexity of this cost evaluation model, we propose a approximated method. We separate the process of computing skyline using space partitioning schema into two phases. Firstly, we maximize the balance of subspaces, *i.e.* the difference of subspaces' size is as small as possible. And then we strive to weak the influence among subspaces with comparable relationship, in other word, reduce the times of dominance tests among points in those subspaces which are comparable.

4.2   Virtual Median Point

The balanced degree of partitioned subspaces depends solely on the position of partitioning point. We now discuss where the partitioning point is can maximize the balance of subspaces.

Theorem 2. *Given a data set S, suppose that all points are distributed uniformly and independently [2, 7] in d-dimensional $[0, m]^d$ space, the point p whose every attribute value is $\frac{m}{2}$ can maximize the balance of partitioned 2^d subspaces.*

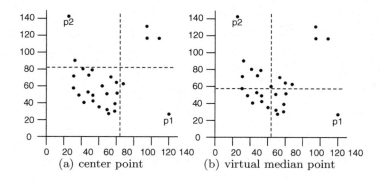

Fig. 1. Selecting partitioning point

Proof. The partitioning point p divides S into 2^d subspaces. Under uniform and independent condition, they have the same number of points since p is the center point of $[0, m]^d$ space.   □

Based on Theorem 2, the center point of space can maximize the balance of subspaces under uniform and independent condition. While in some real environment, it is difficult to get exact distributed region, for example in Fig. 1(a), $p1(140,25)$ and $p2(28,141)$ decide the boundary, if we regard straightly the data set as a uniform and independent distribution, the center of the boundary is $\alpha(84,83)$ and all points are partitioned in a extremely unbalanced method. To solve this problem, We construct the *Virtual Median Point*(VMP) as the partitioning point. The ith attribute value of VMP is the median value of all points in ith dimension. The VMP is $(65,57)$ in Fig. 1(b) and the partitioned subspaces are maximum balance. Besides, VMP is infinitely close to the center point when data set is distributed uniformly and independently.

4.3   Recursive Search Tree

The computations in subspaces based on the framework of space partitioning are non-independent as we have stated before. And we only need to consider the influence among subspaces which are comparable. Now we discuss how to reduce the dominance tests among comparable subspaces as much as possible.

As aforementioned descriptions, for two comparable subspaces V and W, it is either $V \in NIT_W$ or $W \in NIT_V$. So we aim to reduce the dominance tests between any partitioned subspace V and these subspaces which are in NIT_V.

The naive method of judging whether a point p in V is dominated by any point in space set NIT_V such that eliminating p is to do the dominance tests sequentially with p one by one until the occurrence of point q which can dominate p [4]. This method is extremely expensive since p will compare with all points in space set NIT_V if p is a local skyline point. In order to raise the efficiency of NIT tests of p, we introduce a structure recursive search tree(RST) which is similar to *skyline tree* in [3]. The root and inner nodes of RST are the virtual

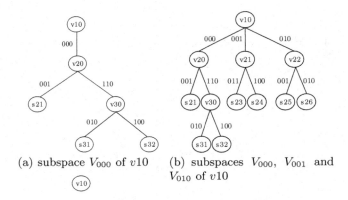

(a) subspace V_{000} of $v10$ (b) subspaces V_{000}, V_{001} and V_{010} of $v10$

Fig. 2. An example of recursive search tree

median points constructed by respective subspaces. And the leaf nodes are local skyline sets of their lying subspaces. The partitioning is recursively conducted, it is necessary to terminate it when space size is less than or equal to a threshold β.

Now we describe the detailed process of maintaining the RST by an example. As well as discuss how to reduce the compared times using it. Given a 3D data set S which contains 32 points and set threshold $\beta = 2$. We construct the first level VMP $v10$ and these points are partitioned into 8 subspaces, V_{000}:{$p1$, $p2$, $p3$, $p4$}, V_{001}:{$p5$, $p6$, $p7$, $p8$}, V_{010}:{$p9$, $p10$, $p11$, $p12$}, V_{011}:{$p13$, $p14$, $p15$, $p16$}, V_{100}:{$p17$, $p18$, $p19$, $p20$}, V_{101}:{$p21$, $p22$, $p23$, $p24$}, V_{110}:{$p25$, $p26$, $p27$, $p28$} and V_{111}:{$p29$, $p30$, $p31$, $p32$}. Then we recursively partition the new produced 8 subspaces in depth-first by their address values(That is the address value presents that $|V_{000}|=0$, $|V_{001}|=1$, $|V_{010}|=2$, and so on. The sorted order is V_{000}, V_{001}, V_{010}, V_{011}, V_{100}, V_{101}, V_{110}, and V_{111}). We recursively divide the subspace V_{000} first until the sizes of partitioned subspaces are all not more than β. The result of partitioning space V_{000} is shown in Fig. 2(a), $v20$ is the VMP of subset {$p1$, $p2$, $p3$, $p4$}. Suppose that $s21$={$p1$}, $s31$={$p2$}, $s32$={$p3$, $p4$}, terminate the partitioning of $s1$ since $1 < \beta$. Then we continue to partition the subspace V_{110} of $v20$, $v30$ is constructed and the other 3 points are put into subspace V_{010} and V_{100} of $v30$. Here, V_{010} and V_{100} of $v30$ both satisfy threshold β. Then we partition the subspace V_{001} of $v10$.

The aforementioned descriptions mainly introduce the process of partitioning. Now we discuss how to reduce the dominance tests using RST. The status of RST now is reported in Fig. 2(b), and suppose that $s21$={$p1$}, $s31$={$p2$}, $s32$={$p3$, $p4$}, $s23$={$p5$, $p6$}, $s24$={$p7$, $p8$}, $s25$={$p9$, $p10$}, $s26$={$p11$, $p12$}.We start to partition the subspace V_{011} of $v10$. Before partition these points, we first eliminate those which are dominate by some points in space set $NIT_{V_{011}}$, $i.e.$, subspaces V_{000}, V_{001}, and V_{010} of $v10$. We suppose p is lied in V_{011} of $v10$, now we check whether p is dominated. First, we compare p with $v20$, suppose that p lies in subspace V_{011} of $v20$. Then p only needs to compare with point $p1$ in $s21$ and skips all the points in subspace V_{110} of $v20$. If p is not dominate by $p1$, we continue this tests in subspace V_{001} and V_{010} of $v10$. Suppose that p lies in subspace V_{010} of $v21$ and in V_{001} of $v22$, p only needs to compare with points in $s25$ and save 6 times dominance tests with points in $s23$, $s24$ and $s26$. If p is not dominated by points in $s25$, keep it and continue this tests for next point in subspace of V_{011} of $v10$. Otherwise, drop p immediately. Finally, we collect all the leaf nodes as the results of skyline computation.

Based on above introduction about RST, it is easy to observe that the more balanced the partitioned subspaces, the less average dominance tests the RST conducts.

5   Algorithms

This section proposes our algorithm VMPSP which adopts constructed virtual median point as the *partitioning point* and utilizes a RST structure to reduce the dominance tests among these subspaces which are comparable. VMPSP strives to keep the balance of partitioned subspaces which can reduce the cost of partitioning and the cost of comparisons in RST. VMPSP algorithm is given in Algorithm 1.

Algorithm 1. VMPSP(S)

Input: A d-dimensional positive numerical data set S
Output: The RST of S

1: create a recursive search tree node RST.
2: $\widehat{p} \leftarrow$ **ConstructVirtualMedianPoint**(S).
3: $max \leftarrow 2^d - 1$
4: create Space Subsets $\{S_0, \cdots, S_{max}\}$ as well as all are *empty*.
5: $RST.value \leftarrow \widehat{p}$. // *Assign partitioning point $\widehat{p}$ to current RST node*
6: **for** $\forall p \in S$ **do**
7:     $i \leftarrow$ **Compare**($p, \widehat{p}$). // *i represents the address of p locating subspace w.r.t. $\widehat{p}$*
8:     Add p into S_i. // *S_i contains the ith space subset's points.*
9: **for** $i \leftarrow 0$ to max **do**
10:    **if** $|S_i| > 0$ **then**
11:        **for** $\forall j < i, (j \mid i) = i$ **do**
12:            **for** $\forall p \in S_i$ **do**
13:                $flag \leftarrow$ **CheckByRST**($p, RST[j]$).
14:                **if** $flag == true$ **then**
15:                    $S_i.discard(p)$
16:        **if** $|S_i| > \beta$ **then**
17:            $T_i \leftarrow$ **VMPSP**(S_i). // *Recursively conduct partition.*
18:            $RST[i].add(T_i)$. // *Add T_i as the RST's ith subtree.*
19:        **else if** $|S_i| > 0$ **then**
20:            $localSkyline \leftarrow$ BNL(S_i).
21:            $RST[i].skyline \leftarrow localSkyline$.
22: **return** RST

Algorithm 1 depicts the pseudo code of VMPSP. We first construct a virtual median point $\hat{p}$ as partitioning point by **ConstructVirtualMedianPoint** function as shown in line 2. And then in lines 3-8 partition all data points into 2^d subsets $\{S_0, \cdots, S_{max}\}$ corresponding to 2^d subspaces $\{V_0, \cdots, V_{max}\}$, and the subscript i of S_i represents the address of subpartitions. **Compare** function in line 7 computes the subspace in which p lies w.r.t $\hat{p}$. For any non-empty subset S_i belonging to V_i, we can immediately skip subsets in IT_{V_i} and solely check remaining subsets in NIT_{V_i} as described in line 11, all subspaces V_j are in NIT_{V_i} if $(j \mid i) = i$. Next, for the dominance tests in NIT_{V_i}, we only check a small part of them using RST maintained by yet found local skyline sets and constructed virtual median points and eliminate dominated points in S_i. After above checking, if the remaining points in S_i is still larger than threshold β, then recursively partition S_i until it is not more than β. Otherwise, we could compute the local skyline sets for the remaining points using basic method BNL [1], it is not inefficient since the threshold β is small. We create a RST at the beginning of the algorithm, the root and inner nodes are the virtual median points constructed by respective subspaces and the leaf nodes are local skyline points of their lying subspaces.

Algorithm 2. CheckByRST(p,RST)

Input: A d-dimensional positive numerical data set S
Output: Whether point p is dominated by any point in recursive search tree RST or not

1: **if** $RST.isLeaf()$ **then**
2:     **if DOMINATE**(p,$RST.skyline$) **then**
3:         return *ture*
4: **else**
5:     $i \leftarrow$ **Compare**(p,RST)
6:     **for** $\forall T \in RST.child, (T.address \mid i) = i$ **do**
7:         **if CheckByRST**(p,T) **then**
8:             return *true*
9: return *false*

RST is used to reduce the dominance tests among these subspaces which are comparable as described in Algorithm 2. First, we need to judge whether current RST node is a leaf. If it is true, check p with local skyline set in it. Otherwise, compare p with it and recursively recall function until check out whether p is dominated by some points. Finally return the result of tests.

6   Experiments

In this section, we evaluate the performance of our proposed algorithm VMPSP in dimensionality and cardinality in detail, by comparing it with state-of-the-art main-memory algorithms SFS [2], OSPS [3], and BSkyTree [4] using both synthetic and real data sets.

6.1   Experimental Settings

We generate two types of synthetic data sets that are Independent(IND) and Anti-correlated(ANT) according to the instructions in [1], respectively. The synthetic data sets dimensionality ranges from 6 to 16 and the data sets cardinality ranges from 200 K to 1M. All attributes values are positive real number in [0,1000]. Specifically, we claim here that the default dimensionality and cardinality are 12 and 200K, respectively. We also collect two real data sets ColorMoments[1] and IPUMS[1]. ColorMoments has 68404 tuples with 9 attributes. It describes the image features extracted from an image collection. IPUMS has 74954 tuples with 23 attributes. It describes unweighted PUMS census data from the Los Angeles and Long Beach areas for the year 1980. All algorithms are implemented by C++ languages and run on Intel Core-Q8300 CPU at 2.5GHz, with 4GB of RAM. Assuming that memory can contain all the origin data sets.

We compare our algorithm VMPSP with three state-of-the-art main-memory algorithms SFS [2], OSPS [3] and BSkyTree [4]. The monotone function of SFS adopts the *sum function* that the sum of all the attributes values. OSPS has two versions and we adopt the better OSPSOnPartitioningFirst method as OSPS and BSkyTree has two versions and we adopt the better BSkyTree-P method as BSkyTree here. We propose a complete algorithm of skyline computation using VMP named VMPSP which also utilizes our designed structure RST. It is better than state-of-the-art algorithms by extensive experiments.

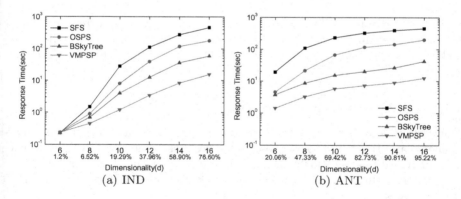

Fig. 3. Performance over dimensionality variation

6.2   Scalability

Varing Dimensionality. This section describes the performance of above algorithms with the dimensionality variation. Figure 3 reports the response time

---

[1] The data set is collected from https://kdd.ics.uci.edu.

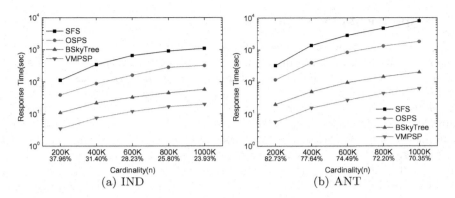

Fig. 4. Performance over cardinality variation

of the four algorithms. The two types distributions data sets that IND and ANT with dimension from 6 to 16 and 200k cardinality are used to conduct the experiments. VMPSP outperforms the other three algorithms in every dimensionality. Experiments validate space-based algorithms is better than sort-based algorithms and our algorithm is the best in existing space-based algorithms. Specifically, VMPSP is about faster 3 to 5 times over state-of-the-art space-partitioning algorithm BSkyTree, and outperforms SFS by up to one to two orders of magnitude.

Varing Cardinality. This section mainly describes the performance of algorithms with cardinality variation. We also generate two types data sets like above with cardinality from 200 K to 1M and 12 dimensionality. As Fig. 4 reports, our proposed algorithm VMPSP is always better than others.

6.3 The Performance on Real Data Sets

This section describes the performance of all algorithms on real data sets ColorMoments and IPUMS. Table 1 presents the response time which it takes to execute all algorithms in ColorMoments. We observe that our algorithm VMPSP is the best algorithm. The ColorMoments has 68040 tuples, 9 attributes, and 1533 of them are skyline results.

Table 1. Performance on real data set.

Algorithm	ColorMoments	IPUMS
	n=68,040; d=9	n= 74,954; d=23
	skyline=1533; 2.25s%	skyline=23190; 30.94 %
SFS	0.17s	35.17s
OSP	0.16s	1.92s
BSkyTree	0.15s	1.37s
VMPSP	0.08s	0.90s

7   Conclusion

In this paper, we illustrated the development of existing non-index main-memory algorithms of skyline computation. Compared two popular framework of being based on sorting and based on space partitioning, respectively. And concluded that space-partitioning-based schema is better. Then, the defections of existing algorithms based on it were introduced which they do not consider the importance of balance in partitioning. To overcome this limitation, we studied the process of partitioning and develop a cost evaluation model of space partitioning in skyline computation. Also, we analyzed the importance of balance in partitioning and proposed an approach to construct the balanced partitioning point VMP. In addition, a structure RST was designed to accelerate dominance tests among comparable subspaces. Finally, the evaluation indicates that our algorithm is faster several times than existing state-of-the-art algorithms.

References

1. Borzsony, S., Kossmann, D., Stocker, K.: The skyline operator. In: ICDE (2001)
2. Chomicki, J., Godfrey, P., Gryz, J., et al.: Skyline with presorting. In: ICDE (2003)
3. Zhang, S., Mamoulis, N., Cheung, D.W.: Scalable skyline computation using object-based space partitioning. In: SIGMOD (2009)
4. Lee, J., Hwang, S.: BSkyTree: Scalable skyline computation using a balanced pivot selection. In: EDBT (2010)
5. Bøgh, K.S., Chester, S.: Assent I. Work-efficient parallel skyline computation for the GPU. In: VLDB (2015)
6. Bartolini, I., Ciaccia, P., Patella, M.: Efficient sort-based skyline evaluation. ACM TODS **33**(4), 1–45 (2008)
7. Godfrey, P., Shipley, R., Gryz, J.: Maximal vector computation in large data sets. In: VLDB (2005)
8. Kossmann, D., Ramsak, F., Rost, S.: Shooting stars in the sky: An online algorithm for skyline queries. In: VLDB (2002)
9. Tan, K.L., Eng, P.K., Ooi, B.C.: Efficient progressive skyline computation. In: VLDB (2001)
10. Lee, K.C.K., Zheng, B., Li, H., et al.: Approaching the skyline in Z order. In: VLDB (2007)
11. Papadias, D., Tao, Y., Fu, G., et al.: Progressive skyline computation in database systems. ACM TODS **30**(1), 41–82 (2005)
12. Lee, J., Hwang, S.: QSkycube: Efficient skycube computation using point-based space partitioning. In: VLDB (2010)

Real-Time Event Detection with Water Sensor Networks Using a Spatio-Temporal Model

Yingchi Mao$^{(\boxtimes)}$, Xiaoli Chen, and Zhuoming Xu

College of Computer and Information, Hohai University, Nanjing 210098, China
{yingchimao,xiaolichen,zmxu}@hhu.edu.cn

Abstract. Event detection with the spatio-temporal correlation is one of the most popular applications of wireless sensor networks. This kind of task trends to be a difficult problem of big data analysis due to the massive data generated from large-scale sensor networks like water sensor networks, especially in the context of real-time analysis. To reduce the computational cost of abnormal event detection and improve the response time, sensor node selection is needed to cut down the amount of data for the spatio-temporal correlation analysis. In this paper, a connected dominated set (CDS) approach is introduced to select backbone nodes from the sensor network. Furthermore, a spatio-temporal model is proposed to achieve the spatio-temporal correlation analysis, where Markov chain is adopted to model the temporal dependency among the different sensor nodes, and Bayesian Network (BN) is used to model the spatial dependency. The proposed approach and model have been applied to the real-time detection of urgent events (e.g. water pollution incidents) with water sensor networks. Preliminary experimental results on simulated data indicate that our solution can achieve better performance in terms of response time and scalability, compared to the simple threshold algorithm and the BN-only algorithm.

Keywords: Event detection with sensor networks · Big data analysis · Spatial-temporal model · Connected dominating set · Probabilistic graphical models

1 Introduction

With the rapid development of computing and sensing technologies, numerous sensors in wireless sensor networks are widely applied to ecological observation, environmental monitoring, and disaster warning, etc. Real-time detection of abnormal event is essential for the practical application [1]. A Water Sensor Network (WSN) is one kind of typical applications of wireless sensor networks. The main objective of WSNs is to monitor various substances, like concentration of free chlorine, in real time way. When a water pollution incident occurs, water quality substantially deteriorates from the upstream to downstream. We should provide the event detection using WSNs in the context of real-time analysis [2,13]. Due to large-scale sensor nodes deployment, the amount of data from

© Springer International Publishing Switzerland 2016
H. Gao et al. (Eds.): DASFAA 2016 Workshops, LNCS 9645, pp. 194–208, 2016.
DOI: 10.1007/978-3-319-32055-7_17

dissimilar sensors has increased tremendously. For example, there are 128, 291 hydrometric sensor nodes deployed in China. They generate the hydrologic data about 60TB in one year. To reduce the computational cost of event detection and improve the response time, sensor node selection is needed to cut down the amount of data for the spatio-temporal correlation analysis. To avoid forwarding and collecting the sensed data from all of the sensors in the WSNs, Connected Dominating Set (CDS) [5,11] is introduced to select some backbone nodes and construct an effective connected structure in a WSN to reduce the amount of transmitted data and improve the response speed for event detection. In this paper, a distributed and effective algorithm to construct CDS called Rule K is adopted to select backbone nodes and forward the sensed data.

Meanwhile, real-time event detection (e.g., water pollution incidences) in the WSNs should consider both spatial and temporal correlation and dynamically record the propagation path of the water pollution with WSNs. This paper focuses on the real-time event detection with a spatio-temporal correlation model in a river to achieve high accuracy and low false alarm rate. Due to the non-linearity and differences of the sensed data, spatio-temporal data analysis faces many challenges. One is to model and analyze the correlation of spatial and temporal data, the other is to make full use of the network topology to reduce the amount of transmitted data. Probabilistic Graphical Models (PGMs) [3] can integrate probabilistic theory and graph theory to deal with the uncertainty and complexity of modeling the real world. A Markov Chain graph is assumed as an overlay graph in a WSN. All of the sensors can individually go through a Markov process and predict the future state. Markov chains can be used to model temporal states. A Bayesian Network (BN) is a directed acyclic graph. Each sensor node is regarded as a node of the BN, and the flow direction is regarded as the direction of each edge. BN enables the incorporation of knowledge and the encoding of uncertainty relationships from different sources. BN can be used to model spatial relationship among different sensor nodes. In this paper, a PGMs-based spatio-temporal model is proposed to achieve the spatio-temporal correlation analysis, where Markov chain is adopted to model the temporal dependency among the different sensor nodes through a learning-based approach, and BN is used to model the spatial dependency [10]. Different from previous work, the proposed event detection approach can effectively detect both temporal anomalies and spatial anomalies and track the propagation path of events.

Preliminary experimental results on the simulation data indicate that the proposed decentralized spatio-temporal abnormal detection approach can achieve better performance than the simple thresholds algorithm and BN-only [9] in terms of the detection precision and efficiency in the same experiment settings.

The rest of this paper is organized as follows. Related work is discussed in Sect. 2. CDS-based nodes selection strategy is proposed in Sect. 3. Section 4 presents the overall framework of online event detection model. In Sect. 5, the real-time spatial-temporal event detection approach is proposed. The performance evaluation is given in Sect. 6. The final section concludes the work.

2   Related Work

Event detection plays an important role in various applications of WSNs. A variety of techniques have been proposed for event detection, like threshold-based, pattern-based, and learning-based approaches.

In the threshold-based approaches, the threshold values are defined through statistical models or experts' experiences. They consider the abnormal event occurrence when the values of sensed data exceed the threshold or is in a range of threshold values. Simplicity is the main advantage of these approaches since raw data can be directly processed in the sensor nodes. Two thresholds were used in [6]. The higher threshold value can reduce the rate of false positives, and the lower threshold can ensure accuracy. All of the nodes were divided into three subsets $R1$, $R2$, and S, with two thresholds. The threshold values in the nodes set $R1$ are larger than the higher threshold; the values in the $R2$ are greater than the lower threshold and smaller than the higher one; and S is the set of the remaining nodes. Furthermore, the sliding windows scheme was introduced to effectively reduce the side effects due to its transient error.

Pattern matching approaches have also been exploited to solve the problem of events detection. In the pattern matching approaches, a specific pattern is usually defined by experts and then is verified with sensor readings to detect events. The integrated pattern-based approach with the in-network sensor query processing framework was proposed [7]. Different from the threshold-based approaches, pattern matching approaches abstract the events into different patterns based on the sensed data and convert the problem of event detection into pattern matching problems.

Learning-based methods can make inferences of the possibility of event occurrence based on the temporal or spatial correlation of the sensed data. The spatial and temporal correlation model is critical to improve the accuracy of event detection, because the spatial and temporal correlation among the sensor readings can effectively deal with the inherent uncertainty and reduce the false alarm rate. It is promising to make full use of the spatial and temporal correlation to detect abnormal events. Probabilistic Graph Models were applied to detect abnormal event. BN and Markov Random Fields (MRFs) are two common models in the WSNs due to the spatial and temporal correlation of the sensors in the high density network.

MRFs were adopted to model spatial context and stochastic interaction among observable quantities [8]. BN was considered as a means for unsupervised learning and anomaly detection in gas monitoring sensor networks for underground coal mines [9]. BN model can learn cyclical baselines for gas concentrations, and reduce the false alarms usually caused by flat line thresholds. Further, the system can learn dependencies among the changes of concentration in different gas concentrations and multiple locations.

Based on the above analysis, most approaches for the event detection have been discussed in the spatial or temporal dimension separately. In the real applications, sensor nodes are deployed in a certain space to collect the sensed data in a certain interval. The changes in sensor readings caused by an event usually exhibit a strong spatio-temporal correlation. Thus, spatial and temporal correlation is very critical to improve the accuracy of event detection. In this paper, the spatio-temporal correlation model is established in a decentralized way, and a distributed event detection approach based on the spatio-temporal correlation model is proposed.

3   Backbone Nodes Selection

A water sensor network consists of thousands of sensor nodes to sense, transmit and forward massive sensed data, which brings heavy burdens to further event detection in the context of real-time analysis. To avoid forwarding and collecting all of the sensed data from all sensors in the WSNs, optimal sensor nodes selection is needed to reduce the amount of transmitted data and improve the response time for event detection. CDS is one of optimal nodes selection strategies from Graph Theory [5].

A dominating set (DS) of an undirected graph $G(V, E)$ is a subset of V' of the vertex set V such that every vertex in $V - V'$ is adjacent to a vertex in V'. A dominating set V' is a connected if for any two vertices u and v, a path (u, v_1), (v_1, v_2), ..., (v_n, v), $(v_i \in V', 1 \le i \le n)$ in E exists. Rule K algorithm was proposed in our previous work [11], which is one energy efficient and distributed algorithm for CDS selection. Through selecting the appropriate communication range, a set of CDS can be constructed to meet the requirements of network connectivity and coverage. In each round of selection, the sensor nodes firstly adjust the appropriate communication radius and select the nodes in the CDS as the backbones. The sensor nodes in the CDS execute the tasks of monitoring and data forwarding. After analysis the collected data from the CDS with a WSN, the system can detect the abnormal event in the context of real-time analysis. In this paper, Rule K algorithm [11] is adopted to select some backbone nodes and construct an effective connected structure in a WSN.

Rule K algorithm consists of two parts: Marking Process and Pruning Rule.

Marking Process: Initially, all of the sensor nodes are the non-dominated nodes. Each node u exchanges its neighbors information in its communication radius R_u with all its neighbors. According to two neighbor nodes' information, node u judges whether it can directly communicate with its neighbor nodes v and w. That is $d(u,v) \le R_u$, $d(u,w) \le R_u$, while $d(w,v) > \min\{R_w, R_v\}$. If two nodes v and w can directly communicate, the node u can be as a candidate dominant node and participate in the construction CDS. Otherwise, node u is controlled by other nodes and is not in the CDS.

Pruning Rule: When the deployment density of nodes is relatively large, there may be other dominant nodes to make communication between two neighboring nodes. Therefore, the number of candidate dominant node selected via marking process is excessive, the candidates dominant nodes can adopt the pruning rule to reduce the size of dominant node set.

The basic idea of pruning rules is that if the k neighbor nodes of candidate dominant node u can make communication with other nodes, then sensor node

u can be deleted from the CDS. In order to avoid multiple candidate dominant nodes simultaneously being removed from the set of CDS, only candidate dominant nodes with maximum priority (e.g., node id, residual energy, node degree, etc.) can be deleted from the CDS.

After selecting backbone nodes, it can keep monitoring several streams from backbone nodes instead of all streams for event detection. Moreover, our strategy of backbone nodes selection can guarantee that unselected nodes can find at least one of the backbone nodes from their neighbors. Therefore, the CDS-based nodes selection strategy can reduce the amount of transmitted data and improve the response time for real-time event detection.

4   Spatio-Temporal Correlation Model

PGMs integrate probabilistic theory and graph theory to deal with the uncertainty and complexity of the real world. A spatio-temporal model is proposed to achieve the spatio-temporal correlation analysis, where a Markov chain is adopted to model the temporal dependency among the different sensor nodes through a learning-based approach, and a BN is used to speculate the occurrence probability of spatial events. To detect the abnormal events in the context of real-time analysis, one distributed processing method is proposed with the spatial and temporal correlation with a WSN.

4.1   Temoral Correlation Model

A first order Markov Chain is a finite or countably infinite sequence of events E_1, E_2, ... over discrete time intervals. At any time, the future behavior of the process is dependent solely on the current state, regardless of the previous states.

Definition 1. *A Markov Chain is a directed graph G with states S and directed edges L. $S = \{N_1, N_2, ..., N_m\}$, $L = \{L_{ij} | i \in 1, 2, ..., m, j \in 1, 2, ..., m\}$, Each edge $L_{ij} = <N_i, N_j>$ is labelled with a transition probability $P_{ij} = P(S_{t+1} = N_j | S_t = N_i) = P(N_j | Ni)$ [4].*

Usually, Markov chains are used to model temporal progress or events. One Markov chain graph is assumed as an overlay communication graph in a WSN. All of the sensors can individually go through a Markov process and predict the future states.

4.2   Spatial Correlation Model

A BN is a directed acyclic graph. Each sensor node is regarded as a node of the BN, and the flow direction is regarded as the direction of each edge. BN can incorporate the knowledge and encode the uncertainty relationships from different sources. BN can be used to model spatial correlation of water sensor networks.

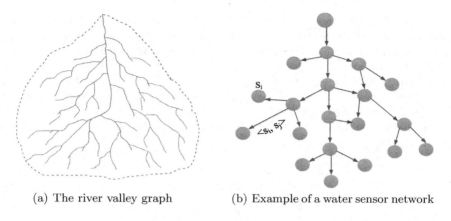

(a) The river valley graph   (b) Example of a water sensor network

Fig. 1. Illustration of water sensor networks.

Definition 2. *BN is a directed acyclic graph. The nodes represent random variables, and the directional edges represent causal relationship between the random variables.*

To clearly illustrate the real-time event detection with spatio-temporal correlation in the water sensor networks, the structure of the water sensor network is shown in Fig. 1. All of sensor nodes are deployed in the water and each sensor can communicate with others via single-hop communication. The topology of the water sensor network can be represented as a graph $G = (V, E)$, where V is the set of sensor nodes. Each directed edge $(S_i, S_j) \in E$ in the graph G can be represented as the single-hop transmission link. In the water sensor network, a set of sensors are deployed to monitor the water quality. Each sensor has a unique ID, from 1 to n. Sensors can sense a variety of attributes, such as the residual chlorine, PH, temperature in the river. The free chlorine is the main contaminants of water, therefore, the free chlorine is selected as the main monitored object for detecting water pollution event.

A BN can model spatial relationship of sensor nodes in the water sensor network G. S_i^t is denoted as the observed value from sensor S_i at the time interval t. One n-tuple $(S_1^t, S_2^t,..., S_n^t)$ can be used as one of the training data cases for learning the parameters of the BN. Thus, it can generate lots of training data cases from the historical sensed data set. The maximum likelihood parameter estimation algorithm is adopted to compute the parameters of the BN.

5   Spatio-Temporal Event Detection

5.1   Temporal Event Detection

The structure of Markov chain graph is defined by states and transition probabilities. The transition probabilities of states, e.g., $P_{(i-j)}$ is the transition probability from the state i to the state j. It can be computed through the learning process by counting the times of each state transition taken over in training data. In addition, the states in a Markov chain graph are determined through a statistical process. It is assumed that all of the sensor nodes are randomly and evenly deployed in the river. The sensed values of sensor nodes have normal distribution. Figure 2 illustrates a 3-state Markov chain and the transition process between three states, which can be represented with a directed graph $G = (S, L)$ with three states. In the graph G, $S = \{Low, Medium, High\}$, $L = \{L_{ij} | <N_i, N_j>, 1 \le i, j \le n\}$. Each arc L_{ij} is labeled with a transition probability $P_{(i-j)} = P(N_j | N_i)$.

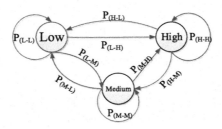

Fig. 2. The free chlorine concentration of Markov process

The concentration of free chlorine monitored by sensors can be classified into three states: *Low, Medium,* and *High*. The states of each sensor node at the time interval N_1 is presented in the Fig. 3, where blue nodes represent a low concentration of free chlorine, yellow nodes represent a moderate concentration, and red ones mean a high concentration. During the period of time interval from N_1 to N_5, the state transition of the free chlorine concentration can be modeled as a Markov chain process, as illustrated in the Fig. 4. It describes their current states and their future states of next time interval.

A deviation from normal temporal behavior of sensors can be interpreted as a noise (as a result of sensors malfunction) or as a temporal event. Since frequent occurrence of deviations rejects the possibility of noise; consecutive occurrence of deviations can be detected as temporal abnormity events. Using the sensed values of nodes, the spatial nodes individually go through the Markov chain process for a length of time τ. A temporal event is recorded when the predicted state through the Markov chain process $\hat{X}(s, t+\tau)$, is different from the normal state $X(s, t+\tau)$.

5.2   Spatial Event Detection

Assume that the historical data are time series $(t_1, t_2, ..., t_n)$ from the interval 1 to n. A BN is used to model causal and spatial relationship between sensor stream and its neighboring sensor streams. As illustrated in Fig. 3, there exists causal relationship between upstream nodes and downstream nodes. The structure of BN is the same as the topological structure of water sensor network.

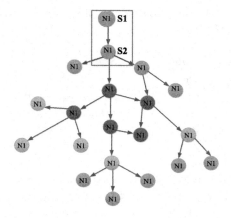

Fig. 3. Status of sensor network at the moment N_1

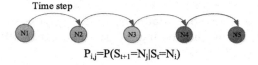

$$P_{i,j}=P(S_{t+1}=N_j|S_t=N_i)$$

Fig. 4. State transitions of free chlorine concentration

There are two sensor nodes S_1 and S_2 to measure the concentration of free chlorine, varying from 0.0 to 1.0. The concentration can be classified into three states: *Low*, *Medium*, and *High*, corresponding to $State_1$, $State_2$, and $State_3$, shown in Fig. 3. Based on the experts' experiences, if the value is smaller than 0.3, the state of sensor node is $State_1$. The value between 0.3 and 0.7 is considered as $State_2$. The node with the value greater than 0.7 belongs to $State_3$. During the training phase, it can compute the parameters of BN by adopting the maximum likelihood parameter estimation algorithm.

Table 1. The probability table of sensor S_1

State	$P(S_1)$	State	$P(S_1)$
$State_1$	0.6	$State_3$	0.1
$State_2$	0.3		

After training phase, the conditional probability table of each sensor can be learned. The conditional probability of sensor S_1 and S_2 are shown in Tables 1 and 2. From Table 1, the probability of three states of sensor S_1 are 0.6, 0.3, and 0.1, respectively. In Table 2, if the state of sensor S_1 is $State_1$, then the conditional probability is 0.7 and the state of sensor S_2 is $State_2$. That is to

Table 2. The conditional probability table of sensor S_2

S_1	S_2	$P(S_2)$	S_1	S_2	$P(S_2)$
$State_1$	$State_1$	0.7	$State_2$	$State_3$	0.2
$State_1$	$State_2$	0.2	$State_3$	$State_1$	0.2
$State_1$	$State_3$	0.1	$State_3$	$State_2$	0.5
$State_2$	$State_3$	0.1	$State_3$	$State_3$	0.4
$State_2$	$State_2$	0.5			

say, the conditional probability is $P(S_2 = State_1|S_1 = State_1)$=0.7. During the reasoning phase, if the state's value of S_1 is 0.24 at a certain interval time, the state of sensor S_1 belongs to $State_1$. It can be estimated that sensor S_2 is still in $State_1$. In the other case, if the state's value of S_1 is 0.8 at other interval time, the state of sensor S_1 is $State_3$. From the conditional probability in Table 2, $P(S_2 = State_2|S_1 = State_3)$=0.5, the state of sensor S_2 should be $State_2$. If the sensed value of sensor S_2 is not within the $State_2$, it can be considered that the sensed value deviates from the obtained spatial relationship. It can be thought that one abnormal event occurred.

5.3   Event Detection Based on Spatio-Temporal Correlation

Before real-time detecting abnormal events, the spatio-temporal correlation is modeled based on the normal historical data. For a large-scale water sensor network, it is hard to simultaneously model spatial and temporal relationship due to the complexity of the data. Furthermore, how to efficiently represent the spatio-temporal correlation is still an open problem. In this paper, spatial and temporal relationship are separately modeled. The conditional probability of BN is used to represent the spatial relationship and Markov chain is used to model the temporal state transition.

Detecting events and tracking the diffusion process of free chlorine is a difficult problem. In the proposed algorithm, the selected nodes in the CDS monitor the event occurrence. Firstly, if the backbone nodes detect the temporal abnormal event occurrence with water sensor network, the corresponding backbone nodes have spatial abnormal states in the BN. It can determine whether the abnormal events will occur based on the spatio-temporal correlation model. If the states of corresponding sensors deviate from the learned spatial and temporal relationship, the system consider that an abnormal event occurrence.

The abnormal event detection algorithm based on spatio-temporal correlation model is as follows:

(1) The nodes in the CDS collect the sensed data. If the temporal abnormal event occurrence, the nodes individually go through the Markov chain process for a period of time τ. A temporal event is recorded when the predicted state goes through the Markov chain process $\hat{X}(s, t+\tau)$, different from the observed state $X(s, t+\tau)$.

(2) If the temporal exception occurs in these backbone nodes, a candidate set are constructed for each abnormal node and the children nodes of the abnormal node are as the corresponding candidate nodes. If there is no temporal exception, these backbone nodes wait for sensing in the next interval time.

(3) The reasoning ability of BN can be used to estimate the expected state SE of the backbone nodes which detected temporal abnormality. Each backbone node compares the expected state SE with the observed value S_i^t to determine whether the value of sensed data deviates from the spatial relationship.

(4) If the observed value S_i^t is different from the expected state SE, it can be considered that there is a spatial deviation. The value of counter storing the spatio-temporal exception increases one.

(5) In the following time interval, the scope of monitoring is expanded, including backbone nodes and candidate nodes. They continue to detect whether there is temporal or spatial anomaly occurs among them.

(6) If the number of nodes with exception values in the CDS is equal to or greater than the threshold θ, it can be considered that an abnormal event occurrence and the corresponding warning is broadcasted.

In our approach, each backbone node has two system states: *IDLE* and *TMPE*. Initially, all of the backbone nodes simultaneously enter into *IDLE* system state. At each time interval t, each backbone node executes Markov chain process and predicts the transition probability in the next time interval $P_{(L|X)}$, $P_{(M|X)}$, $P_{(H|X)}$. For each backbone node, its initial transition probability of states can be estimated from a training data set and recorded in a 3×3 matrix M. When the water sensor network works, its matrix **M** is updated in each time interval. The updated values are stored in the corresponding backbone with an Update Matrix **UM**. At each time interval, each backbone node compares its states in its **UM** with the predicted states, and decides whether there is a deviation occurrence. If the number of deviation in the backbone node is greater than a length of time sequencer τ, an abnormal temporal event occurs. Thus, a corresponding warning message is broadcasted to its neighbor nodes. The system state of the backbone node is changed to *TMPE*.

In *TMPE* state, each backbone node calculates its expected state SE from a learning method. Through comparing the expected state SE with the observed value S_i^t, the backbone node determines whether the observed value deviates from the expected state SE. If there is a spatial deviation, the value of counter recording the spatio-temporal exception increases one. If the number of nodes with spatial and temporal deviation in the CDS is equal to or greater than the threshold θ, we consider that an abnormal spatial-temporal event has occurred.

6 Performance Evaluation

6.1 Experiments Setup

In this section, we evaluate the detection performance if the abnormal event occurs based on the simulations. We use EPANET 2.0 [12] to simulate the river network topology, as shown in Fig. 1. A contaminant simulator using MatLab and EPANET toolkit is also established to simulate events in the network. The proposed decentralized spatio-temporal event detection approach is executed in the simulation platform to evaluate its performance in terms of accuracy, scalability, and response time, compared with the simple threshold algorithm [6], Bayesian Network-based event detection algorithm [9]. Two metrics are used to evaluate the performance of the event detection algorithms.

Accuracy. To measure the accuracy, Precision, Recall, and F1_score are adopted. True Positive (TP) is defined as the number of sensor nodes that can correctly predict the event occurrence. False Positive (FP) is defined as the number of sensor nodes that generate the error detection. False Negative (FN) is defined as the number of sensors that are failure to predict the event while then abnormal event occurred. Thus, the accuracy metrics are defined as:

$$Precision = \frac{TP}{TP+FP}; \; Recall = \frac{TP}{TP+FN}; \; F1\_score = \frac{2 \times Precision \times Recall}{Precision+Recall}$$

Efficiency. The efficiency performance can be evaluated in terms of delay and scalability. The average delay is the average time from the event occurrence to the event being detected. The shorter the average delay, the higher efficiency.

6.2 Experimental Results Analysis

Accuracy. The performance of the proposed approach is compared with the simple threshold and BN-only approaches in the same experimental settings. The threshold-based approach is the most basic one for event detection with WSNs. It has the lowest computational complexity. The BN-only algorithm only considers the spatial relationship among the sensor nodes without the temporal relationship among nodes.

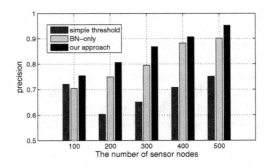

Fig. 5. The accuracy performance - precision in the different node density

Figure 5, 6 and 7 shows that the accuracy performance on the precision, recall rate and F1_score of the proposed approach, the simple threshold and BN-only algorithm, respectively. The number of nodes varies from 100 to 500.

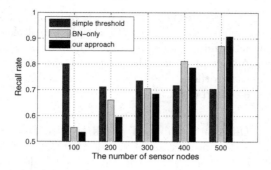

Fig. 6. The accuracy performance - recall in the different node density

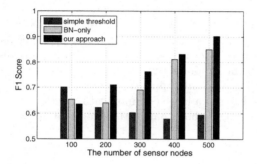

Fig. 7. The accuracy performance - F1_score in the different node density

From Fig. 5, it can find the precision of the proposed scheme is consistently higher than other approaches in the different nodes density. The reason is that the proposed approach can detect the abnormal events considering the spatial and temporal correlation and judge the events occurrence based on the sensed data over a period of time interval. In contrast, the simple threshold-based approach instantaneously reports an event occurrence solely based on the static threshold without taking the spatial and temporal relationship into account.

Meanwhile, the proposed approach has smaller recall rates than the threshold-based approach in the low density of network, as shown in Fig. 6. However, with the increase of network density, the proposed approach can obtain better performance on the recall rate than other two approaches. The simple threshold-based approach recognizes most of temporal deviations as an abnormal event, while some of these deviations are noises which cannot be considered as an abnormal event. Therefore, the recall rate decreases with the increase of the number of nodes in the simple threshold-based approach. To further evaluate the performance of the three event detection approaches, F1_Score, a weighted measure integrating both the detection precision and recall rate, should be considered.

Furthermore, Fig. 7 illustrates the performance in terms of F1_score for three event detection approaches. From Fig. 7, the simple threshold-based approach can obtain the best performance with the small number of sensor nodes. With the increase of the number of nodes, the proposed approach can achieve better performance than the others, due to considering the spatial and temporal correlation among the sensed data.

Detection Delay. Different number of nodes are selected in the same area to execute the proposed approach and BN-only one. As shown in Fig. 8, with the increase of the number of sensor nodes, the average delay time of the two approaches significantly increase, but the average delay time of proposed approach is smaller than that of BN-only one. The results show that the communication overhead increases with the increase of the number of nodes. However, the proposed approach exhibits a slower increase than BN-only one. The reason is that Rule K algorithm is adopted to select part of nodes as backbone nodes for forwarding the sensed data, which can greatly mitigate the communication overhead. Therefore, the proposed approach can achieve shorter average delay of event detections. While BN-only approach needs to collect the sensed data from all of the nodes, resulting in a large communication overhead. The response time of event detection is longer than that of the proposed approach.

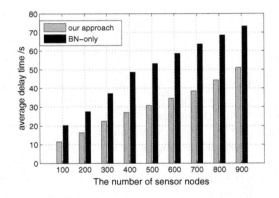

Fig. 8. The average delay time in the different node density

Scalability. The communication complexity of the proposed approach depends on the local transition of the sensed values. Detecting temporal events in the *IDLE* state requires no communication overhead, because the nodes run the algorithm individually. Whenever a temporal event is detected by the backbone nodes, they broadcast the corresponding messages and collect the sensed data. In *TMPE* state, the process communication overhead and the message transmission increase when the number of sensor nodes increases. From Fig. 9, the proposed approach can achieve better performance on data transmission. The reason is that Rule K algorithm can optimally reduce the number of sensor nodes, resulting in a low communication overhead. On the contrary, without optimal nodes selection process, BN-only approach exhibits the linear increase of communication overhead when the number of sensor nodes increases.

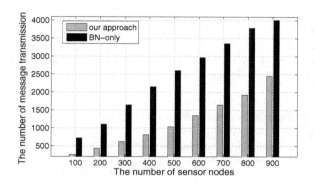

Fig. 9. Scalability with the increase number of nodes

Furthermore, the scale of network also causes the side effect on the efficiency performance. As the number of sensor nodes increases, the average delay time of the proposed approach also grow. However, the increase speed is slower than that of BN-only approach. The weak linear correlation of the number of sensor nodes and communication overhead shows good scalability of the proposed approach.

7 Conclusion

Event detection with the spatio-temporal correlation trends to be a difficult problem of big data analysis due to the massive data generated from large-scale sensor networks like water sensor networks, especially in the context of real-time analysis. To reduce the computational cost of abnormal event detection and improve the response time, sensor node selection is needed to cut down the amount of data for the spatio-temporal correlation analysis. In this paper, a novel algorithm called Rule K is introduced to select backbone nodes from the sensor network. Furthermore, a spatio-temporal model is proposed to achieve the spatio-temporal correlation analysis, where Markov chain is adopted to model the temporal dependency among the different sensor nodes, and Bayesian Network is used to model the spatial dependency. The proposed approach and model have been applied to the real-time detection of composite events with water sensor networks. Preliminary experimental results on simulated data indicate that the proposed decentralized spatio-temporal abnormal detection approach can achieve better performance than the simple thresholds algorithm and BN-only algorithm in terms of the detection precision and efficiency.

Acknowledgments. This research is partially supported by the National Key Technology Research and Development Program of China under Grant No. 2013BAB06B04; Key Technology Project of China Huaneng Group under Grant No. HNKJ13-H17-04; Science and Technology Program of Yunnan Province under Grant No. 2014GA007; the Fundamental Research Funds for the Central Universities under Grant No. 2015B22214; NSF-China and Guangdong Province Joint Project: U1301252.

References

1. Heidemann, J., Stojanovic, M., Zorzi, M.: Underwater sensor networks: applications, advances and challenges. Philos. Trans. Roy. Soc. A: Math. Phys. Eng. Sci. **370**, 158–175 (2012)
2. Eliades, D.G., Lambrou, T.P., Panayiotou, C.G., Polycarpou, M.M.: Contamination event detection in water distribution systems using a model-based approach. Procedia Eng. **89**, 1089–1096 (2014)
3. Koller, D., Friedman, N.: Probabilistic Graphical Models: Principles and Techniques. The MIT Press, Cambridge (2009)
4. Karlin, S.: A First Course in Stochastic Processes. Academic Press, Cambridge (2014)
5. Chandra, A., Tarasia, N., Kumari, A., Swain, A.R.: A distributed connected dominating set using adjustable sensing range. In: Proceedings of 2014 International Conference on Advanced Communication Control and Computing Technologies, pp. 868–871. IEEE Press, New York (2014)
6. Yim, S., Choi, Y.: Fault-tolerant event detection using two thresholds in wireless sensor networks. In: Proceedings of 15th IEEE Pacific Rim International Symposium on Dependable Computing, pp. 331–335. IEEE Press, New York (2009)
7. Xue, W., Luo, Q., Wu, H.: Pattern-based event detection in sensor networks. Distrib. Parallel Databases **30**(1), 27–62 (2012)
8. Piao, D., Menon, P.G., Mengshoel, O.J.: Computing probabilistic optical flow using markov random fields. In: Zhang, Y.J., Tavares, J.M.R.S. (eds.) CompIMAGE 2014. LNCS, vol. 8641, pp. 241–247. Springer, Heidelberg (2014)
9. Wang, X.R., Lizier, J.T., Obst, O., Prokopenko, M., Wang, P.: Spatiotemporal anomaly detection in gas monitoring sensor networks. In: Verdone, R. (ed.) EWSN 2008. LNCS, vol. 4913, pp. 90–105. Springer, Heidelberg (2008)
10. Huang, T., Ma, X., Ji, X., Tang, S.: Online detecting spreading events with the spatio-temporal relationship in water distribution networks. In: Motoda, H., Wu, Z., Cao, L., Zaiane, O., Yao, M., Wang, W. (eds.) ADMA 2013, Part I. LNCS, vol. 8346, pp. 145–156. Springer, Heidelberg (2013)
11. Mao, Y.-C., Xu, Z., Liang, Y.: An energy efficient connected coverage protocol in wireless sensor networks. In: Dong, G., Lin, X., Wang, W., Yang, Y., Yu, J.X. (eds.) APWeb/WAIM 2007. LNCS, vol. 4505, pp. 382–394. Springer, Heidelberg (2007)
12. Rossman, L.A.: EPANET2 user's manual. National Risk Management Research Laboratory: U.S. Environment Protection Agency (2012)
13. Arad, J., Housh, M., Perelman, L., Ostfeld, A.: A dynamic threshold scheme for contaminanat event detection in water distribution systems. Water Res. **47**, 1899–1908 (2014)

Bayesian Network Structure Learning from Big Data: A Reservoir Sampling Based Ensemble Method

Yan Tang, Zhuoming Xu[⊠], and Yuanhang Zhuang

College of Computer and Information, Hohai University, Nanjing 210098, China
{tangyan,zmxu,yuanhangzhuang}@hhu.edu.cn

Abstract. Bayesian network (BN) learning from big datasets is potentially more valuable than learning from conventional small datasets as big data contain more comprehensive probability distributions and richer causal relationships. However, learning BNs from big datasets requires high computational cost and easily ends in failure, especially when the learning task is performed on a conventional computation platform. This paper addresses the issue of BN structure learning from a big dataset on a conventional computation platform, and proposes a reservoir sampling based ensemble method (RSEM). In RSEM, a greedy algorithm is used to determine an appropriate size of sub datasets to be extracted from the big dataset. A fast reservoir sampling method is then adopted to efficiently extract sub datasets in one pass. Lastly, a weighted adjacent matrix based ensemble method is employed to produce the final BN structure. Experimental results on both synthetic and real-world big datasets show that RSEM can perform BN structure learning in an accurate and efficient way.

Keywords: Bayesian network structure learning · Reservoir sampling · Ensemble method · Probabilistic approximation · Big data

1 Introduction

A Bayesian network (BN) [1] is a probabilistic directed acyclic graphical model for representing multivariate probability distributions. BNs have been widely applied to various forms of reasoning in many domains such as Health Care, Finance and Transportation [2–4]. With the increasing availability of big datasets in science, government and business, BN learning from big datasets is potentially more valuable than learning from conventional, small datasets as big data contain more comprehensive probability distributions and richer causal relationships. However, learning BNs from big datasets requires high computational cost [5], easily ending in failure. Facing this challenge, one roadmap is performing the learning task on a big data processing platform using Hadoop or Spark, such as the MapReduce based method proposed by Fang et al. [6] and our previous work [7]. But such a platform is not affordable for all institutions. Therefore, an

© Springer International Publishing Switzerland 2016
H. Gao et al. (Eds.): DASFAA 2016 Workshops, LNCS 9645, pp. 209–222, 2016.
DOI: 10.1007/978-3-319-32055-7_18

alternative is first sampling sub datasets from the big dataset using probabilistic approximation and then learning a BN from the sampled, small sub datasets on a conventional computation platform. This study adopts the second roadmap. Since the most important and challenging step of BN learning is finding the network structure, this paper addresses the issue of sampling-based BN structure learning from a big dataset on a conventional computation platform.

We argue that to achieve Bayesian network structure learning from big data using a conventional computation platform, a big dataset needs to be appropriately sampled into several sub datasets with much smaller sizes, and an ensemble method is necessary for effectively combining the BN structures learned from the sub datasets. Hence a reservoir sampling based ensemble method, called RSEM, is proposed in this paper. The main ideas of RSEM are as follows. We introduce a minimal sample size (MSS) for sub dataset extraction, which can keep DAG-faithfulness [8] of the sub datasets, and design a greedy algorithm for calculating MSS, aimed at achieving a trade-off between learning accuracy and computational efficiency. According to the calculated MSS, we adopt a fast reservoir sampling method based on our proposed notion data reservoir index (DRI) to efficiently extract sub datasets in one pass. Lastly, we employ an ensemble method using a BDeu score [9] based weighted adjacent matrix to combine the BN structures learned from the sub datasets and produce the final BN structure in an approximate but sufficiently accurate way.

Our proposed method has been implemented using R software environment on a conventional computation platform. To validate the effectiveness of the method, we conducted experiments on both three synthetic big datasets and one real-world big dataset. The experimental results show that RSEM can sample appropriate sub datasets from big datasets by means of the calculated MSS, and perform Bayesian network structure learning from big datasets in an accurate and efficient way.

The rest of the paper is organized as follows: Sect. 2 is related work. The proposed method including algorithms is presented in Sect. 3. After giving experimental results and discussion in Sect. 4, we conclude this work in the final section.

2   Related Works

The notion DAG-faithful is the introduced is the work of TPDA algorithm [8]. A dataset is DAG-faithful if its underlying probabilistic model is DAG structured. This condition makes a dataset suitable for BN learning. The fundamental assumption of this research is that given a sufficiently large DAG-faithful dataset, its DAG-faithful sub datasets can be used to approximate the learning on the whole dataset.

In a Bayesian network, the Markov blanket (MB) of a node includes its parents, its children and the children's parents [10]. The MB of a node contains all the variables that shield the node from the rest of the network and is the only knowledge needed to predict the behavior of the node. Many algorithms like MMHC [11] were proposed to learn BN structure. An important property of a BN is its Average Markov blanket size, denoted as AMBS, which is defined as Eq. (1).

$$AMBS = \sum_{i=1..N} MBS_i/N \qquad (1)$$

where is the Markov blanket size of node i and is the total number of nodes in the network. AMBS can measure the complexity of a BN.

Structure Hamming distance (SHD), a metric introduced by Tsamardinos et al. [11], is defined as the number of the following operators required to make the network match: add or delete an undirected edge, and add, remove, or reverse the orientation of an edge [11]. It has become a widely used metric for measuring structure difference between two networks and evaluating the quality of the learned network. Small SHD indicates high learning accuracy. The number of correctly identified edges is equal to the total number of edges in the known BN minus SHD. This paper, therefore, uses SHD to evaluate the accuracy of the learning method.

Jiang et al. [12] studied the sampling of the datasets and applied the sampled datasets to different Bayesian network classifiers to achieve better classification accuracy, which validates the effectiveness of data sampling methods for BN learning. Reservoir sampling [13] is a widely used randomized algorithm for randomly choosing samples from a big dataset which doesn't fit into main memory. We leverage reservoir sampling to efficiently sample sub datasets from a big dataset.

In machine learning, ensemble methods [14] use multiple learning methods to obtain better predictive performance than learning from any of the constituent methods. Hasna and Salma [15] proposed a weighted ensemble Bayesian network learning method for gene regulatory networks. Our previous work [16] achieved higher accuracy of BN structure learning through ensemble methods. In this paper, we continue to adopt ensemble methods for achieve better learning accuracy.

In the field of BN learning from big data, Chickering et al. [17] showed that identifying high-scoring BN from large dataset is NP-hard. Yoo et al. [18] reviewed bioinformatics and statistical methods and concluded that Bayesian networks are suitable in analyzing big datasets from clinical, genomic, and environmental domains.

Furthermore, Fang et al. [6] proposed a Map-Reduce based method for learning BN from massive datasets. Our previous work [7] adopted distributed data-parallelism techniques and scientific workflow for BN learning from big datasets to achieve better scalability and accuracy. To the best of our knowledge, most existing studies applied data parallelization techniques to whole big dataset learning, much less work had used data sampling to learn BN from big datasets for reducing learning complexity.

3   The Proposed Method

3.1   Overview of the Method

Figure 1 is the overview of our proposed reservoir sampling based ensemble method (abbreviated as RSEM) for Bayesian network structure learning from big data, which consists of the following three key steps.

Firstly, RSEM takes the big dataset as the input and uses a greedy algorithm to calculate the minimal sampling size (MSS) of extracted sub datasets for a specific learning task in the BN learning procedure.

Secondly, a fast reservoir sampling algorithm is designed to sample sub datasets with the size of MSS from the big dataset. This sampling algorithm only requires one iteration over the entire dataset.

Lastly, an ensemble algorithm (by means of a BDeu score based, weighted adjacent matrix) is adopted to merge the Bayesian networks (BNs) learned from all the sub datasets and then produce a final BN as the output.

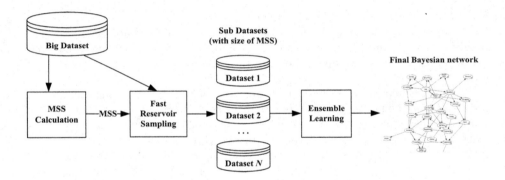

Fig. 1. Overview of the RSEM method

3.2   Calculation of MSS

Given a DAG-faithful (big) dataset with a sufficiently large size, it is reasonable to learn a BN from its sub datasets instead of the whole dataset. Learning on the sub datasets could achieve high computation efficiency and approximate the whole data learning without loss of generality. The key challenge here is the selection of sub dataset size. If the size is too small, then a poorly structured BN will be learned, otherwise, low computation efficiency as well as overfitting will occur. Thus, we introduce a novel concept called the minimal sampling size (MSS), as defined below.

Definition 1. *(minimal sampling size)Given a DAG-faithful and independent identically distributed (iid) dataset D, its minimal sampling size (MSS_D) is the* *minimal size of sub dataset that maintains DAG-faithful. MSS_D is defined in Eq. (2):*

$$MSS_D = N_{attr} * AMBS * sampleCoef_D \qquad (2)$$

where, N_{attr} is the number of attributes in the dataset (i.e. the number of nodes in the underlying Bayesian network). AMBS is the average Markov blanket size of the Bayesian network. And $sampleCoef_D$ is a data sampling coefficient required to maintain the DAG-faithfulness of extracted sub datasets.

Algorithm 1. CalculateMSS

Input:
    D: Dataset;
    ϵ: Threshold;
    $mstep$: Maximum loop steps.

Output:
    $AMBS$: Average Markov blanket size;
    MSS: Minimal sampling size.

1: $bestAMBS = 1$; step $= 0$;
2: sliceSize $= 100$ * number of attributes in D;
3: $D_{sliced} = \text{readData}(D, nrows = sliceSize)$; //nrows is the number of rows to read.
4: $BN_{DS} = \text{LearnBNStructure}(D_{sliced})$;
5: $currentAMBS = $ average Markov Blanket size of BN_{DS};
6: **while** $(currentAMBS > bestAMBS \&\& step \leq mstep \&\& ((currentAMBS - bestAMBS) > bestAMBS * \epsilon)$ **do**
7:     $sliceSize = sliceSize * 2$;
8:     $bestAMBS = currentAMBS$;
9:     $D_{sliced} = \text{readData}(D, nrows = sliceSize)$;
10:     $BD_{DS} = \text{learnBNStructure}(D_{sliced})$;
11:     $currentAMBS = $ average Markov Blanket size of BN_{DS};
12:     $step = step + 1$;
13: **end while**
14: $MSS = $ number of records in D_{sliced};
15: **return** $bestAMBS$ and MSS.

Theorem 1. *Given a DAG-faithful distribution P, there exists two datasets D_{MSS} and D_{S2} drawn from P with sizes MSS and S2 (MSS < S2) respectively so that the difference of the average Markov blanket size between the Bayesian networks learned from D_{MSS} and D_{S2}, denoted as $Diff_{AMBS}(D_{MSS}, D_{S2})$ is zero. This theorem can be formalized as follows:*

$$\forall P, \exists D_{MSS}, D_{S2}, MSS < S2 | Diff_{AMBS}(D_{MSS}, D_{S2}) = 0 \qquad (3)$$

Proof. By Definition 1, D_{MSS} is DAG-faithful. Since $MSS < S2$, D_{S2} is also DAG-faithful. Every DAG-faithful distribution has a unique essential graph [8]. Since D_{MSS} and D_{S2} are drawn from the same distribution P, then the essential graphs of D_{MSS} and D_{S2} are identical. The only difference between an essential graph and a Bayesian network is the edge direction, but the change of edge direction will not affect the sum of the sizes of Markov blankets. Thus, $Diff_{AMBS}(D_{MSS}, D_{S2}) = 0$.

Based on Eq. (2), to calculate MSS_D, both $AMBS$ and $sampleCoef_D$ are required. But in real life, the network structure is unknown. The only way to estimate $AMBS$ is through learning and obtaining the BN structure. And $sampleCoef_D$ is a varying coefficient dependent on each specific dataset instead of a constant number. To conquer this challenge, in light of Theorem 1, we propose a greedy algorithm called CalculateMSS (Algorithm 1) to calculate MSS.

Algorithm 1 starts with small sub dataset D_{sliced}. It learns the BN from D_{sliced} (Step 4) and obtains average Markov blanket size $AMBS$ (Step 5). Since D_{sliced} may not be DAG-faithful, consequently, BN structure learning algorithms will miss many edges, resulting in small $AMBS$. In order to make D_{sliced} DAG-faithful, the loop in the algorithm (Steps 6-13) doubles $sliceSize$ at each iteration, and stops when $AMBS$ becomes relatively stable. This indicates, based on Theorem 1, that the sub dataset size reaches MSS (Step 14). Algorithm 1 obtains both $AMBS$ and MSS, making $sampleCoef_D$ straightforward to compute using Eq. (2). Section 4.2 will show the experimental results of MSS on three datasets and validate the effectiveness of the algorithm.

3.3 Fast Reservoir Sampling

To reduce the scale of learning task, sub datasets need to be drawn from the whole big dataset for BN learning. To make the sampling more efficient, a novel concept, data reservoir index, is introduced in Definition 2.

Algorithm 2. GetdataReservoirIndex

Input:
$numSubDataset_{MSS}$: Number of sub datasets of size MSS in the whole dataset;
K: Number of sub datasets to be extracted.
Output:
dri: Data reservoir index.
1: Initialize dri as an empty array;
2: **for** i = $1..numSubDataset_{MSS}$ **do**
3:   **if** $i \leq K$ **then**
4:     $dri[i] = i$;
5:   **else**
6:     $removedEntry = \text{random}(1..i)$;
7:     **if** $removedEnry \leq K$ **then**
8:       $dri[removedEntry] = i$;
9:     **end if**
10:   **end if**
11: **end for**
12: **return** dri.

Definition 2. *(data reservoir index). A data reservoir index, denoted as dri, is an array that contains K elements, and is produced by reservoir sampling of K integers from one to $numSubDataset_{MSS}$ where $numSubDataset_{MSS}$ is the total number of sub datasets of size MSS in the whole dataset.*

Based on Definition 2, an algorithm named *GetdataReservoirIndex* (Algorithm 2) is proposed. It uses reservoir sampling to obtain *dri*. Since it operates in integer domain up to $numSubDataset_{MSS}$, the computation is very efficient.

After obtaining *dri* and sorting it, K sub datasets can be drawn efficiently from the whole dataset in one pass by extracting data records starting from $dri[i] * MSS$ and ending at $dri[i] * (MSS + 1)$, $i = 1, 2, ..., K$.

3.4 Ensumble Learning

After obtaining the sub datasets, RSEM calls the final procedure *Ensemblelearning* (Algorithm 3) to produce the final BN structure from the big dataset.

Algorithm 3. Ensemblelearning

Input:
    D: Dataset ;
    D_{sub}: Sub datasets sampled by fast reservoir sampling;
    ϵ:Threshold.
Output:
    BN_{final}: Final network structure.
1: $BN_{local}[i] = $ LearnBNStructure$(D_{sub}[i])$;
2: Obtain the Adjacent Matrix $\mathbf{AM}_i$ from $BN_{local}[i]$;
3: Weight each $BN_{local}[i]$ by BDeu score and transform $BN_{local}[i]$ into a Weighted Adjacent Matrix $\mathbf{WAM}_i$;
4: Sum all $\mathbf{WAM}_i$ using Equation (4) to get the final weighted adjacent matrix $\mathbf{FWAM}$;
5: **if** $\mathbf{FWAM}[i, j] > \epsilon$ **then**
6:     Set $BN_{final}[i, j] = 1$;
7: **end if**
8: **return** BN_{final}.

The algorithm invokes a BN learning algorithm (e.g. hill climbing) to learn local BN structure for each sub dataset (Step 1). Then, it uses BDeu score [9] to weight these local structures and transform them into weighted adjacent matrix $\mathbf{WAM}_i$, $i = 1, 2, ..., K$ (Step 3). Next, the algorithm sums all $\mathbf{WAM}_i$ using Eq. (4) to obtain the final weighted adjacent matrix $\mathbf{FWAM}$ (Step 4).

$$\mathbf{FWAM} = \sum_{i=1..K} \mathbf{WAM}_i \tag{4}$$

If an edge exists between node i and node j in majority of local structures, then $\mathbf{FWAM}[i, j]$ should be larger than a threshold ϵ. Therefore, Algorithm 3 adds an edge between i and j in the final network, transforming **FWAM** into the final network structure (Step 5-7).

4   Experiments and Discussion

4.1   Experimental Setup and Datasets

To validate the effectiveness of our proposed method, two experiments were conducted. The first experiment used three synthetic big datasets to confirm the effectiveness of MSS calculation as well as to evaluate the learning accuracy and the computation efficiency of RSEM. The second applied RSEM to a real-world big dataset, in order to show that the method can effectively model causal relationships.

The experiment environment is as follows. The computer is Dell PowerEdge R710, with Intel(R) Xeon(R) CPU E5640, 2.66 GHz, 12 M Cache, and Memory 16 GB (82 GB), 1066 MHz, running the operating system of Windows Server 2008 R2 Enterprise 64-bit, Service Pack 1.

The experiments were run in the R environment (version 3.1.1). Hill climbing and MMHC algorithms [11] in the Bnlearn R Package [19] were used to learn BN structures. The number of sampled sub datasets is 10.

Table 1 lists the datasets (CSV files) used in our experiments. Three synthetic datasets with large data volumes were generated using the data simulation module of the SamIam tool (http://reasoning.cs.ucla.edu/samiam/) from three known Bayesian networks: Child [20], Alarm [21] and HEPAR2 [22]. These known networks provide ground truth for the comparison of average Markov blanket size (AMBS) between the learned networks and the original ones, and for the resulting structural hamming distance (SHD). In Table 1, HMDALAR [23] is a real-world dataset from the Data.gov portal, representing 2009 Home Mortgage Disclosure Act (HMDA) Loan Application Register (LAR) Data.

Table 1. Experimental datasets

Name	#Rows(million)	Size(GB)	#Attributes	Domain
Child	10	1.2	20	Medical
Alarm	10	1.9	37	Weather
HEPAR2	10	4.9	70	Medical
HMDALAR	5.8	3.8	45	Finance

4.2   MSS Cacluation Results

Table 2 shows the computation results for minimal sampling size (MSS) and comparison between calculated AMBS and actual AMBS.

Table 2. MSS and AMBS comparison

Dataset	Calculated MSS	Calculated AMBS	Actual AMBS
Child	4,000	3.00	3.00
Alarm	14,800	3.30	3.51
HEPAR2	224,000	4.29	4.51
HMDALAR	40,000	8.00	n/a

From the first three lines in the table, we observe that the calculated AMBS by Algorithm 1 is close to the actual AMBS, indicating an accurate estimation of the BN complexity.

With the purpose of verifying the correctness of calculated MSS, letting the calculated MSSs (Table 2) be reference values, we used the hill climbing algorithm to perform BN learning on the synthetic datasets by doubly decreasing and increasing the values of MSS, and recorded the resulting SHDs. Figure 2 shows SHD trends over varying MSS on three synthetic datasets.

From the curves in Fig. 2, we observe that SHD rises sharply with the decrease of MSS starting from the reference value, while starting from the calculated MSS (second column in Table 2), SHD becomes stable with the growth of MSS. In other words, the calculated MSS by our algorithm (Algorithm 1) is a reasonable tradeoff between learning accuracy and computational efficiency.

In short, the above experimental results (Table 2 and Fig. 2) confirm the effectiveness of MSS calculation in the proposed RSEM method.

4.3   Results on the Synthetic Datasets

Table 3 shows the comparison of structural hamming distances (SHDs) and computation time for learning BN structures from the datasets between our method (RSEM) and whole dataset learning (WDL) using hill climbing algorithm. When applying RSEM, the threshold ε of the ensemble learning procedure is 0.667.

Table 3. SHD and computation time

Dataset	Total # of edges in original BNs	SHD		Computation time (minute)	
		RSEM	WDL	RSEM	WDL
Child	25	0	0	1.5	10.1
Alarm	46	13	13	3.20	22.9
HEPAR2	123	17	Failure	34.10	Failure (insufficient memory)
HMDALAR	n/a	n/a	Failure	19.60	Failure (insufficient memory)

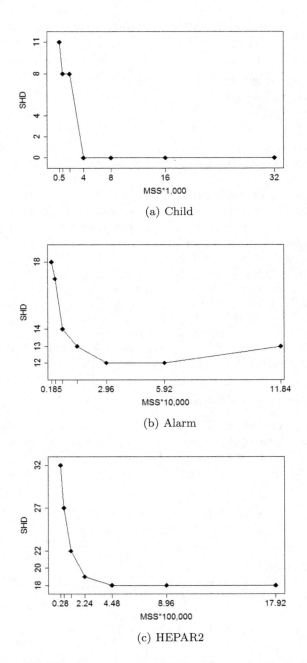

(a) Child

(b) Alarm

(c) HEPAR2

Fig. 2. SHD trends over varying MSS on the synthetic datasets.

From the third column of Table 3, we can find that RSEM achieves the same SHD compared with whole dataset learning (WDL) for the Child and Alarm datasets. In particular, RSEM found the correct network for the Child dataset (SHD=0). For the HEPAR2 dataset, RSEM identified over 86 % of the correct edges while WDL failed due to insufficient memory. These results indicate a high learning accuracy of our proposed RSEM.

Regarding the comparison of computation time (the last column in Table 3), it is observed that RSEM achieves nearly an order of magnitude improvement in computation time on the Child and Alarm datasets compared with WDL. Meanwhile, the HEPAR2 and HMDALAR datasets are too big to learn the BN structure from the whole dataset, resulting in computation failure caused by insufficient memory. But our method finished successfully within an hour for both big datasets.

The above experimental results confirm high learning accuracy and good computation efficiency of RSEM.

4.4 Results on the Real-World Dataset

For the HMDALAR dataset, there is no ground truth for the comparison of average Markov blanket size (AMBS) between the learned networks and the original ones, and for the resulting structural hamming distance (SHD). Nonetheless, the following results (cf. Figs. 3 and 4) on the real-world dataset show that our

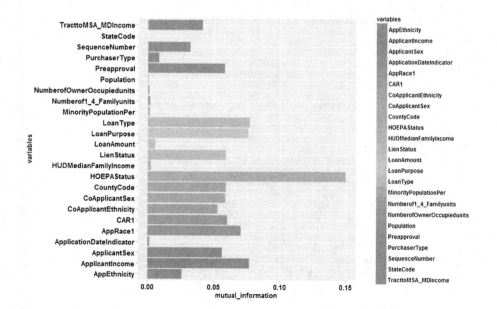

Fig. 3. MI of class variable ActionType and other variables in a sub dataset of HMDALAR

method (RSEM) can sample appropriate sub datasets from the big dataset as well as effectively model causal relationships between the data attributes.

After analyzing the HMDALAR dataset, we found that the ActionType attribute of the data is a class variable. Based on the calculated MSS (40,000) in Table 2, ten sub datasets were sampled from the big dataset. Figure 3 shows the mutual information (MI) of class variable ActionType and other variables in one of the sub datasets. In Fig. 3, we can find that variables HOEPAStatus (Home Owners Equity Protection Act Status), LoanType, ApplicantIncome, LoanPurpose, and AppRace1 (the race of the first applicant) have the top five MI values with the class variable. This is reasonable because from the perspective of loan approval, these variables indeed have a major impact on the approval decision. On the other hand, Fig. 3 indicates that state code, population, numberOfOwnerOccupiedUnits (number of units occupied by the owner), MinorityPolulationPer (Percentage of minority population) have the lowest MI values with the class variable.

As for the modeling of causal relationships between the data attributes, we applied RSEM to the ten sub datasets and produced the final BN. Figure 4 shows the Markov blanket of node ActionType in the Bayesian network. Observing the Markov blanket in Fig. 4, we find that the variables (Preapproval, PurchaserType, HOEPAStatus, and TractooMSA_MDincome) that have direct causal relationships with the class variable are modeled in the Markov blanket. Furthermore, variable Preapproval has six parents including LoanAmount, LoanPurpose, ActionType, Numberof1_4_Familyunits, HUDMedianFamilyIncome, and PurchaserType, which are truly important decision-making factors in loan pre-approval. On the other hand, most variables that have a low

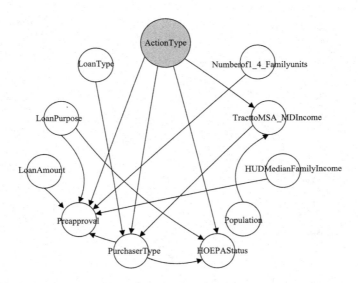

Fig. 4. Markov Blanket of node ActionType in the learned BN from HMDALAR

MI value are not in the Markov blanket of the node ActionType. This shows the effectiveness of RESM in modeling causal relationship for the real world dataset.

5 Conclusion

In this paper, we have proposed a reservoir sampling based ensemble method for Bayesian network structure learning from big data. We have demonstrated through experiments that our method can sample appropriate sub datasets from big datasets using the probabilistic approximation technique, and perform Bayesian network structure learning from big datasets in an accurate and efficient way. This method allows Bayesian network structure learning from big data using a conventional computation platform rather than a big data processing platform. Our future work focuses on enhancing the ensemble method to obtain higher learning accuracy.

Acknowledgments. This work was supported by the Natural Science Foundation of Jiangsu Province, China (Grant No. BK20141420 and Grant No. BK20140857) and the "Six Talent Peaks Program" of Jiangsu Province, China (Grant No. 2008135).

References

1. Ben-Gal, I.: Bayesian Networks. Encyclopedia of Statistics in Quality and Reliability. Wiley, New York (2007)
2. Zhang, Y., Zhang, Y., Swears, N., et al.: Modeling temporal interactions with interval temporal bayesian networks for complex activity recognition. IEEE Trans. Pattern Anal. Mach. Intell. **35**(10), 2468–2483 (2013)
3. Fenton, N.E., Neil, M.: A critique of software defect prediction models. IEEE Trans. Softw. Eng. **25**(5), 675–689 (1999)
4. Sun, S., Zhang, C., Yu, G.: A bayesian network approach to traffic flow forecasting. IEEE Trans. Intell. Trans. Syst. **7**(1), 124–132 (2006)
5. Al-Jarrah, O., Yoo, P., et al.: Efficient machine learning for big data: A review. Big Data Res. **2**(3), 87–93 (2015)
6. Fang, Q., Yue, K., Fu, X., Wu, H., Liu, W.: A mapreduce-based method for learning bayesian network from massive data. In: Ishikawa, Y., Li, J., Wang, W., Zhang, R., Zhang, W. (eds.) APWeb 2013. LNCS, vol. 7808, pp. 697–708. Springer, Heidelberg (2013)
7. Wang, J., Tang, Y., Nguyen, M., Altintas, I.: A scalable data science workflow ap-proach for big data bayesian network learning. In: Proceedings of the 2014 IEEE/ACM International Symposium on Big Data Computing (BDC 2014), pp. 16–25 (2014)
8. Cheng, J., Greiner, R., Kelly, J., Bell, D., Liu, W.: Learning bayesian networks from data: An information-theory based approach. Artif. Intell. **137**(1–2), 43–90 (2002)
9. Heckerman, D., Geiger, D., Chickering, D.: Learning bayesian networks: The combination of knowledge and statistical data. Mach. Learn. **20**, 197–243 (1995)

10. Pearl, J.: Probabilistic Reasoning in Intelligent Systems: Networks of Plausible Inference. Series in Representation and Reasoning. Morgan Kaufmann, San Mateo (1988)
11. Tsamardinos, I., Brown, L.E., Aliferis, C.F.: The max-min hill-climbing bayesian network structure learning algorithm. Mach. Learn. **65**(1), 31–78 (2006)
12. Jiang, L., Li, C., Cai, Z., Zhang, H.: Sampled bayesian network classifiers for class-imbalance and cost-sensitive learning. In: Proceedings of the IEEE 25th International Conference on Tools with Artificial Intelligence (ICTAI), pp. 512–517 (2013)
13. Vitter, J.S.: Random sampling with a reservoir. ACM Trans. Math. Softw. **11**(1), 37–57 (1985)
14. Rokach, L.: Ensemble-based classifiers. Artif. Intell. Rev. **33**(1–2), 1–39 (2010)
15. Hasna, N.J.S.: Weighted ensemble learning of bayesian network for gene regulatory networks. Neurocomputing **150**((B)), 404–416 (2015)
16. Tang, Y., Wang, Y., Cooper, K., Li, L.: Towards big data bayesian network learning - an ensemble learning based approach. In: Proceedings of the IEEE International Congress on Big Data (BigData Congress), pp. 355–357 (2014)
17. Chickering, D., Heckerman, D., Meek, C.: Large-sample learning of bayesian networks is np-hard. J. Mach. Learn. Res. **5**, 1287–1330 (2004)
18. Yoo, C., Ramirez, L., Liuzzi, J.: Big data analysis using modern statistical and machine learning methods in medicine. Int. Neurourol. J. **18**(2), 50–57 (2014)
19. Scutari, M.: Learning bayesian networks with the bnlearn r package. J. Statist. Softw. **35**(3), 1–22 (2010)
20. Spiegelhalter, D., Cowell, R.: Learning in probabilistic expert systems. Bayesian Statistics, 4. Clarendon Press, Oxford (1992)
21. Beinlich, I., Suermondt, H., Chavez, R., Cooper, G.: The alarm monitoring system: A case study with two probabilistic inference techniques for belief networks. In: Proceedings of the 2nd European Conference on Artificial Intelligence in Medicine, pp. 247–256 (1989)
22. Onisko, A.: Probabilistic Causal Models in Medicine: Application to Diagnosis of Liver Disorders. Ph.D. thesis, Institute of Biocybernetics and Biomedical Engineering, Polish Academy of Science, Warsaw (2003)
23. Data.gov - the U.S. Government Open Data: 2009 Home Mortgage Disclosure act (HMDA) Loan Application Register (LAR) Data, Accessed December 15, 2015. http://catalog.data.gov/dataset/2009-home-mortgage-disclosure-act-hmda-loan-application-register-lar-data

Correlation Feature of Big Data in Smart Cities

Yi Zhang[1,2,3], Xiaolan Tang[1,2,3], Bowen Du[1,2,3], Weilin Liu[1,2,3],
Juhua Pu[1,2,3($\boxtimes$)], and Yujun Chen[1,2,3]

[1] State Key Laboratory of Software Development Environment,
Beihang University, Beijing 100191, China
pujh@buaa.edu.cn
[2] Research Institute of Beihang University in Shenzhen, Shenzhen 518057, China
[3] College of Information Engineering, Capital Normal University,
Beijing 100048, China

Abstract. Smart cities are constantly faced with the generated data resources. To effectively manage and utilize the big city data, data vitalization technology is proposed. Considering the complex and diverse relationships among the big data, data correlation is very important for data vitalization. This paper presents a framework for data correlation and depicts the discovery, representation and growth of data correlation. In particular, this paper proposes an innovative representation of data correlation, namely the data correlation diagram. Based on the basic and the multi-stage data relations, we optimize the data correlation diagrams according to the transitive rules. We also design dynamic data diagrams to support data and relation changes, reducing the response time to data changes and enabling the autonomous growth of the vitalized data and the relations. Finally an instance of smart behaviors is introduced which verifies the feasibility and efficiency of the data relation diagram.

Keywords: Smart cities · Big data · Data vitalization · Data correlation

1 Introduction

Currently smart city is a hot spot, which monitors, analyzes and integrates all sorts of data in cities dynamically. There are many smart applications in economy, traffic, energy, security, administration, service, culture, etc. They improve the operation and management efficiency, modify public services, strengthen the emergency response capability, etc. [1]. Nowadays, the rapid development of smart city technology has drawn the extensive concern of academy, industry and government at home and abroad. Many countries are launching the research programs to develop smart city.

Data is the strategic resource related to urban economy and social development. Smart city construction prompts the explosive increase of urban data scale, redundancy and isolation of original data, which considerably restrain the full utilization of data values. Take Beijing for an example; various divisions assemble about 500,000 surveillance cameras in the city, which produce 3PB

© Springer International Publishing Switzerland 2016
H. Gao et al. (Eds.): DASFAA 2016 Workshops, LNCS 9645, pp. 223–237, 2016.
DOI: 10.1007/978-3-319-32055-7_19

video data each day. Nevertheless, these data belongs to and can be used only by its owner. As a consequence, it is very common that several cameras are assembled in the same spot by different divisions. It leads to waste of resources and expansion of data. Furthermore, there exists a large number of RFID, sensors, and on-board GPS devices, collecting various data. How to make fuller and smarter use of these massive and multi-source heterogeneous data (called big data) is becoming the core issue in prompting the development of smart city gradually [2]. However the statics and isolation of data make this a big challenge.

Xiong proposed data vitalization (DV) concept, which consists of data representation, organization, and utilization. In DV's viewpoint, data has the necessity and ability of self-recognition, self-study and automatic growth if the correlation among data are fully considered and used [3]. Thus, DV concentrates on describing, handling and utilizing the data in the way of live entity, which creates a brand new data organization and management mode. Therefore, how to compose live data entity of abundant isolated data via potential correlation, how to build the correlation among data entities, and how to analyze the principle of autonomous growth of data correlation are challenging in DV theory [4].

City data is extensively isomerous, and data applications are complex and varied. Thus it is hard to mine, display and grow the correlation among data, and there still lacks the systematic research study. This paper focuses on the big data in smart city. First, we propose a framework for studying data correlation which helps recognize data correlation features. Second, from the perspective of graph theory, data relation diagram (DRD) is illustrated to describe data correlation. DRD also helps to discover the potential data association and improves the efficiency of autonomous growth of data entity's correlation.

2   Related Work

Presently big data is ubiquitous in smart city. In the field of energy, manufacturing, transportation, etc., massive data in the level of TB, PB, even EB is accumulated. These data contains immense value which has significant strategic impact on society, economy and science research. For instance, the well-known Walmart needs to process over 1,000,000 requests every hour and its database contains more than 2.5 PB data. Compared to conventional massive data and very large data, big data in smart city has the following features: (1) massiveness in data scale; (2) variety in data diversity, including the structured, the semi-structured and the unstructured data and the greater proportion of the semi-structured and unstructured data; (3) difficulty in determining data schema beforehand which could only be determined after the emergence of data and is constantly changing with the growth of the data size; (4) data is treated as a way of assisting to solve the problems in other fields; (5) varieties in processing tools and there is no way capable of processing all the data [5,6].

In order to make full use of big data in smart city, DV comes into existence. The structure of DV includes data description and recognition, data maintenance and management, and data correlation and growth, etc.

Traditionally, the models of data and correlation representation include Entity-Relation model (ER), UML, data dictionary, etc. The main data description languages include DML, DTD, XML, RDF, OWL, etc., while the main description tools are database, data warehouse, OWL, etc. They support identification, extraction and retrieval data characteristics, and construct multi-granularity cross-media data mapping of data characteristics and high level semantics. However, they do not consider the relevance among different data correlations, which commonly exists in the real world. Besides, the data are static and dead in their approaches, ignoring the life cycle of the data entities. In this paper, we propose DRD to represent data and relations, which classifies different relations according to the life cycle of data. In this way, DRD adapts to the data and relation variations well.

The research on data correlation is significant for DV, which aims at identifying and mining data correlation and promote the correlation of data entities to realize their automatic growth. Nowadays, a great number of industries have already concentrated on the data mining in correlation. Take business application as an example; shopping basket analysis utilizes data correlation to discover the connection in product purchase information to increase the total sales and profits. For instance, the analysis of shopping basket finds that 15 % of customers who buy scarf also purchase gloves, and 60 % of customers who buy bread take milk [7]. The typical algorithms of correlation mining include AIS [8], SETM [9], Apriori [10], FP-Growth [11], etc. These algorithms seek the correlations through detecting frequent item sets and the largest frequent item set. However, they do not explore the relation features to discover new correlations, such as transitivity. In our proposal, we utilize the relation transitivity and the transmission constraint to optimize DRD, in order to convert indirect relations into direct relations. Therefore, DRD shortens the response time when data or correlations change, and further supports autonomous growth of data entities.

In spite of this, the framework of big data correlation in smart city remains ambiguous, and the modeling method and its further utilization still need to be studied further.

3   Data Correlation Framework

In DV opinion, big data in smart city consists of living data entities based on correlation and attributes, such as spatial and temporal correlation attributes and other correlation attributes related to figure, event, object, etc. Multi-dimensional and all-around data correlation are attempted to extract through the analysis on representation characteristics, standard dimension, temporal relation, spatial characteristics of data, etc. The research on the correlation of big data mainly includes the mining, the representation and the growth of data correlations, which will be illustrated in detail as follows.

3.1   The Mining of Data Correlations

The mining of data correlations is to discover the correlation among two or more sets of data. Its main challenges lie in two aspects: (1) The scale of data is huge, resulting in an intensely increasing complexity to catch the correlations among all the data. (2) The data has different owners and features, thus the correlation is very complex. To solve these problems, the layering and partitioning mechanism is used. When mining the data correlations, it gives higher priority to partial and tight coupling data, and then builds models for cross-layer and cross-partition correlations.

Currently, there are two ways to mine the correlations of big city data. On the one hand, because data entities usually contain the temporal and spatial information, it is possible to use the existing mining technology and take advantages of content or semantic discovery to identify the correlations among data. For instance, the flourishing type of business near some infrastructure could be found based on the registration records of business institutions. Thereby these information can be used to instruct the commercial layout. On the other hand, correlations among some regular and common data could be discovered from the perspective of the applications. Take the data entity shop for an example; the related social management department includes industry and commerce, taxation, urban management, environment protection, etc. Therefore, there is correlation between shops and each department. For the applications, it is likely to detect some primary and obvious correlations with the two ways mentioned above, while the mining of potential and complex correlations requires further study.

3.2   The Representation of Data Correlations

The representation of correlations means using mathematical model to show the correlations among data. The main challenges are: (1) Data correlations can be classified into many kinds according to the number of variables, triggering method, functioning phase, semantic logic and other classifying conditions. Therefore, how to create a unified approach to present correlations of different categories is one of the difficult tasks to tackle. (2) Different correlations may be not independent and share some variables. The independent variables of one correlation might be the dependent variables of another correlation. How to describe the relation among correlations is also one of the tasks in the correlation representation.

Generally, a data correlation consists of the following basic elements, which is presented by $R = (A_1, A_2, \ldots, A_n, B, rule, type, duration, enzyme)$. Here A_1, A_2, ..., A_n represent independent variables; B represents dependent variables; $rule, type, duration, enzyme$ represent correlated rules, the way of correlations taking effect, the time when the correlations taking effect, and data enzyme (the indispensable conditions of correlations taking effect). Similar to the description of data, a typical representation method of correlation is XML. XML shows the variation of dependent variables under the influence of independent variables, which is suitable for the correlation display of single data. Nevertheless, each correlation has independent XML display and the relation among several correlations are not displayed. Thus it is impossible to present the widely existing correlations among big data in smart city comprehensively and synthetically.

In this paper, we design a data relation graph to illustrate the relation among correlations clearly, named DRD. In DRD, its vertices and directed edges represent the data and the impacts on dependent variables. Specially, the existence of directed edge is affected by the way of data functioning, the time of data functioning and data enzyme. When the correlations among data come into existence, there is a directed edge between independent and dependent variables. Therefore, DRD is a kind of dynamic way to illustrate data correlations comprehensively. Its graph structure can be optimized with the help of graph theory, thereby improving the validity of data automatic growth according to the correlations.

3.3   The Growth of Data Correlations

The growth of correlations means the process of the produce and destroy of correlations with time passing, which is the dynamic characteristics based on the mining and representation of correlations and one of the key parts in the correlations research. When the vitalized data grows automatically according to the correlations, the new requests may create new correlations in the data entities. When the conditions of multidimensional data change, the correlations among data also change constantly. When the conditions of the correlations fail to be satisfied, the correlations disappear instantly. In the light of correlation phases from creation, maintenance to deletion, its growth process mainly includes the creation of new correlations, the update of existing correlations and the deletion of invalid correlations. As for a new correlation, the problems to be solved include the setting of generating conditions, the evaluation of new correlation elements and the assignment of correlation tags, etc. As for the deletion of an invalid correlation, the issues to be tackled include the setting of deletion conditions, the recycling of correlation tags, etc. Comparatively speaking, the correlation update tends to be more complicated, which requires the smart applications to set the conditions and procedures of update.

The challenges of the correlation growth research include: (1) The growth conditions are diverse. Every smart application requires its own conditions during its correlation creation, maintenance and deletion processes. Besides, the method of discovering and displaying the conditions is still one of the difficulties to cope with. (2) The correlation growth exerts complex influence on the data. The growth usually affects its related data, and the independent or dependent variables may change. How to erase and create the relation between data and correlations also remains to be handled. In general, there are many problems remained to be solved in the growth of correlations. Particularly, the growth of correlations is based on the mining and the correlation representation. The research on these two issues are helpful for that on correlation growth.

Currently, the research on data correlations concentrates on correlation mining, while the correlation representation is omitted. How to present data correlations is the prerequisite to further study, such as the dynamic growth of correlations. Therefore, this paper will introduce the method of correlation representation so as to provide the support for the further study on other characteristics of correlations. In detail, this paper proposes the data relation diagram to present data correlations. The efficiency of automatic growth of big data could be improved by optimizing the data relation graph.

4   Data Relation Diagram

Data relation diagram is the graph structure which represents the correlations among data. In detail, DRD takes data entities as vertices and the impacts of independent variables on dependent variables as the directed edges (the arrowhead points to dependent variables and the nock points to independent variables). Considering a data entity contains its carrier, temporal and spatial information and other attributes, the variations of different attributes are distinct. In order to make it clear, the correlation features in DRD refers to the relation of contents, and the relation of other attributes is handled in similar way.

Because DRD discussed in this paper only explores one-to-one data relation, namely the impact of one independent variable on another. Thus, the data relation needs to be separated and aggregated beforehand. As for the data relation with multiple independent variables, the data is separated first. In other words, the impact of each independent variable that functions independently is extracted and the correlations of one multiple independent variables turn to those of multiple single independent variable. As for the relation of multiple independent variables joining together, if these independent variables always function together in the relation, they will be aggregated. They are represented by single data vertices in the data relation diagram. Furthermore, for the data relations of multiple independent variables that function together but could not be aggregated, a new way of representation in the data relation diagram is requested. This work remains to be studied in the future and this paper is not involved in it. This paper aims at utilizing graph theory to optimize the data relation diagram and discover the potential correlations. Moreover, the dynamic update of data relation diagram is used to improve the efficiency of autonomous growth of data entities.

4.1   Data Relation

Data relation is the foundation of data relation diagram. Because the vitalized data entities are alive, its life cycle includes data generation, update and death stages. To simplify the analysis, the data relations which only take effect in one stage is called basic data relations. Considering practical applications usually having multi-stage data relations, we extend the basic relations to multi-stage data relations.

(1) **Basic Data Relations.** According to the life stage of dependent variables when the data relation takes effect, the basic data relations are classified into the relations of data generation, data update and data deletion. Specifically, the relations of data generation include the replica relation, etc., and the relations

Table 1. Basic data relations in DRD

Name of relation	Representation of relation	Name of data	Life stage of data	Meaning of data
Replica relation	$D \overset{>r}{-} D_r$	D: source data, D_r: replica data	Generation stage	D_r is the replica of D
Interference relation	$I_s \overset{Ifunc(I_s,I_t)}{\longrightarrow} I_t$	I_s: interference source data, I_t: interference acceptor data	Update stage	After I_s changes, I_t changes according to $IFunc(I_s, I_t)$; but I_t change does not affect I_s
Conjunction relation	$C_a \overset{CFunc(C_a,C_b)}{\underset{CFunc(C_b,C_a)}{\longleftrightarrow}} C_b$	For $CFunc(C_a, C_b)$, C_a: agentive data, C_b: passive data; For $CFunc(C_b, C_a)$, C_b: agentive data, C_a: passive data	Update stage	After C_a changes, C_b changes by $CFunc(C_a, C_b)$; after C_b changes, C_a changes by $CFunc(C_b, C_a)$
Dependency relation	$D \overset{>d}{-} D_d$	D: native data, D_d: following data	Death stage	The deletion of D leads to that of D_d

of data update include the interference relation, the conjunction relation, etc., and the relation of data deletion includes the dependency relation, etc. The representation of these typical basic data relations and its instructions are shown in Table 1. In the case of shop mentioned in the Sect. 3.1, the relation between the regulatory data of industrial and commercial departments and the operating data of shops is interference relation.

Although the four basic data relation cannot cover all the data relation in smart cities, they are able to explain much data relation and representative. The following part includes the examples of data relation diagram consisting of replica relation, interference relation, conjunction relation and dependency relation in detail. As for the other data relations, the way to depict the four relations above could be helpful and it is easy to be integrated into the data relation diagram.

Because the relevance is the basic characteristics of data, in the replica relation, its newly-generated replica data requires the application to create data relation. According to the way how the replica data builds its relation, the replica relation are classified into four categories, i.e., replica-function partition relation, replica-function extension relation, replica-new function relation and replica-non-function relation, which are called the application-oriented replica relations. Each relation has certain effect based on their characteristics to satisfy the application demand of data load balance, relation backup, data backup, new task backup, etc. The application-oriented replica relations in smart cities are shown in Table 2.

Table 2. Application-oriented replica relations

Name of relation	Representation of relation	Meaning of relation	Task and function
Replica-function partition relation	$D\{R\} \overset{>r}{-} D_r\{R_r\}, R_r \subseteq R$	Transfer part of data relation of D to replica D_r	Decrease communication volume of source data D, load balancing
Replica-function extension relation	$D\{R\} \overset{>r}{-} D_r\{R\}$	Replica data D_r inherits all relations from source data D	Backup of relations, improving security of relations and decreasing the time span of update
Replica-new function relation	$D\{R\} \overset{>r}{-} D_r\{R'\}, R' \subsetneq R$	Assign new data relations to replica data D_r	Assign new tasks
Replica-non-function relation	$D\{R\} \overset{>r}{-} D_r\{\varnothing\}$	Replica data D_r have no data relations.	Only for data back up, not involved in tasks

(2) Multi-stage Data Relations. In the applications of smart cities, in order to display the changing process of each stage in the life circle of data, the data relations of different stages are aggregated to obtain the multi-stage data relations. Take the replica relation for an example; 2-stage data relations are obtained by integrating the interference relations and conjunction relations in data update stage, including replica-interference relation, replica-anti-interference relation and replica-conjunction relations, as shown in Table 3. Furthermore, these data relations could be integrated with dependency relation in data death stage, and 3-stage data relations are obtained. Take replica-conjunction for an example, 3-stage data relations contain replica-conjunction-dependency relation, replica-conjunction-anti-dependency relation and replica-conjunction-equal-dependence relation, is shown in Table 3. Similarly, replica-interference relation and replica-anti-interference relation could also be integrated with dependency relation to obtain the 3-stage data relations.

4.2   The Optimization of Data Relation Diagram

The optimization of data relation diagram refers to making use of transitivity of data relations to transfer some indirect data relations to direct data relations under the constraint of relation transmission, which aims to increase the response speed of data growth.

(1) Relation Transmission Rule. In the data relation diagram, some data relations can be transmitted among multiple data. The first data to transmit is

Table 3. Instances of 2-stage and 3-stage data relations

Name of relation	Representation of relation	Life stage of data	Meaning of relation
Replica-interference relation	$D \overset{\geq r}{\rightarrow} D_r$	Generation stage, update stage	The change of source data D affects replica data D_r, not vice versa
Replica-anti-interference relation	$D \overset{\geq r}{\leftarrow} D_r$	Generation stage, update stage	The change of replica data D_r affects source data D, not vice versa
Replica-conjunction relation	$D \overset{\geq r}{\leftrightarrow} D_r$	Generation stage, update stage	Replica data D_r and source data D affect each other
Replica-conjunction-dependency relation	$D \overset{>r,>d}{\longleftrightarrow} D_r$	Generation stage, update stage and death stage	The deletion of source data D leads to the deletion of conjunction replica D_r, not vice versa
Replica-conjunction-anti-dependency relation	$D \overset{>r,<d}{\longleftrightarrow} D_r$	Generation stage, update stage and death stage	The deletion of conjunction replica D_r leads to the deletion of source data D, not vice versa.
Replica-conjunction-equal-dependency relation	$D \overset{>r,>d,<d}{\longleftrightarrow} D_r$	Generation stage, update stage and death stage	The deletion of source data D leads to the deletion of conjunction replica D_r, vice versa

called the initial data and the last data to transmit is called the terminal data. Particularly, a k-step transmission relation refers to the transmissions from initial data S to terminal data T through k transfers, which are shown as $\{S - Ii_1\} \cap \{Ii_1 - Ii_2\} \cap ... \cap \{Ii_{k-1} - T\} \Rightarrow \{S \overset{(k)}{\rightarrow} T\}$. Specially, in order to avoid the transfer loop, each data relation only appears once in the transmission process.

A 2-step transmission is taken for an instance. According to the categories of data relations finally obtained, a 2-step transmission may be 2-step replica relation, 2-step interference relation, 2-step conjunction relation and 2-step dependency relation.

The 2-step replica relation is the new replica relation on the basis of two replica relations, represented by $\{D \overset{\geq r}{\rightarrow} Dr_1\} \cap \{Dr_1 \overset{\geq r}{\underset{*}{\rightarrow}} Dr_2\} \Rightarrow \{D \overset{\geq r,(2)}{\rightarrow} Dr_2\}$. Similarly, the 2-step conjunction relation is $\{Ca_1 \overset{CFunc(Ca_1,Cb_1)}{\underset{CFunc(Cb_1,Ca_1)}{\longleftrightarrow}} Cb_1\} \cap$ $\{Cb_1 \overset{CFunc(Cb_1,Cb_2)}{\underset{CFunc(Cb_2,Cb_1)}{\longleftrightarrow}} Cb_2\} \Rightarrow \{Ca_1 \overset{CFunc(Ca_1,Cb_2),(2)}{\underset{CFunc(Cb_2,Ca_1),(2)}{\longleftrightarrow}} Cb_2\}$, and the

2-step dependency relation is $\{D \xrightarrow{\geq d} Dd_1\} \cap \{Dd_1 \xrightarrow{\geq d} Dd_2\} \Rightarrow \{D \xrightarrow{\geq d,(2)} Dd_2\}$. 2-step interference relations are classified into interference-conjunction relation, conjunction-interference relation and pure-interference relation, in terms of the transmission categories of its two components. Specifically, the interference-conjunction relation is the new interference relation based on an interference relation and a conjunction relation, as $\{Is \xrightarrow{IFunc(Is,It)} It\} \cap \{It \underset{CFunc(Cb,It)}{\overset{CFunc(It,Cb)}{\longleftrightarrow}} Cb\} \Rightarrow \{Is \xrightarrow{IFunc(Is,Cb),(2)} Cb\}$. The conjunction-interference relation is the new interference relation based on a conjunction relation and an interference relation, $\{Cb \underset{CFunc(Is,Ca)}{\overset{CFunc(Ca,Is)}{\longleftrightarrow}} Is\} \cap \{Is \xrightarrow{IFunc(Is,It)} It\} \Rightarrow Ca \xrightarrow{IFunc(Ca,It),(2)} It$. The pure-interference relation is the new interference relation based on two interference relations, $\{Is_1 \xrightarrow{IFunc(Is_1,It_1)} It_1\} \cap \{It_1 \xrightarrow{IFunc(It_1,It_2)} It_2\} \Rightarrow \{Is_1 \xrightarrow{IFunc(Is_1,It_2),(2)} It_2\}$.

Afterwards, 3-step transmission relation is obtained by integrating a 2-step transmission relation and a new data relation. In a similar way, the k-step transmission relations can be calculated by $\{D_1 \xrightarrow{(m)} D_2\} \cap \{D_2 \xrightarrow{(n)} D_3\} \Rightarrow \{D_1 \xrightarrow{(k)} D_3\}$, where $k = m + n$.

(2) Data Relation Constraint. From the perspective of data security, the effect of some data relations are only permitted to be obtained from initial data, which means that the relation transmission is unallowed or only allowed within several steps. This is named the data relation constraint. Those relations meeting the constraint are named constraint transmission relations. k-step constraint implies that the valid relation transmission has no more than k steps. The set of valid relations which satisfy the k-step transmission constraint is called k-step constraint relation set, denoted by $\bigcup_{i=1}^{k} \{S \xrightarrow{(i)} T\}$.

The k-step constraint interference relation is $\bigcup_{i=1}^{k} \{S \xrightarrow{(i)} T\}$, meaning that the step size from interference actor to acceptor is no more than k steps. It is notable that the inter relations may be mixed by the interference-conjunction relations, the conjunction-interference relations and the pure-interference relations in the process of interference relation transmissions.

The set of k-step constraint conjunction relations is $\bigcup_{i=1}^{k} \{S \xrightarrow{(i)} T\}$, which means that the number of steps from conjunction actor S (or T) to conjunction acceptor T (or S) is less than or equals k. Similarly, there also exist k-step constraint replica relations and k-step constraint dependency relations.

The number of constraint steps varies with the categories of data relations. Take the four categories as example, i.e., replica relations, interference relations, conjunction relations and dependency relations; their valid data relation sets are represented by $\bigcup_{step=1}^{kr} \{S \xrightarrow[(step)]{>r,} T\}$, $\bigcup_{step=1}^{ki} \{S \xrightarrow{(step)} T\}$, $\bigcup_{step=1}^{kc} \{S \xleftrightarrow{(step)} T\}$, $\bigcup_{step=1}^{kd} \{S \xrightarrow[(step)]{>d,} T\}$ respectively, among which S and T are random distinct data nodes, kr, ki, kc and kd are the values of constraint step parameters of replica relations, interference relations, conjunction relations and dependency relations, respectively.

(3) k-step Optimization of Data Relation Diagram. k-step optimization of data relations means that i-step constraint relation sets are permitted with $i \leq k$, and the indirect relations in DRD are converted into direct relations. Besides, full-step optimization or k_{MAX}-step optimization means that all the indirect relations are converted into direct relations, where k_{MAX} is the biggest number of transmissions in DRD. After optimization, the data relations having different steps are allowed in a pair of certain data, and the data relations having the same step is merged together. Take a DRD shown in Fig. 1(a) as an instance; its 2-step optimized DRD is shown in Fig. 1(b), while its full-step optimized DRD is displayed in Fig. 1(c).

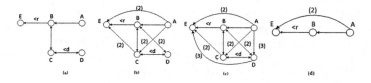

Fig. 1. The optimization of DRD.

4.3   Dynamic Data Relation Diagram

It is apparent that the dynamic variations in DRD mainly include data variation and relation variation. These two kinds of variations will be illustrated in detail as follows.

(1) Data Variation. Data variation consists of data insertion, data modification and data deletion. (a) Insertion of data D. Firstly, after a data node D is inserted into DRD, D waits for its related data relations to join. If D is a replica data generated by replica relation, the relation is a kind in Table 2. Then set data relations for D in reference to the data relations of source data and the task requirements. If the new data D is assigned with data relations, it is handled according to relation insertion introduced in the following part. (b) Modification of data D. After D is modified in DRD, the values of interference acceptors whose interference source is D are changed according to the interference functions. Meanwhile, the passive data whose agentive data is D adjust their values according to the conjunction functions. (c) Deletion of data D. In DRD, D as well as the relations between D and the other data is deleted. Then the transmission relations including D in the optimized DRD are decided to be deleted or not according to the original DRD. Moreover, if there exists a dependency relation where D is precursor data, the successor data is deleted. Take Fig. 1 for an example; the full-step optimized DRD after the deletion of D is shown in Fig. 1(d). Because there is a dependency relation between C and D, C is also deleted. The update network only contains data A, B and E.

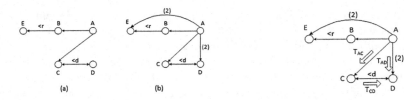

Fig. 2. Effects of relation changes   **Fig. 3.** Transaction scheduling instance.

(2) Relation Variation. Similar to data variation, relation variation is classified into relation insertion and relation deletion. Due to the relative independence of relations, relation modification can be considered as the deletion of original relation and the insertion of new relation. (a) Relation insertion. After a new data relation is inserted, the constraint transmission relations related to this relation are created in the optimized DRD according to the transmission constraint. (b) Relation deletion. After the deletion of one data relation, the transmission constraint relations it produces are removed from the optimized DRD. Take Fig. 1(a) as an example; delete the conjunction relation $B \leftrightarrow C$, and insert the interference relation $A \rightarrow C$. The updated DRD is Fig. 2(a), while its corresponding full-step optimized DRD is Fig. 2(b).

4.4   Transaction Scheduling

In DRD, a transaction implies a series of data update resulted from the change of independent variables according to the data relations. In the actual applications, it is normal that there exist multiple concurrent transactions between two data. In order to avoid the data mistakes made by the multiple update of the same data at the same time, this paper sets an order for the execution of concurrent transactions. Specifically, when the data relations with a greater transmission step update the terminal data, they also have impacts on intermediate data and override the original changes of these data. Therefore, the transactions of these data relations should not be taken the priority. In other words, the transactions are with smaller step size should be set in a high priority. This also indicates that in the process of data relation transmission, the data relations having fewer steps should be set higher priority. The output of former step serves as the input of next step, and the data update is transmitted step by step backwards.

In Fig. 3, two requests of data D update occur simultaneously, i.e., T_{AD} : $A \xrightarrow{(2)} D$ and $T_{CD} : C \leftrightarrow D$. If $T_{AD} : A \xrightarrow{(2)} D$ runs first, data A and data D are updated. Then $T_{AC} : A \rightarrow C$ starts, which leads to the loss of the original value of C. After that, $T_{CD} : C \rightarrow D$ is executed. The effect of A on C in $T_{AC} : A \rightarrow C$ is transmitted to D, which means that the effect of A on D is calculated twice and the original data C' s effect on D does not be calculated. Therefore, it leads to a wrong value of D update. To avoid this issue, $T_{CD} : C \leftrightarrow D$ with fewer transmission steps should be executed first, while $T_{AD} : A \xrightarrow{(2)} D$ with more transmission steps should be executed later.

4.5   Instance Analysis of Smart Behaviors

From the microscopic view, smart applications in modern cities intellectualize the human behaviors, which helps to promote the security, comfort and efficiency of human activities on the basis of smart behaviors. In order to realize smart behaviors, data correlations need to be built among big data in smart cities, and they are adjusted according to data diagram. Some stock dealers transact in the Stock Exchange, which is an example to illustrate the function of DRD. The process of data transmission is shown in Fig. 4.

When the vehicle leaves for the destination (the Stock Exchange), the report of destination is sent to the traffic control center, which figures out the optimized driving route (data 1) and sends it back to the vehicle. The vehicle drives under the instructions. Apparently, the expected driving route (data 2) is the replica of the optimized route data calculated by the traffic computing center, and it changes as data 1 does. There is a replica-interference relation between data 1 and data 2.

When the request of stock information is released, abundant stock information is transmitted by the communication patterns in vehicular networks. Therefore, there is a replica-interference relation between the stock data cache in the roadside infrastructures (data 3) and that in this vehicle (data 4). Considering inter-vehicle content sharing, data 4 could also be obtained through other vehicles. Because the stock data collected by the other vehicles (data 5) varies in time, each vehicle updates the latest information. Thus, there is a replica-conjunction relation between data 5 and data 4.

According to the hypothesis above, the stock dealer arrives at the Stock Exchange through the optimized route and obtains the latest transaction data. The stock dealer drives back after the transaction is handled.

On the way, the weather changes suddenly. The meteorological departments calculates and sends the warning of rainstorm (data 6) to traffic departments. The warning of rainstorm in traffic departments is data 7, and it is sent to the news media (such as traffic radio and television broadcast). The warning of rainstorm in news media is data 8, and then it is sent to the driver by news media. The warning of rainstorm recorded by the driver is data 9. Above all, data 6 and data 7, data 7 and data 8, data 8 and data 9 are all replica-interference relations. In this way, it takes three transmissions for the user to get the weather warning, and any mistake among these data replicas will lead to the danger of driving. In order to improve the traffic security and intelligence, the 3-step optimization of DRD discussed above can establish direct replica-interference relation between data 6 and data 9, thereby sending the warning of rainstorm to the driver promptly.

Based on DRD, different divisions in smart cities coordinate with each other, and are not separated any more. When the storm comes, the government is informed of the hydrops on the roads and all the manhole cover status (data 10) by the sensors deployed. Then data 10 is delivered through traffic division (data 11) and the news media (data 12) to the vehicles and pedestrian (data 13). Using the 3-step optimized DRD, the indirect relations are converted into direct

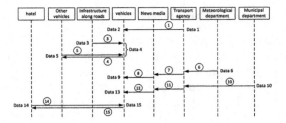

Fig. 4. Sketch map of the data transmission process in the instance of smart behaviors.

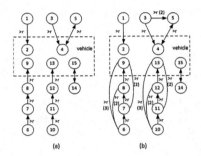

Fig. 5. DRD in the instance of smart behaviors.

relations, which accelerate the update speed of data. The locations of severe hydrops and that of the dangerous manhole covers are sent to the vehicles and pedestrian at once. Meanwhile, the hotels release the renting information (data 14) to the vehicles and pedestrian nearby to accommodate the pedestrian and avoid the traffic accidents. The users can order the rooms (data 15) based on data 14, which indicates that there is a conjunction relation between data 14 and data 15.

In the instance above, the original DRD is shown in Fig. 5(a) and the DRD after 3-step optimization is shown in Fig. 5(b). Thus, the research on correlations among data and the use of DRD to represent the complex data relations is helpful for the potential data relation discovery. In addition, the optimization of transmission relation diagram accelerates the response speed of data variation, improves the efficiency of smart behavior decision and provides all-dimensional and high-quality smart services for the daily life of citizens in smart cities.

5   Conclusion

The correlations of big data is a key technology in DV. Through the research on the framework of data correlations, this paper illustrates the mining, representation and growth of the data correlations separately, including their connotation, challenges and progress. Specially, this paper proposes a representation method based on data relations diagram to display the data correlations. First, we define the basic data relations and multi-stage data relations in DRD. Second, four typical data relations are illustrated in detail. Third, by the transitivity of data relations, the optimization method of DRD is introduced. The DRD optimization can speed the responding to the variation and improve DRD's efficiency. Furthermore, this paper designs the dynamic DRD to cope with the complicated change of data and their relations. This can satisfy the growing needs of vitalized data and its relations. Finally, this paper provides an instance of smart behaviors. In this instance, the data relation characteristics are analyzed and its DRD is illustrated, which can be used to instruct the decisions of smart behaviors. This instance testifies the feasibility and validity of DRD solution proposed in this paper.

Acknowledgement. We gratefully acknowledge the support from National High Technology Research and Development Program of China (2013AA01A601), National Natural Science Foundation of China (61173009, 61502320), the Science Foundation of Shenzhen City in China (JCYJ20140509150917445), Science & Technology Project of Beijing Municipal Commission of Education in China (KM201410028015), and the State Key Laboratory of Software Development Environment (SKLSDE-2015ZX-25).

References

1. Grady, M., Hare, G.: How smart is your city? Science **335**(6076), 1581–1582 (2012)
2. Hancke, G.P., Silva, B.D.E., Hancke, G.P.: The role of advanced sensing in smart cities. Sensors **13**(1), 393–425 (2013)
3. Xiong, Z., Luo, W., Chen, L., Ni, L.M.: Data Vitalization: a new paradigm for large-scale dataset analysis. In: IEEE 16th International Conference on Parallel and Distributed Systems (ICPADS), pp. 251–258. Shanghai, 8–10 December 2010
4. Tian, X.: The Technology Research of Multi-sensor Data Association and Track Fusion. Harbin Engineering University, Harbin (2012)
5. Zhu, H.: An institution theory of formal meta-modelling in graphically extended BNF. Front. Comput. Sci. **6**(1), 40C56 (2012)
6. Zemke, F.: Whats new in SQL. SIGMOD **41**(1), 67–73 (2012)
7. Du, Y.: An improved algorithm for mining association rules. Xidian University (2012)
8. Agrawal, R., Imieliski, T., Swami, A.: Mining association rules between sets of items in large databases. ACM SIGMOD Record **22**(2), 207–216 (1993)
9. Agrawal, R., Ghosh, S., Imielinski, T., Iyer, B., Swami, A.: An interval classifier for database mining applications. In: International Conference on Very Large Data Bases (VLDB), pp. 560–573 (1992)
10. Agrawal, R., Srikant, R.: Fast algorithms for mining association rules. In: International Conference on Very Large Data Bases (VLDB), pp. 487–499 (1994)
11. Wang, K., Liu, T., Han, J.: Mining frequent patterns using support constraints. In: International Conference on Very Large Data Bases (VLDB), pp. 43–52 (2000)

Nearly Optimal Probabilistic Coverage for Roadside Data Dissemination in Urban VANETs

Yawei Hu[1,2], Mingjun Xiao[1,2(✉)], An Liu[3], Ruhong Cheng[1,2], and Hualin Mao[1,2]

[1] School of Computer Science and Technology,
University of Science and Technology of China, Hefei, People's Republic of China
xiaomj@ustc.edu.cn
[2] Suzhou Institute for Advanced Study,
University of Science and Technology of China, Suzhou, People's Republic of China
[3] School of Computer Science and Technology, Soochow University,
Suzhou, People's Republic of China

Abstract. Data disseminations based on Roadside Access Points (RAPs) in vehicular ad-hoc networks attract lots of attentions and have a promising prospect. In this paper, we focus on a roadside data dissemination, including three basic elements: RAP Service Provider (RSP), mobile vehicles and requesters. The RSP has deployed many RAPs at different locations in a city. A requester wants to rent some RAPs, which can disseminate their data to vehicles with some probabilities. Then, it tries to select the minimal number of RAPs to finish the data dissemination, in order to save the expenses. Meanwhile, the selected RAPs need to ensure that the probability of each vehicle receiving data successfully is no less than a threshold. We prove that this RAP selection problem is NP-hard, since it's a meaningful extension of the classic Set Cover problem. To solve this problem, we propose a greedy algorithm and give its approximation ratio. Moreover, we conduct extensive simulations on real world data to prove its good performance.

Keywords: Vehicular ad-hoc network · Data dissemination · Roadside access points selection

1 Introduction

Vehicular ad-hoc networks (VANETs) as a new paradigm of mobile ad-hoc networks, can offer convenient data service for passengers and have attracted attentions of many researchers. Further, with the development of intelligent transportation system, some cities have deployed many RAPs at different locations, such as bus stations, taxi pickup points and intersections. These RAPs usually have good capacities of storage and communication. When a vehicle enters a RAP's communication range, the RAP can deliver the information stored in its

© Springer International Publishing Switzerland 2016
H. Gao et al. (Eds.): DASFAA 2016 Workshops, LNCS 9645, pp. 238–253, 2016.
DOI: 10.1007/978-3-319-32055-7_20

memory to the vehicle via WiFi or other protocols of short-range wireless communication. Obviously, the introduction of RAPs can improve the connectivity of VANETs. Moreover, by using these RAPs in urban areas, some large-size data can be disseminated to passengers in vehicles at a low cost. Due to this advantage, much research ([1–3]) has studied the new form of data dissemination based on RAPs in VANETs.

Consider a scenario of roadside data dissemination. There is a RAP Service Provider (RSP), for instance, certain government agency, who has deployed many RAPs at different locations in a city. If a requester hopes to disseminate its data, it can rent parts of the RAPs from the RSP to finish the dissemination. For example, a shop can be a requester, which wants to disseminate advertisements of its new commodities. On the one hand, they need to select as less RAPs as possible for this requester to conduct the data dissemination, in order to save the requester's expenses. On the other hand, they also need to let these RAPs cover adequate vehicles, so that the data can be received by as many passengers as possible. Moreover, as the communication connections between RAPs and mobile vehicles are unstable and intermittent, the selected RAPs also need to ensure that the probability of each vehicle receiving the data successfully is no less than a reasonable threshold.

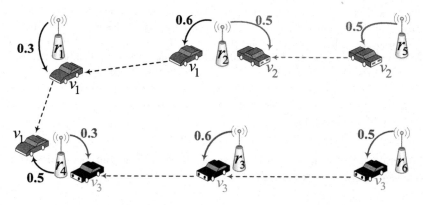

Fig. 1. A example of the roadside data dissemination. (Three vehicles pass six RAPs, during which the vehicles can receive data with some probabilities.)

Let's take Fig. 1 as an example. There are 6 RAPs $(r_1, \cdots, r_6)$ and 3 mobile vehicles (v_1, v_2, v_3). Suppose that a requester hopes to disseminate its data to the three vehicles, and expects that the probability of each vehicle receiving the data successfully is not less than the threshold 0.7. In this example, we assume that the requester selects $\{r_2, r_3, r_4, r_5\}$ to perform the data dissemination. Although each vehicle cannot receive the data successfully from each single RAP with the probability no less than 0.7, their joint probability exceeds the threshold. For instance, v_1's probabilities of receiving the data successfully from r_2 and r_4 are 0.6 and 0.5, respectively. Their joint probability is $1 - (1 - 0.6) * (1 - 0.5) = 0.8$, beyond the threshold 0.7.

For the schemes of data dissemination in VANETs, most of them pay attentions to the delivery ratio or delay of messages. However, our roadside data dissemination focuses on how to select the minimal number of RAPs to conduct the dissemination. Unlike other schemes of data dissemination, our scheme can ensure that the probability of each vehicle receiving data successfully is no less than a predefined threshold. For example, in [2], the authors select several seed vehicles to disseminate advertisement data to other vehicles, and the seed vehicle selection is based on the degree centrality [4] of each vehicle. Nevertheless, the solution cannot guarantee that each vehicle's probability of receiving advertisement data successfully is large enough. From another perspective, the data dissemination can be regarded as the problem of coverage. For example, in [2], the dissemination is achieved by selecting some seed vehicles to cover as many vehicles as possible. Their "cover" means that there is at least one social contact between a single seed vehicle and each vehicle. In contrast, our problem is to ensure the joint dissemination probability from all selected RAPs no less than the threshold. Actually, our problem will lead to a combining probabilistic set cover mixed by non-linear programming, which is also different from the classic Set Cover problem.

In this paper, we first describe the moving pattern of a mobile vehicle, by using a set of RAPs which this vehicle often passes by. The mobile vehicles can receive data successfully from these passed RAPs with some probabilities. Then, we define the RAP selection problem for roadside data dissemination and prove its NP-hardness. The problem is to select the minimal number of RAPs to cover all of the vehicles, while ensuring that each vehicle can receive data successfully from the RAPs with probability no less than a threshold. Finally, we propose a greedy algorithm to solve the RAP selection problem, analyze the complexity and approximation ratio of this algorithm, and conduct extensive simulations to prove its good performance.

We highlight our main contributions as follows:

1. We propose a scheme for roadside data dissemination based on probabilistic coverage of some selected RAPs.
2. We introduce a new optimization problem, which is the RAP selection problem for roadside data dissemination. We prove that this problem is NP-hard.
3. We propose a greedy approximation algorithm to solve the RAP selection problem, give the corresponding approximation ratio of this algorithm and conduct extensive simulations to prove its good performance.

The remainder of this paper is organized as follows. Section 2 presents the related work. In Sect. 3, the network model, the definition of RAP selection problem and the proof of the NP-hardness of this problem are described. We propose a greedy algorithm, analyze its complexity and approximation ratio in Sect. 4. In Sect. 5, the evaluation of this algorithm is showed, followed by the conclusion of this paper in Sect. 6. Partially complex proofs are moved to the Appendix.

2   Related Work

The scenario of data dissemination based on RAPs in VANETs is different from data dissemination in Delay Tolerant Networks or Mobile Social Networks [5–7]. In VANETs, J. Qin et al. [2] considered to select seed vehicles to diffuse advertisement data to others. They analyzed the sociality of the vehicular network and proved the dynamic and temporal correlations of sociality. Based on the analysis, they proposed a greedy scheme to solve it. Z. Li et al. [3] proposed a scheme for advertisement data diffusion with an incentive-centered architecture. The architecture encourages the advertisement providers to trade off the effect and cost of their advertisements messages, aiming to avoid unnecessary distractions to drivers and message storms in VANETs. In [1], H. Zheng et al. designed a system, which can disseminate advertisement data via some placed RAPs. When a driver receives a shop's advertisements from RAPs, he or she may detour to the shop and the detour probability depends on the detour distance. The RAP placement needs to balance the tradeoff between the traffic density and the detour probability. S.-B. Lee et al. [8] proposed a secure incentive framework to avoid the noncooperative behavior of selfish or malicious vehicle nodes in the advertisement data dissemination. Unlike these schemes, our paper uses the RAPs to disseminate data to the vehicles passing by.

Our roadside data dissemination is relevant to the data dissemination based on roadside units (RSUs) in VANETs. For example, in [9], J. Jeong et al. considered to forward data from stationary APs to moving vehicles. The forwarding scheme took into account the AP's location and the destination of the vehicle's trajectory, and selected a target point as packet-and-vehicle-rendezvous-point. M. Sardari et al. in [10] proposed a message dissemination paradigm, in which each RSU encoded a huge message into k data packets and forwarded them to vehicles, then the vehicles can decode a specific RSU's message by collecting sufficient packets. K. Liu et al. [11] focused on accessing information stored at RSUs and paid attention to the channel division. Comparing with these papers, we consider how to select the minimal RAPs to finish the data dissemination from RAPs to vehicles.

Our problem, i.e., the RAP selection, is related to the deployment of RSUs in VANETs. In [12], T. Wu et al. studied the RSU placement problem for vehicular networks in a highway-like scenario. Their placement strategy maximized the aggregate throughput in the network, taking into account the impact of wireless interference, vehicle population distribution, and vehicle speeds in the formulation. In [13], B. Aslam et al. focused on the placement problem in the scenario of urban vehicular network environment. They proposed two methods aiming to minimize the reporting time of event for RSUs, with incorporating the density and speed of vehicle, as well as the occurrence likelihood of an event in urban. By comparison, our problem is based on the joint probabilistic coverage of RAPs.

Therefore, our roadside data dissemination is different from other studies of advertisement or data dissemination, and cannot be solved directly by existing solutions.

3   Network Model and Problem Definition

In this section, we set up the network model, define the RAP selection problem for roadside data dissemination and analyze the hardness of this problem.

3.1   Network Model

We first describe the set of i vehicles, that is, $V = \{v_1, v_2, \cdots, v_i\}$. We assume that the vehicles have been equipped with wireless communication devices. Consequently, they can communicate with the RAPs. Then, we use the set $R = \{r_1, r_2, \cdots, r_j\}$ to denote all the RAPs which are deployed by the RSP at different locations in a city. As we know, a vehicle often visits some locations frequently due to the sociality of the vehicle (actually, the driver or the passenger). For simplicity, we describe the *moving pattern* of each vehicle with a set of locations where the vehicle passes and the RAPs are placed. Namely, we use a subset of R, i.e., $R(v_i) = \{r_{i_1}, r_{i_2}, \cdots, r_{i_k}, \cdots\}$ to describe the move of the vehicle v_i. Next, for $v_i \in V$ and $r_j \in R(v_i)$, we use $p_j^i (0 < p_j^i < 1)$ to indicate the probability that v_i successfully receives data from r_j. Although a vehicle v_i may pass by certain RAP twice or more, we only consider the vehicle passing by each RAP in $R(v_i)$ once. This assumption is reasonable. If v_i passes by r_j n times, we can use the $1-(1-p_j^i)^n$ to replace the p_j^i, and meanwhile, treat it as one-time passing. In addition, we assume that the value of p_j^i can be derived from the history records.

We construct a graph G to describe the model. The vertex set of G is $V \cup R$. For an arbitrary pair of v_i and r_j, if $r_j \in R(v_i)$, we add an edge (v_i, r_j) into the edge set of G, and attach p_j^i as the weight of (v_i, r_j). Let's take Fig. 2(a) as an example. The vehicle node set is $V = \{v_1, v_2, \cdots, v_5\}$, the RAP node set is $R = \{r_1, r_2, r_3, r_4\}$. v_1 passes by the RAPs in $R(v_1) = \{r_1, r_2\}$, v_2 passes by the RAPs in $R(v_2) = \{r_2, r_3\}$, and so on. By the way, we denote $V(\cdot)$ as the set of neighbor vehicle nodes of one RAP, denote $R(\cdot)$ as the set of neighbor RAP nodes of one vehicle, denote $deg(\cdot)$ as the degree of a node. In Fig. 2(a), $V(r_1) = \{v_1, v_3\}$, $R(v_3) = \{r_1, r_3, r_4\}$, $deg(v_1) = 2$.

3.2   Problem Definition

In our scheme of roadside data dissemination, a requester hopes that the vehicles in V can successfully receive its data from the RAPs with probability. Therefore, it's necessary to select as less RAPs as possible to conduct the data dissemination to these vehicles, and ensure that each vehicle's probability of receiving data successfully is not less than a threshold. We use τ to denote this threshold $(0 < \tau < 1)$, use S to denote the set of all selected RAPs from R. Moreover, for a vehicle v_i, it may receive data from each RAP in $S \cap R(v_i)$, and we denote $S(v_i) = S \cap R(v_i)$. As a result, given a selected RAP set S, v_i's probability of receiving data successfully from the RAPs in $S(v_i)$ is $Pr(v_i|S) = 1 - \prod_{r_j \in S(v_i)}(1 - p_j^i)$. Especially, if $S(v_i) = \emptyset$, $Pr(v_i|S) = 0$.

In this scheme, the problem we need to solve is how to select minimal number of RAPs and ensure that each vehicle's probability of receiving data successfully is not less than the threshold τ. In summary, we illustrate this problem of RAP selection as follow.

$$
\begin{aligned}
&minimize : |S| \\
&subject\ to : \bigcup_{r_j \in S} V(r_j) = V; \\
&\qquad\qquad Pr(v_i|S) \geq \tau\ for\ \forall v_i \in V; \\
&\qquad\qquad Pr(v_i|S) = 1 - \prod_{r_j \in S(v_i)}(1 - p_j^i); \\
&\qquad\qquad S \subseteq R.
\end{aligned} \tag{1}
$$

Here, we assume that $Pr(v_i|R) \geq \tau$ for $\forall v_i \in V$. If there is a vehicle v_k satisfying $Pr(v_k|R) < \tau$, the v_k cannot receive data successfully with the probability no less than τ, even if all RAPs in R are selected. Actually, there is no feasible solution to this case. In this paper, we only study the cases which exist feasible solutions.

3.3   Problem Hardness Analysis

Theorem 1. *The RAP selection problem for roadside data dissemination is NP-hard.*

Proof. We simplify the RAP selection problem by assuming $p_j^i = 1$ for $\forall v_i \in V$ and $\forall r_j \in R$. Hence, $Pr(v_i|S) \geq \tau$ for $\forall v_i \in V$ always holds. Therefore, the simplified problem is to select the minimal number of RAPs from R to cover all vehicles in V. It is obvious that $\bigcup_{r_j \in R} V(r_j) = V$ and $V(r_j) \subseteq V$. Let's denote $V_R = \{V(r_1), V(r_2), \cdots, V(r_j)\}$, i.e., V_R is a collection of V's subsets. For the RAP selection problem, if to make the selection of the RAP r_k corresponds to the selection of V's subset $V(r_k)$, this problem finally can be illustrated as the following (2).

$$
\begin{aligned}
&minimize : |V_S| \\
&subject\ to : \bigcup_{V(r_k) \in V_S} V(r_k) = V; \\
&\qquad\qquad V_S \subseteq V_R.
\end{aligned} \tag{2}
$$

Based on [14], we can conclude that Eq. 2 is same as the problem of Set Cover. Therefore, the RAP selection problem is NP-hard. Moreover, we can find that, the RAP selection problem is different from the Set Cover problem and is a meaningful extension of it.

4   Greedy Algorithm and Performance Analysis

In this section, we propose a greedy algorithm to solve the RAP selection problem, and analyze the correctness, complexity and approximation ratio of this algorithm.

4.1 Greedy Algorithm

Before the algorithm, we define a utility function $f(\cdot) : 2^R \to Q^+$, which is a mapping from a RAP set to a real utility value. It indicates the probabilistic coverage utility of a given RAP set, in other words, the sum of probabilities that each vehicle in V successfully receives data from the given set of RAPs. Concretely, the utility is defined as follows.

$$f(S) = \theta * \sum_{v_i \in V} min\{Pr(v_i|S), \tau\} \tag{3}$$

where $\theta = max\{\frac{1}{\theta_1}, \frac{1}{\tau - \theta_2}, \frac{1}{\theta_3}\}$, $\theta_1 = p_{min}*(1-p_{max})^{d_{max}}$, $\theta_2 = max\{Pr(v_i|S) \mid \forall v_i \in V, S \subseteq R, Pr(v_i|S) < \tau\}$, $\theta_3 = \frac{\tau*|V|}{|R|}$, $d_{max} = max\{deg(v_i) \mid \forall v_i \in V\}$, $p_{max} = max\{p_j^i \mid \forall v_i \in V, \forall r_j \in R\}$, $p_{min} = min\{p_j^i \mid \forall v_i \in V, \forall r_j \in R\}$.

In Eq. 3, $min\{Pr(v_i|S), \tau\}$ is the probabilistic coverage utility of S to the vehicle v_i, i.e., the (joint) probability that v_i successfully receives data from the RAPs in S. Moreover, along with the increase of the probability, the utility value $min\{Pr(v_i|S), \tau\}$ will become larger and larger. When the probability exceeds the threshold τ, the utility value will not change any more. θ is a parameter that we defined for the approximation ratio analysis in the following part.

Let S be the selected RAP set. When a RAP r_k is added to S, some vehicles which haven't been covered will be covered by r_k, or some vehicles' probabilities of receiving data successfully will increase. Both of them result in the growth of the utility. Then, the *basic idea* of our algorithm is to select the RAP r_k which maximizes the growth of the utility, and add it to the set S in each round. The concrete algorithm is showed in Algorithm 1, where S is initialized to be $\emptyset$. In each round, the RAP r_k which maximize $f(S+\{r_k\})$ is selected and added into S. The algorithm terminates when $f(S) = \theta*\tau*|V|$.

Algorithm 1. Greedy Algorithm for RAP Selection

1: **Input:** V, R, S, τ, p_j^i for $\forall v_i$ and $\forall r_j$;
2: **Start:**
3:   $S = \emptyset$;
4: **while** $f(S) < \theta*\tau*|V|$ **do**
5:     choose the RAP $r_k \in R \backslash S$ which maximize $f(S+\{r_k\})$);
6:     $S = S \cup \{r_k\}$;
7: **End**
8: **Output:** S

We use the example in Fig. 2(a) to illustrate Algorithm 1 and let $\tau = 0.6$. In addition, there is a small trick in Algorithm 1. Since θ is a constant and only be used in the theoretical analysis of approximation ratio, we can simply let $\theta = 1$ in the real implementation of Algorithm 1. That will not change the final result. Note that, we denote $f_i(S) = min\{Pr(v_i|S), \tau\}$. Figure 2(b)~(d) show the corresponding results.

1. First round: $f(\{r_1\}) = 0.9$, $f(\{r_2\}) = 1.6, f(\{r_3\}) = 1.2$, $f(\{r_4\}) = 1.2$. Hence, we select r_2 and add it to S. After this round, $Pr(v_1|S) = 0.4$, $Pr(v_2|S) = 0.6, Pr(v_3|S) = 0$, $Pr(v_4|S) = 0.4$, $Pr(v_5|S) = 0.2$. That is to say, v_2 is covered and its probability of receiving data successfully is not less than 0.6, v_3 isn't covered, v_1, v_4, v_5 are covered with some probabilities of receiving data successfully less than 0.6.

2. Second round: $f(S \cup \{r_1\}) = 2.3$, $f(S \cup \{r_3\}) = 2.2$, $f(S \cup \{r_4\}) = 2.5$. Therefore, r_4 is selected and added into S. After this round, both v_4 and v_5 can receive data successfully with the probability no less than 0.6.

3. Third round: $f(S \cup \{r_1\}) = 3$, $f(S \cup \{r_3\}) = 2.8$. Therefore, r_1 is selected and added into S.

   After this round, we have $f(S) = 3 = \theta * \tau * |V|$. Therefore, Algorithm 1 terminates. As $f_i(S) = 0.6$ for $\forall v_i \in V$, $S = \{r_1, r_2, r_4\}$ is a feasible solution.

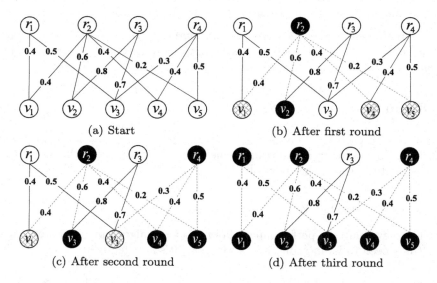

(a) Start   (b) After first round

(c) After second round   (d) After third round

Fig. 2. The results of Algorithm 1 being conducted on the example in Fig. 2(a). (The white RAP nodes are not selected and the black RAP nodes are selected. The white vehicle nodes are not covered, the black vehicle nodes are covered and their probabilities of receiving data successfully are not less than 0.6, the grid vehicle nodes are covered but their probabilities of receiving data successfully are less than 0.6.)

4.2 Correctness, Complexity and Approximation Ratio

We first show the correctness of Algorithm 1 by theoretical analysis. On the one hand, in Algorithm 1, only one RAP is added into S in each round. In the worst case, all of RAPs in R are added into S after $|R|$-th round. For each vehicle, if all of RAPs are selected, its probability of receiving data successfully is not less than τ (assumed before). In this case, $f(S)$ must be $\theta*\tau*|V|$, since $min\{Pr(v_i|S), \tau\} = \tau$ for $\forall v_i \in V$. Therefore, Algorithm 1 terminates for sure. On the other hand, if $f(S) = \theta*\tau*|V|$, $min\{Pr(v_i|S), \tau\}$ must equal to τ for $\forall v_i \in V$. Hence, each vehicle in V can receive data successfully from the RAPs in S with the probability no less than τ, namely, S is a feasible solution of the RAP selection problem. In turn, if S is a feasible solution, $f(S)$ must be $\theta*\tau*|V|$ after each RAP in S is selected. In summary, our Algorithm 1 is correct.

Then, we analyze the complexity of Algorithm 1. In Algorithm 1, the loop body will run $|R|$ times in the worst case. During each round of Algorithm 1, in order to choose the maximal $f(S+\{r_k\})$ for $\forall r_k \in R\backslash S$, the algorithm needs to test all of $|R\backslash S|$ cases. Moreover, Algorithm 1 also need to compute $f(S+\{r_k\})$, so as to compare the values of $f(S+\{r_k\})$ for $\forall r_k \in R\backslash S$. The complexity of compute $f(S)$ is $O(|V|*|R|)$, according to Eq. 3. In sum, the complexity of Algorithm 1 is $O(|V|*|R|^3)$.

Finally, we give the approximation ratio of Algorithm 1, via the following Theorem 2, and its detail proofs are showed in appendix.

Theorem 2. *Our Algorithm 1 produces an approximation solution with a ratio of $1 + ln(\frac{\theta*\tau*|V|}{opt})$ from the optimal, where opt is the number of selected RAPs produced by the optimal solution.*

5 Evaluation

We conduct extensive simulations to evaluate the performance of our proposed algorithm. The compared algorithms, the traces that we used, the simulation settings, and the results are presented as follows.

5.1 Compared Algorithms and Traces

In order to evaluate the performance of our proposed algorithm, some compared algorithms need to be introduced. As we know, it's an effective method to solve the problem of coverage, by using the degree centrality. In [2], the authors also took into account the degree centrality of each vehicle, in order to select the seeds. Based on this fact, we design two compared algorithms: MSCC (Minimum Selection for Covering Completely) and MSPE(Minimum Selection with Probability Ensuring). Differing from our Algorithm 1, which selects the RAP r_k from $R\backslash S$ to maximize $f(S+\{r_k\})$ in each round until all of the vehicles are covered with some probabilities of receiving data successfully no less than the threshold, MSCC selects the RAP r_k with maximum degree, i.e., the RAP with maximum $deg(r_k)$ for $\forall r_k \in R\backslash S$ in each round, until all of the vehicles are covered. Note that, MSCC do not ensure that each vehicle can receive data successfully with probability no less than the threshold. Obviously, it's the minimal guarantee that a vehicle can receive data. MSPE is similar with MSCC, which also select the RAP with maximum degree in each round, and terminates when each vehicle can receive data successfully with enough probability, i.e., no less than the threshold. Finally, we also design RS (Random Selection), which randomly selects a RAP in each round until each vehicle in V is covered with probability of receiving data successfully no less than the threshold. Moreover, in order to improve the performance of RS, its values showed in Figs. 3, 4 and 5 are the optimal among 10 running results.

We evaluate the metric of these four schemes, by using the real data of bus systems in two cities, Hefei and Shanghai in China. The data of Hefei includes 125 bus lines and 899 bus stations, and the data of Shanghai includes 503 bus lines and 3850 stations. We regard each bus line as a mobile vehicle, each station as a location where a RAP is deployed. Therefore, the moving pattern of a vehicle is the set of the stations passed by corresponding bus line. Further, the probabilities that each vehicle receives data successfully from the RAPs are generated randomly.

5.2 Evaluation Metrics, Methods and Results

First, we use the metric *Selection Ratio* to compare these algorithms. Selection ratio refers to the ratio of the number of the selected RAPs to the total number of all RAPs. We compare their performances with different threshold τ, and θ is set to 1.0. The detail results are showed in Fig. 3.

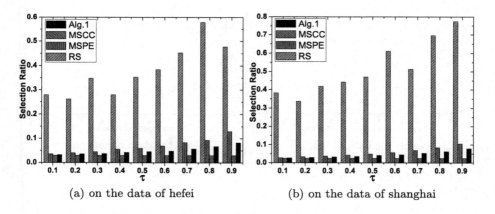

(a) on the data of hefei   (b) on the data of shanghai

Fig. 3. Performance comparison: selection ratio vs. τ

In both Fig. 3(a) and (b), we can see that Algorithm 1 achieves good performance and precedes MSPE and RS, with $\tau = \{0.1, \cdots, 0.9\}$. MSCC selects the minimal number of RAPs, about 3 % in Fig. 3(a) and 2.4 % in Fig. 3(b), without ensuring the probabilities. And its values don't change with the increment of τ, therefore can be regard as benchmark. RS has the worst performance. Its ratio much exceeds the others and fluctuates strongly. MSPE has close performance to Algorithm 1 when τ is rather small, and becomes worse and worse with τ increasing. When $\tau = 0.9$, the results of Algorithm 1 showed in Fig. 3(a) and

(b) are about 36 % and 25 % smaller than MSPE, respectively. The reason is that, when τ is small, most of vehicles can receive data successfully with some probabilities no less than τ from one single RAP. Along with τ increases, more and more vehicles need to be covered jointly by two or more RAPs, and at this moment, taking into account only degree without the probability attached by the edge is partial. However, for Algorithm 1, its RAP selection in each round considers the coverage utility to all of vehicles, therefore, it achieves the best performance.

We also conclude from Fig. 3, that the selection ratio of Algorithm 1 doesn't rise severely as the threshold τ increases. Therefore, an appropriate and large value of τ can be chose without worrying about the consequent bad performance.

Second, we also define the metric of *average receiving probability* to compare Algorithm 1, MSPE, and RS. Here, we don't care that, whether or not the probability of each vehicle of receiving data successfully is less than the threshold τ. And we only select parts of RAPs ($1\%, \cdots, 9\%$), and compute the corresponding average probability of all of vehicles receiving data successfully. The results are depicted in Fig. 4(a) and (b). From Fig. 4(a) and (b), we can find

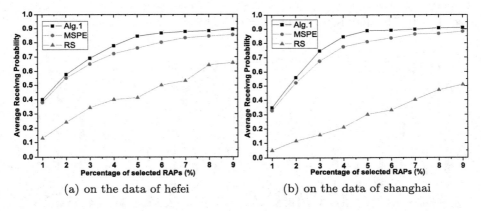

(a) on the data of hefei   (b) on the data of shanghai

Fig. 4. Performance comparison: average receiving probability vs. percentage of selected RAPs

that, with percentage of selected RAPs increasing, the three values of average receiving probabilities increase. Among them, RS increases slowly and show fluctuation with the lowest average probability. However, the average probabilities of Algorithm 1 and MSPE increases fast when selected RAPs is less than roughly 5 %, then approach slowly to about 0.9. in spite of this, Algorithm 1 is still better than MSPE since its average probability reaches to 0.9 faster. It's because that Algorithm 1 considers the coverage utility of each RAP to all vehicles in each round, instead of just the degree of a RAP.

Finally, we also examine the metric of selection ratio of these schemes, by randomly selecting the probability between any pair of vehicle and RAP, i.e., p_j^i

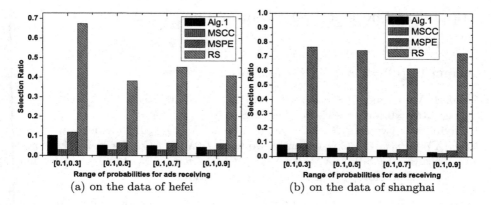

Fig. 5. Performance comparison: selection ratio vs. range of probabilities for data receiving

for $\forall v_i \in V$ and $\forall r_j \in R(v_i)$, from $[0.1, 0.3]$, $[0.1, 0.5]$, $[0.1, 0.7]$ $[0.1, 0.9]$. Here, $\tau = 0.5$, $\theta = 1.0$. Such settings have practical significance, because it simply describe the environment of VANETs with different levels of connectivity. Figure 5(a) and (b) plot the results, from which we still can get that Algorithm 1 achieves the best performance.

6 Conclusion

In this paper, we focus on the roadside data dissemination in VANETs. Its scenario includes three basic elements: RSP, mobile vehicles and requesters. We first introduce the model of the roadside data dissemination. Based on the model, we propose a RAP selection problem for roadside data dissemination, which is to select the minimal number of RAPs to cover all vehicles in model, and ensure that each vehicle can receive data successfully with probability no less than a threshold. Then we prove that the RAP selection problem is NP-hard by converting it to the Set Cover problem. Next, we propose a greedy approximation algorithm to solve this problem, analyze the complexity and approximation ratio of this proposed algorithm. Finally, we conduct extensive simulations to show the good performance of our greedy algorithm. The results indicate that our algorithm can select the minimum number of RAPs to finish the roadside data dissemination, than other compared schemes.

Acknowledgment. This research was supported in part by the National Natural Science Foundation of China (NSFC) (Grant No. 61572457, 61572336, 61502261, 61379132), and the Natural Science Foundation of Jiangsu Province in China (Grant No. BK20131174, BK2009150).

Appendix: The Detailed Proof of Theorem 2

We first give two important properties of our utility function $f(\cdot)$.

Lemma 1. $f(\emptyset) = 0$ and $f(\cdot)$ is an increasing function.

Proof. 1. We define $f_i(S) = min\{Pr(v_i|S), \tau\}$. If $S = \emptyset$, $Pr(v_i|S) = 0$ for $\forall v_i \in V$. Hence, $f_i(S) = 0$ for $\forall v_i \in V$. Further, $f(S) = \sum_{v_i \in V} f_i(S) = 0$.

2. Giving any two sets X and Y, and suppose $X \subseteq Y \subseteq A$, obviously we have $Pr(v_i|X) \leq Pr(v_i|Y)$. Consequently, we have $f_i(X) \leq f_i(Y)$ for $\forall v_i \in V$. Moreover, because of $\theta > 0$, $f(X) = \theta * \sum_{v_i \in V} f_i(X) \leq \theta * \sum_{v_i \in V} f_i(Y) = f(Y)$ when $X \subseteq Y \subseteq R$. Therefore, $f(\cdot)$ is an increasing function.

Then, we prove that our utility function $f(\cdot)$ is submodular by giving the below Lemma 2, since we all know that, submodular function relates closely to greedy algorithm.

Lemma 2. $f(\cdot)$ is a submodular function.

Proof. Given $X \subseteq Y \subseteq R$, $\forall r_k \in R\backslash Y$, if $\Delta_{r_k} f(X) = f(X + \{r_k\}) - f(X) \geq \Delta_{r_k} f(Y) = f(Y + \{r_k\}) - f(Y)$, $f(\cdot)$ is a submodular function. Before proving this conclusion, we first compare the size of $\Delta_{r_k} f_i(X) = f_i(X + \{r_k\}) - f_i(X)$ and $\Delta_{r_k} f_i(Y) = f_i(Y + \{r_k\}) - f_i(Y)$, by dividing all possibilities into the following six cases. Note that, $Pr(v_i|(X + \{r_k\})) \geq Pr(v_i|X)$, $Pr(v_i|(Y + \{r_k\})) \geq Pr(v_i|Y)$, $Pr(v_i|Y) \geq Pr(v_i|X)$ and $Pr(v_i|(Y + \{r_k\})) \geq Pr(v_i|(X + \{r_k\}))$.

1. $\tau < Pr(v_i|X)$, $\tau < Pr(v_i|Y)$. Hence, $\Delta_{r_k} f_i(X) - \Delta_{r_k} f_i(Y) = [min\{Pr(v_i|(X + \{r_k\})), \tau\} - min\{Pr(v_i|X), \tau\}] - [min\{Pr(v_i|(Y + \{r_k\})), \tau\} - min\{Pr(v_i|Y), \tau\}] = (\tau - \tau) - (\tau - \tau) = 0$;

2. $Pr(v_i|X) \leq \tau < Pr(v_i|(X + \{r_k\}))$, $\tau < Pr(v_i|Y)$. Hence, $\Delta_{r_k} f_i(X) - \Delta_{r_k} f_i(Y) = (\tau - Pr(v_i|X)) - (\tau - \tau) \geq 0$;

3. $Pr(v_i|X) \leq \tau < Pr(v_i|(X + \{r_k\}))$, $Pr(v_i|Y) \leq \tau < Pr(v_i|(X + \{r_k\}))$. Hence, $\Delta_{r_k} f_i(X) - \Delta_{r_k} f_i(Y) = (\tau - Pr(v_i|X)) - (\tau - Pr(v_i|Y)) = Pr(v_i|Y) - Pr(v_i|X) \geq 0$;

4. $Pr(v_i|(X + \{r_k\})) \leq \tau$, $\tau < Pr(v_i|Y)$. Hence, $\Delta_{r_k} f_i(X) - \Delta_{r_k} f_i(Y) = [Pr(v_i|(X + \{r_k\})) - Pr(v_i|X)] - (\tau - \tau) \geq 0$;

5. $Pr(v_i|(X + \{r_k\})) \leq \tau$, $Pr(v_i|(Y + \{r_k\})) \leq \tau$. Hence, $\Delta_{r_k} f_i(X) - \Delta_{r_k} f_i(Y) = [Pr(v_i|(X + \{r_k\})) - Pr(v_i|X)] - [Pr(v_i|(Y + \{r_k\})) - Pr(v_i|Y)]$;

6. $Pr(v_i|(X + \{r_k\})) \leq \tau$, $Pr(v_i|Y) \leq \tau < Pr(v_i|(Y + \{r_k\}))$. Hence, $\Delta_{r_k} f_i(X) - \Delta_{r_k} f_i(Y) = [Pr(v_i|(X + \{r_k\})) - Pr(v_i|X)] - [\tau - Pr(v_i|Y)] \geq [Pr(v_i|(X + \{r_k\})) - Pr(v_i|X)] - [Pr(v_i|(Y + \{r_k\})) - Pr(v_i|Y)]$.

For the cases 5 and 6 above, we need to continue analyzing the size of $Pr(v_i|(S + \{r_k\})) - Pr(v_i|S)$ for $\forall v_i \in V$ and $\forall r_k \in R\backslash S$. We define $\Delta_{r_k} Pr(v_i|S) = Pr(v_i|(S + \{r_k\})) - Pr(v_i|S)$. In fact, $\Delta_{r_k} Pr(v_i|S)$ for $\forall v_i$ is the increment of its probability of receiving data successfully when a new RAP r_k is added to S. It is obvious that r_k only influences the vehicles in $V(r_k)$. In order to compute $\Delta_{r_k} Pr(v_i|S)$, there exists three possible case below:

1. v_i is covered by r_k and (part or all of) the RAPs in S. Hence, $\Delta_{r_k} Pr(v_i|S)$ $= [1 - \prod_{r_j \in (S + \{r_k\})(v_i)}(1 - p_j^i)] - [1 - \prod_{r_j \in S(v_i)}(1 - p_j^i)] = \prod_{r_j \in S(v_i)}(1 - p_j^i) - (1 - p_k^i)\prod_{r_j \in S(v_i)}(1 - p_j^i) = p_k^i * \prod_{r_j \in S(v_i)}(1 - p_j^i)$;

2. v_i is covered by r_k but isn't covered by any RAPs in S. Hence, $\Delta_{r_k} Pr(v_i|S) = [1 - (1 - p_k^i)] - 0 = p_k^i$;

3. v_i isn't covered by r_k, namely, wether or not r_k is selected has no influence on v_i. Obviously, $\Delta_{r_k} Pr(v_i|S) = 0$.

Based on the analyses above, we can compute $\Delta_{r_k} Pr(v_i|X) - \Delta_{r_k} Pr(v_i|Y)$ by dividing it into following four possible cases. Note that, $X \subseteq Y$ and $X(v_i) \subseteq Y(v_i)$.

1. v_i isn't covered by any RAPs in Y, but is covered by r_k. Hence, $\Delta_{r_k} Pr(v_i|X) - \Delta_{r_k} Pr(v_i|Y) = p_k^i - p_k^i = 0$;

2. v_i isn't covered by any RAPs in X, but r_k and RAPs in $Y \backslash X$ cover v_i. Hence, $\Delta_{r_k} Pr(v_i|X) - \Delta_{r_k} Pr(v_i|Y) = p_k^i - p_k^i * \prod_{r_j \in Y(v_i)}(1 - p_j^i) \geq 0$;

3. r_k and RAPs in X and Y cover v_i. Hence, $\Delta_{r_k} Pr(v_i|X) - \Delta_{r_k} Pr(v_i|Y) = p_k^i * [\prod_{r_j \in X(v_i)}(1 - p_j^i) - \prod_{r_j \in Y(v_i)}(1 - p_j^i)] \geq 0$ since $X(v_i) \subseteq Y(v_i)$;

4. v_i isn't covered by r_k. Hence, $\Delta_{r_k} Pr(v_i|X) - \Delta_{r_k} Pr(v_i|Y) = 0 - 0 = 0$.

In summary, we have $\Delta_{r_k} Pr(v_i|X) - \Delta_{r_k} Pr(v_i|Y) \geq 0$. Finally, we also can conclude that $\Delta_{r_k} f_i(X) - \Delta_{r_k} f_i(Y) \geq 0$ in all six cases. Moreover, as $\theta > 0$, $\Delta_{r_k} f(X) - \Delta_{r_k} f(Y) = \theta * \sum_{v_i \in V}[\Delta_{r_k} f_i(X) - \Delta_{r_k} f_i(Y)] \geq 0$. Therefore, we have the conclusion that $f(\cdot)$ is a submodular function.

According to Lemmas 1 and 2, we know that, $f(\cdot)$ is polymatroid since it's an increasing submodular function with $f(\emptyset) = 0$. Similarly, we can easily prove that the cardinality function $c(X) = |X|$ is polymatroid. Given two polymatroid functions $g(\cdot)$ and $h(\cdot)$ on 2^E, the problem of Minimum Submodular Cover with Submodular Cost (MSC/SC) is defined, which is the minimization problem $min\{h(X)|g(X) = g(E), X \subseteq E\}$[15].

In this paper, for the utility function $f(\cdot)$, given a selected RAP set S, if $f(S) = \theta * \tau * |V|$, S is a feasible solution of the RAP selection problem and $f(S) = f(V) = \theta * \tau * |V|$. Therefore, the RAP selection problem can be described as: $min\{c(S)|f(S) = f(V), S \subseteq V\}$, where $c(\cdot)$ is the cardinality function. That is to say, our selection problem is a MSC/SC problem. Based on this fact, we give the following Theorem 3.

Theorem 3. *[15] Suppose $g(\cdot)$ is a polymatroid function on 2^E, and $g(E) \geq opt$ where opt is the cost of a minimum submodular cover. For a greedy algorithm, if the selected x in each round always satisfies that $\frac{g(X + \{x\}) - g(X)}{c(\{x\})} \geq 1$, then the greedy solution is a $1 + \rho ln(\frac{g(E)}{opt})$-approximation, where $\rho = 1$ if $c(\cdot)$ is modular (i.e., linear).*

Now, we give the following crucial Lemma 3.

Lemma 3. *Give the set $S \subseteq R$, which is the set of RAPs selected by Algorithm 1 after r-th round, and the RAP r_k selected during $(r+1)$-th round, we have $\frac{f(S + \{r_k\}) - f(S)}{c(\{r_k\})} \geq 1$.*

Proof.   1. It is obvious that the cardinality function $c(\cdot)$ is linear and $c(\{r_k\})=1$. 2. In each round of Algorithm 1, if $f(S)<\theta*\tau*|V|$, a new RAP will be selected and added into the set S. In fact, if Algorithm 1 doesn't terminate after r-th round, there must exist a vehicle v_i with $Pr(v_i|S)<\tau$. Otherwise, if $Pr(v_i|S)\geq\tau$ for $\forall v_i\in V$, then $f(S)=\theta*\tau*|V|$, consequently, Algorithm 1 will terminate. Moreover, for this vehicle v_i and the selected RAP r_k, since $f(S+\{r_k\})$ is maximized during $(r+1)$-th round, we can suppose r_k must cover this v_i, i.e., $p_k^i>0$, without loss of generality. That is to say, if the RAP r_k is selected during $(r+1)$-th round, r_k must cover at least one vehicle v_i with $Pr(v_i|S)<\tau$. Otherwise, we have $f(S+\{r_k\})-f(S)=0$ for $\forall r_k\in R\backslash S$, which conflicts with the fact that there must exist a vehicle v_i with $Pr(v_i|S)<\tau$ if Algorithm 1 doesn't terminate. Based on these analyses, we have

$$f(S+\{r_k\})-f(S) = \theta*\sum_{v_i\in V}[min\{Pr(v_i|(S+\{r_k\})),\tau\}-min\{Pr(v_i|S),\tau\}]$$
$$\geq \theta*(min\{Pr(v_i|(S+\{r_k\})),\tau\}-min\{Pr(v_i|S),\tau\}) = \theta*(min\{Pr(v_i|(S+\{r_k\})),\tau\}-Pr(v_i|S)) = \theta*min\{Pr(v_i|(S+\{r_k\}))-Pr(v_i|S),\tau-Pr(v_i|S)\}$$
$$=\theta*min\{p_k^i*\prod_{r_j\in S(v_i)}(1-p_j^i),\tau-Pr(v_i|S)\} \geq min\{\theta*\theta_1,\theta*(\tau-\theta_2)\}\geq 1$$

To sum up, $\frac{f(S+\{r_k\})-f(S)}{c(\{r_k\})}\geq f(S+\{r_k\})-f(S)\geq 1$

Finally, suppose that S_{opt} is an optimal solution of the RAP selection problem and $c(S_{opt})=|S_{opt}|=opt$ which is number of RAPs in S_{opt}. Consequently, we have $f(R)=\theta*\tau*|V|\geq\frac{\tau*|V|}{\theta_3}\geq|R|\geq opt$ since $\theta\geq\frac{1}{\theta_3}$. Therefore, we can conclude that our proposed Algorithm 1 is a $1+\ln(\frac{\theta*\tau*|V|}{opt})$-approximation, by applying Theorem 3.

References

1. Zheng, H., Jie, W.: Optimizing roadside advertisement dissemination in vehicular cyber-physical systems. In: IEEE International Conference on Distributed Computing Systems (ICDCS) (2015)
2. Qin, J., Zhu, H., Zhu, Y., Li, L., Xue, G., Li, M.: Post: Exploiting dynamic sociality for mobile advertising in vehicular networks. In: IEEE INFOCOM (2014)
3. Li, Z., Liu, C., Chigan, C.: On secure vanet-based ad dissemination with pragmatic cost and effect control. IEEE Trans. Intell. Transp. Syst. **14**(1), 124–135 (2013)
4. Freeman, L.C.: Centrality in social networks conceptual clarification. Soc. Networks **1**(3), 215–239 (1979)
5. Guo, L., Zhang, C., Yue, H., Fang, Y.: A privacy-preserving social-assisted mobile content dissemination scheme in DTNs. In: IEEE INFOCOM (2013)
6. Ning, T., Yang, Z., Hongyi, W., Han, Z.: Self-interest-driven incentives for ad dissemination in autonomous mobile social networks. In IEEE INFOCOM (2013)
7. Gao, W., Cao, G.: User-centric data dissemination in disruption tolerant networks. In: IEEE INFOCOM (2011)
8. Lee, S.-B., Park, J.-S., Gerla, M., Songwu, L.: Secure incentives for commercial ad dissemination in vehicular networks. IEEE Trans. Veh. Technol. **61**(6), 2715–2728 (2012)
9. Jeong, J., Guo, S., Gu, Y., He, T., Du, D.H.C.: TSF: Trajectory-based statistical forwarding for infrastructure-to-vehicle data delivery in vehicular networks. In: IEEE International Conference on Distributed Computing Systems (ICDCS) (2010)

10. Sardari, M., Hendessi, F., Fekri, F.: Infocast: A new paradigm for collaborative content distribution from roadside units to vehicular networks. In: IEEE SECON proceedings (2009)
11. Liu, K., Lee, V.C.S.: RSU-based real-time data access in dynamic vehicular networks. In: International IEEE Conference on Intelligent Transportation Systems (ITSC), pp. 1051–1056 (2010)
12. Tsung-Jung, W., Liao, W., Chang, C.-J.: A cost-effective strategy for road-side unit placement in vehicular networks. IEEE Trans. Commun. **60**(8), 2295–2303 (2012)
13. Aslam, B., Amjad, F., Zou, C.C.: Optimal roadside units placement in urban areas for vehicular networks. In: IEEE Symposium on Computers and Communications (ISCC) (2012)
14. Feige, U.: A threshold of ln n for approximating set cover. J. ACM **45**(4), 634–652 (1998)
15. Wan, P., Dingzhu, D., Pardalos, P., Weili, W.: Greedy approximations for minimum submodular cover with submodular cost. Comput. Optim. Appl. **45**(2), 463–474 (2010)

OCC: Opportunistic Crowd Computing in Mobile Social Networks

Hualin Mao[1], Mingjun Xiao[1(✉)], An Liu[2], Jianbo Li[3], and Yawei Hu[1]

[1] School of Computer Science and Technology, Suzhou Institute for Advanced Study, University of Science and Technology of China, Hefei, People's Republic of China
xiaomj@ustc.edu.cn
[2] School of Computer Science and Technology, Soochow University, Suzhou, People's Republic of China
[3] School of Computer Science and Technology, Qingdao University, Qingdao, People's Republic of China

Abstract. Crowd computing is a new paradigm, in which a group of users are coordinated to deal with a huge job or huge amounts of data that one user cannot easily do. In this paper, we design an Opportunistic Crowd Computing system (OCC) for mobile social networks (MSNs). Unlike traditional crowd computing systems, the mobile users in OCC move around and communicate each other by using short-distance wireless communication mechanisms (e.g., WiFi or Bluetooth) when they encounter each other, so as to save communication costs. The key design of OCC is the task assignment scheme. Unlike the traditional crowd computing task assignment problem, the task assignment in OCC must take into consideration the users' mobile behaviors. To solve this problem, we present an optimal user group algorithm (OUGA). It can minimize the total cost, while ensuring the task completion rates. Moreover, we conduct a performance analysis, and prove the optimality of this algorithm. In addition, the simulations show that our algorithm achieves a good performance.

Keywords: Crowd computing · Mobile social network · Task assignment

1 Introduction

Crowd computing is a paradigm, in which a group of users are coordinated to deal with a large job that one user cannot easily do [1–3]. Currently, there have been two typical crowd computing paradigms: crowdsourcing and mobile crowdsensing. In a crowdsourcing, a user can publish a large job onto an online platform, and other users can apply for participating in the work via this platform [1,2,4,5]. For example, a crowd of online users can cooperatively conduct an image labeling job, or perform a joint data analysis job via the platform of Amazon Mechanical Turk. On the other hand, the mobile crowdsensing involves a group of mobile

© Springer International Publishing Switzerland 2016
H. Gao et al. (Eds.): DASFAA 2016 Workshops, LNCS 9645, pp. 254–267, 2016.
DOI: 10.1007/978-3-319-32055-7_21

users who carry smart phones or iPads, which are generally equipped with different sensors [6–8]. These mobile users can cooperatively perform a sensing job by using their carried smart devices, such as urban WiFi characterization, traffic information mapping, and so on [9,10]. Since crowd computing can accomplish a large complex job by exploiting the intelligence or mobility of crowds, it has attracted a lot of attention recently.

In this paper, we focus on the crowd computing in Mobile Social Networks (MSNs). Envision that a user in an MSN has a lot of jobs which need to be finished before a deadline, and it is hard for the user to finish them alone, e.g., park searching, noise detection, social surveys, big data analysis, and so on. Then, the user can assign some tasks to other mobile users in the MSN and ask for their help by launching a crowd computing. In order to save communication costs, the user can adopt the short-distance wireless communication schemes (e.g., WiFi or Bluetooth) to deliver tasks and receive results [11,12]. Since the users encounter or visit WiFi access points with some probabilities, the task assignment in MSNs is opportunistic. In addition, in order to attract other users to complete the tasks, the user needs to pay a reward for each task assignment. To provide the crowd computing service in MSNs, we design an Opportunistic Crowd Computing system (OCC), taking these factors into consideration.

Among the current crowd computing schemes, the crowdsourcing systems are generally online systems [13–15], in which the users participate in the crowdsourcing voluntarily, and they select the tasks on their own initiative, so that there is no task assignment process. On the other hand, although the current crowdsensing systems have the task assignment process, the users in these systems are generally assumed to publish tasks and submit the corresponding results at any time anywhere via a cloud platform [16–18]. However, the users in our system need to proactively assign tasks to other users while taking the efficiency into consideration. Moreover, the task assignment is opportunistically performed. These characteristics make our system different from current crowd computing systems.

The key challenge of our system design is the opportunistic task assignment in MSNs. To solve this problem, we propose an optimal task assignment algorithm. More specifically, the major contributions include:

1. We first present the basic design of the OCC system, including the framework and all models in this system.
2. We propose an opportunistic task assignment algorithm, named OUGA, by which we can assign tasks to the minimum users while making the total usefulness degree of the corresponding task results exceed a given threshold. Furthermore, we prove the optimality of this algorithm.
3. We also conduct extensive simulations to evaluate our proposed algorithm. The results prove that our algorithm has a better performance.

The remainder of the paper is organized as follows. We introduce the background and the design challenges in Sect. 2. The framework is proposed in Sect. 3. The task assignment algorithm is presented in Sect. 4. Simulations are shown in Sect. 5. Related work and conclusion are presented in Sects. 6 and 7, respectively.

2   Background and Challenges

Each user can use our system to publish crowd computing tasks by assigning these tasks to other users. The user can also perform crowd computing tasks which are received from other users. When a user publishes crowd computing tasks, we call this user a publisher. When a user performs crowd computing tasks, we call this user a worker. We design the system while taking the feature of MSNs into consideration. When a publisher and a user encounter, the publisher will send a task list to the encountered user. The user chooses some tasks and sends back the chosen-task list, which includes all tasks that the user wants to perform. Then, the publisher uses the system to assign some crowd computing tasks to this user (worker) according to the chosen-task list. When the worker encounters the publisher next time, the worker can (use the system to) return the tasks results to the publisher if the worker has finished the task. Otherwise, the worker will not return the results. For example, a student has a large language translation task. He/she visits a classroom and encounters his/her classmates. Then, he/she can use the system to publish the task. When his/her classmates have finished the task and they visit the classroom next time, he/she can receive the results.

The assignment strategy of our system is based on the estimated finish time of the task, the encounter probability and the reputation of users. Because of the probabilistic encounter, the result recycling is also probabilistic. Hence, the tasks may be assigned to more than one worker. When two users encounter each other, their systems can adjust the encounter probability and update the reputation of users. All of the conditions should be taken into consideration. The optimization goal of our system is to find minimum users to complete a crowd computing task cooperatively. There are three challenges in designing the system. Firstly, our system is an offline system. It means that the system assigns the task actively. Secondly, it takes some time to complete the tasks. Moreover, the encounter of two users is probabilistic. Hence, the task assignment should be dynamical. Thirdly, the user's information, which includes the encounter probability, the reputation of users, and so on, changes dynamically. The system will adjust the information when two users encounter each other.

3   System Design

As shown in Fig. 1, our system can be divided into five models: user profile model, task generation model, task assignment model, task execution model, and result management model.

3.1   User Profile Model

The User Profile Model (UPM) is designed to record and manage the potential workers' profiles. Each worker's profile contains the user id (UI), the reputation (REP), encounter probability index (EPI), and user name (UN), as shown in Table 1 and Fig. 2(a).

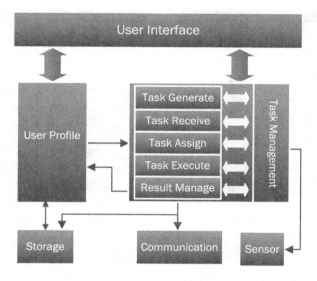

Fig. 1. System framework

The REP of a worker is a probability of the usefulness of the results returned by this worker, which is evaluated by the historical results returned by this worker. For example, the worker has returned 10 task results and three of them have been adopted. Then, the REP value of the work is 0.3. If the encountered user has no historical REP, the initial REP value is 1. Once the worker finishes a new crowd computing task and the result has been returned, the publisher will update the corresponding REP value by the following equation:

$$\xi^{new} = \alpha \cdot \xi' + (1 - \alpha) \cdot p \tag{1}$$

Table 1. Worker information

Worker information
User id (UI)
User name (UN)
Reputation (REP)
Encounter probability index (EPI)

In Eq. 1, ξ^{new} is the new REP value of the worker, and ξ' is the historical REP value of the worker. p is the REP value after the new crowd computing task. α is a coefficient to balance the weights of the historical and new REP values, which can be defined by the user. In our system, we set the value of α as $1/2$.

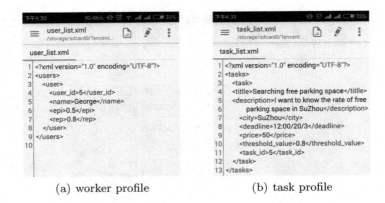

(a) worker profile   (b) task profile

Fig. 2. Worker profile example and task profile example.

Besides, when the publisher encounters another mobile user, the publisher will adjust the REP according to this mobile user's record by using Eq. 2.

$$\xi = \beta \cdot \xi' + (1 - \beta) \cdot \xi'' \qquad (2)$$

In Eq. 2, ξ' is the REP value from the encountered user, ξ'' is the old REP score, ξ is the new REP score, and β is the weight of REP score of the encountered user.

The EPI of a worker is the average frequency for the publisher encountering this worker within the unit time, i.e. the reciprocal of the average inter-meeting time between the publisher and this worker. When the publisher encounters another user, the publisher will update the corresponding EPI value by using the following equation:

$$\frac{1}{\lambda} = \frac{\gamma}{\lambda'} + (1 - \gamma) \cdot T \qquad (3)$$

In Eq. 3, λ is the new EPI value, and λ' is the historical EPI value. T is the new inter-meeting time. The unit of T is hour. γ is the weight of historical inter-meeting time.

3.2 Task Generation Model

Task Generation Model (TGM) is used to generate tasks. This model provides the human-device interface for the user to generate tasks. Each task is expressed by using a xml text, including the properties of task id, title, description, city, deadline, price, threshold, and so on. The title, description, and city describe the basic information of a task. The deadline indicates the time before which the task must be finished. The price is the reward that the publisher is willing to pay for others conducting this task. The threshold is a value, which the total usefulness degree of the results for this task must exceed, meaning that it is a successful and satisfying task performing. The TGM allocates an unique id for each task. An example is shown in Fig. 2(b), in which a user publishes a task of searching free parking space in a city Suzhou. The task must be finished before 12 o'clock on March 20, and the user will pay 50 yuan for each participant. The threshold is 0.8, and the task id is 5.

In addition, we use a task list to manage all tasks, including the tasks generated by the user itself or the tasks assigned by other users. On the other hand, the tasks can be divided into two categories. One is the tasks that this user needs to assign to other users. Another is the tasks that this user must perform by itself.

3.3   Task Assignment Model

Task Assignment Model (TAM) is responsible for making the strategy to assign crowd computing tasks. When two users encounter each other, our system will send the task list to the user from the publisher. The user receives the task list and picks some interested tasks, and then the user sends his interested task list back to the publisher. The system on the publisher side will estimate the result satisfaction degree (RSD) of each task on the interested task list, and the system will assign the task according to the RSD. Here, RSD is defined as follows:

Definition 1. *Result satisfaction degree (RSD) indicates the degree of usefulness of the task result which a worker can return. This value is estimated by the publisher of the task, using the probability of receiving the task result and the historical REP value of the user. Concretely, it is defined by the following equation:*

$$\phi = \xi \cdot p, \tag{4}$$

where p is the probability of the publisher receiving the task result from the user, and ξ is the historical REP value of the user.

The above RSD value is the usefulness degree of the task result from a single user. Actually, it is a probability that the task result might be adopted. If the task is assigned to more than one user, we use the joint probability of the usefulness degrees of task results from all users, called the *total usefulness degree*. If the total usefulness degree is larger than the threshold π, then we regard that the task is worth to assign to the user and the user becomes a worker after he/she receives the task.

3.4   Task Execution Model

Task Execution Model (TEM) handles the received task. There are two kinds of tasks. One is the human intelligence task, and another is the sensing task. The human intelligence tasks need to be done by users themselves, for example, finding animals, questionnaire survey, giving advices, and so on. The sensing tasks can be executed by mobile devices, for example, noise detection, temperature detection, and so on. After the worker executes a task, the model will send the result to the result management model.

3.5 Result Management Model

Result Management Model (RMM) is used to decide the opportunity moment to return the task result when two users encounter each other. The system will store the result, the user id and task id, after the worker finishes the task. Once the worker encounters another user, the system will check the encountered user's id. If it matches the publisher id of the task in the result block, the system will send the result to the encountered user. Otherwise, the worker ignores the encountered user. Besides, if the user receives a result from other users, the system will check the type of the task in task list. If it is forwarding task, the system will store the result until the user encounters another user. Otherwise, it will show the result.

4 Task Assignment

In this section, we introduce the task assignment strategy, including the problem formalization, a greedy task assignment algorithm, and the analysis of the optimality and complexity of the algorithm.

4.1 Problem Formalization

In our system, when a publisher assigns a task to an encountered user, he/she will pay a price to the encountered user. The total cost that the publisher will pay for this task is proportional to the number of users that the task is assigned to. Then, in order to save the total cost, the publisher will assign the task to as less users as possible. On the other hand, the task assignment should satisfy the constraint, i.e., the total usefulness degree of the task result must be larger than the given threshold. Otherwise, the task performing will be regarded as a failure.

More specifically, we assume that the publisher has n potential users which are denoted by $U = \{u_1, \cdots, u_n\}$. There are m tasks chosen by the encountered user, denoted by $L = \{l_1, \cdots, l_m\}$. The REP values of users are denoted by $\xi = \{\xi_1, \cdots, \xi_n\}$. The EPI values are denoted by $\lambda = \{\lambda_1, \cdots, \lambda_n\}$. Here, we assume that the inter-meeting time between each pair of users follows the exponential distribution, where the ratio is the corresponding EPI value. This assumption is reasonable, since many real MSN traces approximately follow the exponential distribution, as shown in [19]. The thresholds of the chosen tasks are denoted by $\pi = \{\pi_1, \cdots, \pi_m\}$. Consider the assignment of task l_j, and let S_j be the set of users that the publisher decides to assign this task to. Then, the problem can be formalized as follows:

$$Minimize \ |S_j| \tag{5}$$

$$Maximize \ 1 - \prod_{u_i \in S_j} (1 - \phi_i) \tag{6}$$

$$s.t. \ 1 - \prod_{u_i \in S_j} (1 - \phi_i) \geq \pi; \tag{7}$$

$$\phi_i = \xi_i \cdot p_i; \tag{8}$$

$$S_j \subseteq U. \tag{9}$$

Here, $|S_j|$ is the size of set S_j, ξ_i is the historical REP of user u_i, p_i is the probability that the publisher receives the result from user u_i, ϕ_i is the RSD value related to user u_i, and $1 - \prod_{u_i \in S_j}(1 - \phi_i)$ is the total usefulness degree of this task to be assigned. Moreover, the optimization objective is first to minimize the size of the set S_j, and then to maximize the total usefulness degree.

4.2   Computing RSD

As we mentioned in TAM, the worker will send the selected task list back to the publisher. For each task chosen by the worker, the system will prepare for the assignment. The preparation process has three steps. First, the system computes the RSD value with the information of the encountered worker. Second, the system computes the RSDs of the rest users in the user list. Third, the system sorts the array of RSDs. The key process is to calculate the probability that the publisher receives the result from the users. There are two cases:

The first case is that the publisher has encountered and assigned the task (e.g. l_j) to the user u_i. In this case, the probability that the publisher receives the result from u_i is the probability that the publisher encounters this user once in the future before the deadline. That is:

$$p_i = \int_0^{\tau - t_0} \lambda_i e^{-\lambda_i t} = 1 - e^{-\lambda_i (\tau - t_0)} \tag{10}$$

Here, τ is the deadline of the task, t_0 is the current time, and λ_i is the EPI of user u_i.

Another case is that the publisher has not assigned the task (e.g. l_j) to the user $u_{i'}$. In this case, the probability that the publisher receives the result from user $u_{i'}$ is the probability that the publisher encounters this user twice before the deadline. One encounter is for assigning the task, and another is for receiving the result. Then, the formula is:

$$p_{i'} = \int_0^{\tau - t_0} \lambda_{i'} e^{-\lambda_{i'} t} \cdot (1 - e^{-\lambda_{i'} (\tau - t_0 - t)}) \tag{11}$$

$$= 1 - (1 + \lambda_{i'} (\tau - t_0)) e^{-\lambda_{i'} (\tau - t_0)} \tag{12}$$

Here, $\lambda_{i'}$ is the EPI of user $u_{i'}$.

According to the probabilities p_i and $p_{i'}$, we can calculate the all RSD values by using Eq. 8.

4.3   The OUGA Algorithm

We use a greedy algorithm, called Optimal User Group Algorithm (OUGA), to find an optimal user group with a smallest group size and then, the largest total usefulness degree. Moreover, this total usefulness degree must be no less than the threshold. First, the system computes the RSDs by using Eq. 8 and sorts these

Algorithm 1. Allocate Procedure

1: **for** $i = 1$ $to TaskList.length()$ **do**
2:   $getInfo(\tau, t_0, \lambda, \pi_i, \xi, TaskList(i), Userlist(id))$;
3:   $\phi \leftarrow \xi \cdot (1 - e^{-\lambda(\tau - t_0)})$;
4:   **for** $j = 1$ to $Userlist.length()$ **do**
5:     **if** $UserList(j) \neq id$ **then**
6:       $getInformation(\lambda, \xi_j, Userlist)$;
7:         $\phi_j \leftarrow \xi_j \cdot (1 - (1 + \lambda_j(\tau - t_0))e^{-\lambda_j(\tau - t_0)})$;
8:   $RSD() \leftarrow Sort(\phi, \phi_j)$;
9:   $sum\phi \leftarrow 0$;
10: **for** $k = 1$ to $RSD.length()$ and $sum\phi < \pi$ **do**
11:    **if** $RSD(k).value == \phi$ **then**
12:      $Send(id, TaskList(i))$
13:      $TaskList(i).threshold \leftarrow 1 - \frac{1-\pi}{1-\phi}$;
14:      $sum\phi \leftarrow 1 - (1 - sum\phi) \cdot (1 - RSD(k).value)$;

RSDs in the descending order. Then, the system can find the maximum one in the ordered RSDs. If this RSD is larger than the threshold of the task, the system will only assign the task to the user related to this RSD. Otherwise, the system will determine a group of users, whose size is smallest and the corresponding total usefulness degree is larger than the threshold. This can be realized by selecting the users in the descending order of their RSDs, one by one. Then, the system will assign the task to any encountered user in this group.

The detailed algorithm is showed in Algorithm 1. Lines 1–3 are used to calculate the RSD, which is related to encountered user. Lines 4–7 are to get RSDs of the rest users. Line 8 sorts the RSDs in the descending order. Lines 9–14 describe the assignment strategy according to the RSDs we compute. The overhead is dominated by Line 8, in which a quick-sort algorithm is adopted to sort the RSDs. Then, the overhead is $O(mlog_2 n)$, where m is the number of tasks, and n is the number of users.

4.4 The Optimality of OUGA

Now we prove that OUGA can find an optimal user group as follows:

Theorem 1. *OUGA can find the minimum number of users whose total usefulness degree is larger than the threshold.*

Proof. Assume that the sorted order of RSDs is $\{\phi_1, \cdots, \phi_k, \cdots, \phi_n\}$ ($\phi_1 > \phi_2 > \cdots > \phi_n$), and the group of users produced by OUGA is $S_j = \{u_1, \cdots, u_k\}$. This means that the total usefulness degree of S_j is larger than the threshold, while the total usefulness degree of any proper subset of S_j is less than the threshold. Now, we prove the optimality of the set S_j as follows:

We assume that another set $S_{j'}$ is the optimal user group. Then, we have $u_1 \in S_{j'}$. Otherwise, if $u_1 \notin S_{j'}$, then we construct a new user set $S_{j''}$ by using u_1 to replace any user in $S_{j'}$. Now, we calculate the total usefulness degrees of

$S_{j'}$ and $S_{j''}$, respectively, and compare them. As a result, we can get that the later value is the larger one:

$$1 - \prod_{u_i \in S_{j''}} (1 - \phi_i) > 1 - \prod_{u_i \in S_{j'}} (1 - \phi_i) \tag{13}$$

This is the contradiction to the optimality of the user set $S_{j'}$. Therefore, $u_1 \in S_{j'}$. In the same way, we can prove that $u_2 \in S_{j'}$, $u_3 \in S_{j'}$ and so on. That is $S_{j'} = \{u_1, u_2, u_3, \cdots u_{|S_{j'}|}\}$. According to OUGA, if $|S_{j'}| < k$, the corresponding total usefulness degree is less than the threshold. Thus, $|S_{j'}| \geq k$. On the other hand, due to the optimality of $S_{j'}$, $|S_{j'}| = k$. Therefore, $S_{j'} = S_j$. That is to say, our theorem is correct.

5   Performance Evaluation

In this section, we present extensive simulations to evaluate the performance of our algorithm under different settings. The compared algorithms, the evaluation methods, settings, and results are presented as follows.

5.1   Algorithms in Comparison

Since there are no other similar algorithms, we design two compared algorithms to analyze the efficiency of OUGA algorithm. Because traditional crowd computing systems are online systems and the systems will assign the task to voluntary users. We set the Encounter-based Assignment algorithm (EA) to be the compared algorithm. EA assigns the encountered user with all tasks picked by this user. The publisher may assign the task to more than one user until the total usefulness degree of the task is larger than the threshold. Moreover, we improve the EA to RSD-based Assignment algorithm (RA). When a publisher encounters another user, the publisher assigns the task with the largest RSD value to this user. Moreover, RA will satisfy the constraint that the total usefulness degree is larger than the threshold.

5.2   Simulation Settings and Metrics

We set different parameters of the simulations, including number of users, number of tasks, EPI, REP, threshold, as shown in Table 2. The number of users is selected from $\{100, \cdots, 500\}$, and the number of tasks is from $\{200, \cdots, 1000\}$. Moreover, we set the average EPI as the value, selected from $\{0.1, \cdots, 1\}$. The REP of each user is selected from $\{0.1, \cdots, 1\}$, and the threshold of each task is selected from $\{0.5, \cdots, 1\}$.

In the simulations, we evaluate our algorithm by two metrics. The first metric is the average number of the users that are selected to perform the crowd computing tasks. Actually, this indicates the average cost that the publishers will pay for their crowd computing tasks. The second metric is the total usefulness degree of the task.

Table 2. Evaluation settings.

Parameter name	Range
Total number of users	100–500
Total number of tasks	200–1000
EPI λ	0.1–1
REPξ	0.1–1
TV π	0.5–1

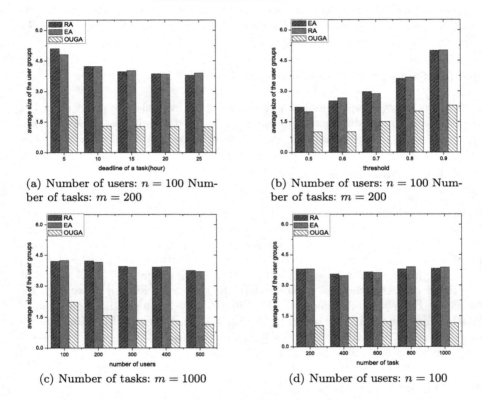

(a) Number of users: $n = 100$ Number of tasks: $m = 200$

(b) Number of users: $n = 100$ Number of tasks: $m = 200$

(c) Number of tasks: $m = 1000$

(d) Number of users: $n = 100$

Fig. 3. Average number of the users

5.3   Evaluation Results

In the first simulation, we change the deadline of the tasks from 5 h to 25 h. The number of users is 100, the total number tasks is 200, and the average threshold value is 0.8. The second simulation changes the threshold of the task from 0.5 to 0.9. The number of users is 100, and the total number tasks is 200. The deadline of each task is 15 h. The third simulation changes the number of users from 100 to 500. The total number of tasks is 1000, the average threshold is 0.8, and

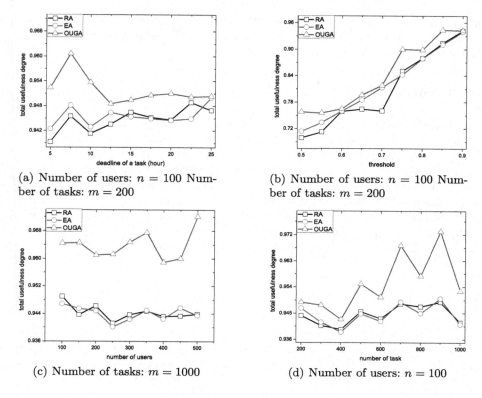

(a) Number of users: $n = 100$ Number of tasks: $m = 200$

(b) Number of users: $n = 100$ Number of tasks: $m = 200$

(c) Number of tasks: $m = 1000$

(d) Number of users: $n = 100$

Fig. 4. Total usefulness degree

the deadline of each task is 15 h. The last one we change the total number of tasks from 200 to 1000. The number of user is 200 while other variables are same as third simulation. We record the average size of the user groups produced by the algorithms, and the average total usefulness degree. The average size of the user groups is associated with the total cost of tasks. It can reflect the performance as well.

As the result shown in Fig. 3(a), the average size of the user groups will decrease, along with the increase of the task deadline. Because the publisher has more time to encounter other users, and will get a higher total usefulness degree, when the deadline delays. Figure 3(b) shows that the threshold of a task will affect the performance, too. It is more difficult to complete the task, if the task has a higher threshold. Accordingly, there will be more users in each user group. As shown in Fig. 3(c), if there are more users, the publisher will encounter more users before the deadline, so that the total usefulness degrees of the tasks become bigger, and the average size of user groups becomes smaller. Figure 3(d) shows that the number of tasks effects little. As shown in Fig. 4, the total usefulness degree is irregular because it is affected by the encounter of users, and the process of encounter is probability. Moreover, the performance of our algorithm is better than the compared algorithms while the total usefulness degree is larger than the threshold.

6   Related Work

There are several research about crowd computing. Most research of crowd computing focused on designing a system for a specific application. For example, Mathur [18] designs a ParkNet system. When vehicle which carried ParkNet system drives by a parking space, the system collects parking space occupancy information and sends the data to a center server. The server can use the data to build a real-time map of parking availability, and can provide the information to client. Other researches such as Medusa [17] focus on the framework of crowd computing system. These crowd computing system are generally online systems [13–15,20,21]. There are no task assignment processes on these system which is different from ours.

7   Conclusion

In this paper, we design a new crowd computing system in MSNs, which is used to publish crowd computing tasks. We present an opportunistic task assignment algorithm (OUGA). OUGA can help to find the optimal user group with a smallest group size and the largest total usefulness degree. We conduct a serious of simulations, and the results show that our algorithm has a better performance. In the future, we will continue to study task assignment problems in our crowd computing system. Besides, we will study the more complex task model, and we will study the problem of resource allocation in the crowd computing system.

Acknowledgment. This research was supported in part by the National Natural Science Foundation of China (NSFC) (Grant No. 61572457, 61572336, 61502261, 61379132), and the Natural Science Foundation of Jiangsu Province in China (Grant No. BK20131174, BK2009150).

References

1. Alonso, O., Rose, D.E., Stewart, B.: Crowdsourcing for relevance evaluation. ACM SIGIR Forum Arch. **42**, 9–15 (2008)
2. Guo, X., Chan, E.C.L., Liu, C., Wu, K., Liu, S., Ni, L.M.: ShopProfiler: profiling shops with crowdsourcing data. In: INFOCOM, pp. 1240–1248 (2014)
3. Murray, D.G., Yoneki, E., Crowcroft, J., Hand, S.: The case for crowd computing. In: MobiHeld 10 Proceedings of the Second ACM SIGCOMM, pp. 39–44 (2010)
4. Hu, X., Li, X., Ngai, E.C., Leung, V.C., Kruchten, P.: Multidimensional context-aware social network architecture for mobile crowdsensing. IEEE Commun. Mag. **52**, 78C–87 (2014)

5. Li, M., Li, P.: Crowdsourcing in cyber-physical systems: stochastic optimization with strong stability. IEEE Trans. Emerg. Top. Comput. **1**(2), 218–231 (2013)
6. Ma, H., Zhao, D., Yuan, P.: Opportunities in mobile crowd sensing. IEEE Commun. Mag. **52**(8), 29–35 (2014)
7. He, S., Shin, D., Zhang, J., Chen, J.: Toward optimal allocation of location dependent tasks in crowdsensing. In: IEEE INFOCOM, pp. 745–753 (2014)
8. Weppner, J., Lukowicz, P.: Bluetooth based collaborative crowd density estimation with mobile phones. In: IEEE PerCom (2013)
9. Krontiris, I., Dimitriou, T.: Privacy-respecting discovery of data providers in crowd-sensing applications. In: IEEE DCOSS (2013)
10. Ganti, R.K., Ye, F., Lei, H.: Mobile crowdsensing: current state and future challenges. IEEE Commun. Mag. **49**, 32C–39 (2011)
11. Li, Y., Jiang, Y., Jin, D., Su, L., Zeng, L., Wu, D.: Energy-efficient optimal opportunistic forwarding for delay-tolerant networks. IEEE Trans. Veh. Technol. **59**, 4500–4512 (2010)
12. Wang, X., Chen, M., Han, Z., Wu, D.O., Kwon, T.T.: TOSS: traffic offloading by social network service-based opportunistic sharing in mobile social networks. In: INFOCOM, pp. 2346–2354 (2014)
13. Demirbas, M., Bayir, M., Akcora, C., Yilmaz, Y.: Crowd-sourced sensing and collaboration using twitter. In: IEEE WoWMoM, pp. 1–9 (2010)
14. Simoens, P., Xiao, Y., Pillai, P.: Scalable crowd-sourcing of video from mobile devices. In: ACM MobiSys, pp. 139–152 (2013)
15. Yang, D., Xue, G., Fang, X.: Crowdsourcing to smartphones: incentive mechanism design for mobile phone sensing. In: ACM MobiSys, pp. 173–184 (2012)
16. Hu, X., Chu, T.H.S., Chan, H.C.B., Leung, V.C.M.: Vita: a crowdsensing-oriented mobile cyber-physical system. IEEE Commun. Mag. **1**, 78–87 (2014)
17. Ra, M., Liu, B., Porta, T., Govindan, R.: Medusa: a programming framework for crowd-sensing applications. In: ACM MobiSys, pp. 337–350 (2012)
18. Mathur, S., Jin, T., Kasturirangan, N.: ParkNet: driveby sensing of road-side parking statistics. In: ACM MobiSys, pp. 123–136 (2010)
19. Xiao, M., Wu, J., Huang, L., Wang, Y., Liu, C.: Multi-task assignment for crowdsensing in mobile social networks. In: IEEE INFOCOM (2015)
20. Ganti, R., Pham, N., Ahmadi, H.: GreenGps: a participatory sensing fuel-efficient maps application. In: ACM MobiSys, pp. 151–164 (2010)
21. Mohan, P., Padmanabhan, V.N., Ramjee, R.: Nericell: rich monitoring of road and traffic conditions using mobile smartphones. In: ACM SenSys, pp. 323–336 (2008)

Forest of Distributed B+Tree
Based on Key-Value Store for Big-Set Problem

Thanh Trung Nguyen[1,2]([✉]) and Minh Hieu Nguyen[2]

[1] Research and Development Department, VNG Corporation,
Hanoi, Vietnam
trungthanhnt@gmail.com, thanhnt@vng.com.vn
[2] Le Quy Don Technical University, Hanoi, Vietnam

Abstract. In many big-data systems, the amount of data is growing
rapidly. Many systems have to store big-sets: the sets with a large number of items. Efficiently storing a large number of big-sets to support
high rate updating and querying is a challenging problem in data storage
systems. Nowadays, distributed key-value stores play important roles in
building large-scale systems with many advantages. They support horizontal scalability, low-latency, high throughput when manipulating small
or medium key-value pairs. Unfortunately, when working with big-set
data structure, they do not work well and most of them are not scalable
with a large number of big sets. In this research, we analyze the difficulty in storing big-sets using key-value stores. An architecture called
"Forest of distributed B^+Tree" and algorithms are proposed to build
NoSql data store for storing big data structures such as set, dictionary.
The big-sets are split into multiple small sets of limited size and stored
in key-value stores. A Multi-level meta-data is also proposed and used
to reduce the complexity in writing operations of big-sets when using
key-value stores from $O(N)$ to $O(log(N))$. This research can store larger
number of items in a set than Cassandra and Google BigTable. Parts of
big set in this research is distributed while a row in Google BigTable only
has a limited size and must be fit in a server. Experiment results show
that proposed system has better read performance than Cassandra. The
proposed architecture may potentially be used in various applications
such as storage system for data from sensors in the Internet of Things
(IoT) systems, commercial transaction storages and social networks.

Keywords: Big set · Forest of distributed B+Tree · Key-value · Big
data structure · Storage

1 Introduction

Many data systems have to store a large number of big sets in which each set contains a large number of unique items. Each item has three main logical attributes:
identification, sorting index and data. Therefore, a set can be represented as a
sorted list of unique elements. Each big-set is identified by a *key*. Additionally,

© Springer International Publishing Switzerland 2016
H. Gao et al. (Eds.): DASFAA 2016 Workshops, LNCS 9645, pp. 268–282, 2016.
DOI: 10.1007/978-3-319-32055-7_22

the sets can be frequently changed and potentially grow to millions or billions items. In distributed environment, beside complexity of in-memory data structure, the complexity of adding or removing item in sets can be estimated by the amount of data transfered and disk I/O. The big-set problem thus is how to build a big-set storage system simultaneously satisfies these requirements: (1) ability to store a large number big-sets, size of a set can be larger than size of main-memory or even larger than size of filesystem in a single server; (2) minimize the complexity of updating the sets; (3) all big-set data are persistent, all writes are durable; (4) low latency, high throughput, frequently accessed items are cached in main memory.

Storing big-sets is an important requirement in many systems such as social networks, e-commerces, massively multi-player online games, data for anomaly detection [16], sensors data for Internet of Things(IoT). Since writing rate to the sets is high and the size of sets gets larger overtime, maintaining high performance, throughput and low latency of the store is difficult. Recently, a number of research has been carried out to enhance the performance of key-value store including SILT [10], our Zing Database [15], Mega-KV [22], Masstree [12], etc. Many open-source systems including Redis [20], LevelDB [7], Kyoto cabinet [8] have been developed that support high performance key-value stores. These works aimed to minimize the memory overhead, latency for key-value store. We also had built a backend storage framework on top of these key-value stores to store data for many high load and data intensity applications. They performed efficiently in systems with small or medium size key-value pairs. However, for large data structure in value such as large string set, large dictionary, using these key-value store directly we did not get high performance and system was not well scalable.

This research is done to efficiently solve problems for storing big data structures such as *Set, Dictionary/Map* based on key-value stores. These are main contributions of this research:

- Defining big-set problem based on key-value store and analyzing complexity of read and write operations.
- Propose "Forest of distributed B^+Tree" based on key-value stores, with multilevel meta-data management system, algorithms for minimize data transfer when updating the big sets.
- Implement a big-set storage system based on key-value stores which has ability for storing large number of big-sets. It can store bigger set than row in Google BigTable and Cassandra due to ability to distribute big-set data(or parts of a row) in distributed key-value store. It is useful for Big Data Management.

The remaining two subsections of this section present our framework to build high performance data storage backend service using key-value stores, then bigset problem and its complexity are presented. After that, the rest of the paper is organized as follow: In Sect. 2 we present related works. Section 3 shows the detail architecture and algorithms of proposed "Forest of distributed B+Tree". Section 4 describes workloads and benchmark and comparison scenario to verify the advantages of proposed work with some discussion. Finally, Sect. 5 concludes the paper.

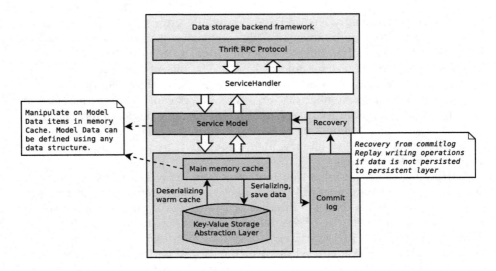

Fig. 1. Storage backend framework overview

1.1 Zing Data Storage Framework

Zing Data Services are used for various products of VNG[1] such as Social Network Media Distribution Platform, Mobile Game Distribution which have millions of Internet users. Zing Data Services require low-latency in every read and write operation. They also require data persistence and fast recovery. We developed a framework for Zing Data Services to satisfy these requirements. This framework is used as a base for building data store service for complex data structures. The framework consists of these components: *Memory Cache, Service Model, Key-Value Store Abstraction,* and *Commit-Log.* The overview of the framework is described in the Fig. 1. This framework uses key-value stores to persist data. Structured data are serialized and stored in key-value stores. In the read path, if an item is missed from *Memory Cache,* its data will be loaded from key-value store and be de-serialized then be warmed into the *Memory Cache.* Caching data-structures are used to cache structured data as a dictionary, mapping from key to value. Zing data service framework can be used to build high-performance, specific back-end storage system to fit the requirements of applications.

Memory Cache. *Memory Cache* is a dictionary data structure for mapping from keys to structured values. It is used to fast access data with low latency. The size of *Memory Cache* is limited, it only store frequently accessed items. When the number of key-value pairs in *Memory Cache* reaches the configured limited size, some items will be selected and evicted using a replacement algorithm. Each *Memory Cache* class has a replacement algorithm: LRU (least recent

---

[1] http://www.vng.com.vn/en/.

used)[17,18], LFU (least frequent used), ARC (Adaptive Replacement Cache) [13,14], etc. Memory Cache with a specific Cache Replacement Policy is configurable to be appropriate to workload characteristic of the each storage backend service. We implemented hash tables with various cache replacement algorithms for caching data and atomic visiting function to visit item in the cache for reading and/or writing. In this framework, caching functions are thread-safe. Lock-free algorithms can be used to implement some type of cache structures.

Key-Value Store Abstraction. Key-value store abstraction is a persistent layer of the framework. This layer store binary key-value data to persistent devices or remote services. It is abstracted and supports multiple implementation of key-value engines and distributed key-value store. This layer currently support ZDB [15], LevelDB [7], Kyoto Cabinet [8], and Remote Key-Value service wrapped by ZDBService described in [15]. With this design, the framework can support any type of key-value store in the backend. *Memory Cache* and *Key-Value Store Abstraction* are combined to create *Storage* component. This component supports serialization and then saves modified data from memory cache to persistent layer and warm-up cache item when it is missed from cache. The process of flushing modified key-value pairs from *Memory Cache* to *Key-Value Store Abstraction* is controlled by *Storage* component. This process can be configured to be done synchronously or asynchronously.

Service Model. *Service Model* is the core logical layer of the framework. It implements all algorithms and functions for reading, atomic manipulating, modifying, removing model data. Model data in each service is defined depending on the application requirements such as User-Profile, Friend-List, E-Commerce Transaction Info. It can be any type of data structure we want to store. The model uses *Storage* component to read and write data items for its logical algorithms.

Commit-Log. Commit-Log is a component that sequentially logs all write operations to files. All running operations with its parameters are serialized using Thrift binary protocol to get binary data then be appended to commit-log files. Commit-log files are used to recover data after a system crash. All writing to commit-log file is append-only, it is very fast. This component is appropriate with in-memory store service.

1.2 Big Set Problem Statement

Problem. Build key-value store that mapping from key SK_i to value SV_i, each value SV_i is a set of items $\{i_1, i_2, ...\}$. Item type can be either as simple as integer, double, or as complex as string, binary, a key - value pair or other data structure. There is a function to get the unique sortable key for an item called $K(item)$. For simple item type such as *Integer*, *Double*, $K(item)$ can be the *item* itself.

For key - value item type, $K(item)$ can be equal to *item.key* , etc. The storage system supports these basic functions:

- add(SK_i, *item*) - add *item* to value SV_i associated with SK_i.
- remove(SK_i, *itemKey*) - remove *item* which have $K(item) = itemKey$ from value SV_i associated with SK_i.
- existed(SK_i, *itemKey*): boolean - check if there is an item with its key equal to *itemKey*.
- get(SK_i): get all items in the set SK_i.
- getSlice(SK_i, *start*, *num*) - retrieve *num* items in value SV_i from $start^{th}$ item.
- rangeQuery(SK_i, *startItemKey*, *endItemKey*) - retrieve items in value SV_i which has key in range of $[startItemKey, endItemKey)$

All data need to be persisted in distributed key-value store.

Complexity. If each set value is stored in a separate data structure such as B-Tree, RB-Tree, the complexity when insert a new item is $O(log(N))$, N is size of the set. The space complexity and the amount serialized data writing to key-value store is $O(N)$. In case of big set, the big set serialization will make large key-value pairs. And storing these pairs into key-value store when the set is changed by writing operations will consume high the network traffic and disk I/O. That decreases the performance of the whole backend services and increase the load of operation system.

2 Related Works

Many works were done for solving problems in distributed storage system and storing big set data in both academic and industrial fields.

Marcos K. et al. presented a paper about distributed B+Tree [1]. In that work, they used Sinfonia to build distributed B+Tree. In the evaluation the work compared it with Berkeley DB [19] and the design of that work is not for rapidly growing B-Tree so the performance of inserting workload is not very good. Based on [1], a later work [21] proposed an in-memory distributed multi-version B-Tree was proposed. It can scale to hundreds of cores and TBs of memory. It supports copy-on-write snapshot and writable clone.

In [11], Witold Litwin et al. presented a family algorithms called RP* for building distributed files on multicomputers. It proposed distributed data structures and used it for file management in networking environment.

BigTable [3] is a distributed storage system for structured data developed by Google. It is a sparse, distributed, persistent multidimensional sorted map. Data Model of BigTable is indexed by row key, column key, timestamp. Each value in the map is an array of byte. ($row : string, column : string, time : int64) \rightarrow string$ [3]. Data of BigTable is persisted in GFS [5]. BigTable ensures that there is at most one active master at a time by using Chubby [2] -a high availability and persistent distributed lock service. BigTable partition data of a table into multiple tablets by row keys. Each tablet is a range of a row key. BigTable consists of one master server and many tablet servers. A tablet server manages a set of tablets. The design of BigTable shows that data of a row is stored in a tablet which is assigned to one server at a time. This design leads to the limitation in total size of a row[2]. This research aim to new design that support distribution of partial data of a row into distributed key-value store. Proposed design will allow to store wider row than BigTable.

Cassandra [9] is an open source column-oriented nosql store developed by Facebook[3]. It support complex data structure in form of columns, each row can consists of multiple columns. Latest version of Cassandra can support up to 2 billions columns per row. Rows in Cassandra belong to a node and can be replicated. This design makes a row in Cassandra never split across nodes and it is impossible to partially cache columns in a row. In production, row cache feature of Cassandra often is disable when using wide row. For Big-Set problem, Cassandra can be used to store big sets of up to 2 billions items. In our benchmarks we have done, Cassandra has good write performance, however, read throughput is lower than the others in the comparison.

Redis [20] is a high performance structured key-value store that supports many data structures. All items in Redis are placed in main memory implemented as a cache with cache-replacement algorithms such as LRU. For storing *Set* data structure, Redis splits items into blocks and stores them in the main memory using skip-list data structure. So, the maximize number of items in a set is limited by the size of main memory and can not be scaled for big-set, each Redis instance can store only a small number of big sets.

3   Forest of Distributed B+Tree for Solving Big-Set Problem

In this section, we show detail idea in our contributions. For storing big set efficiently, we split the big set into multiple limited size small sets and design a multi-level set meta-data to manipulate and query the sets.

We treat a set like a sorted list of unique items. We call l_{max} is the maximum size of each small set. If the size of small set is greater than l_{max}, it will be split into two smaller sets. We designed and implemented a distributed back-end service based on key-value store for storing these small sets. Figure 3 describes the overview of big set components.

Figure 2 shows an example of how big set data are organized. In this example, each small set contains at most 3 items. When the number of items in a small set is greater than 3, it will be split into 2 smaller set. Each item is a key-value pair, sorted in small set by the key. Small sets are stored in *Small Set Service*. There is a multi-level meta-data system to manage the big set. Meta data nodes are stored

---

[2] https://cloud.google.com/bigtable/docs/schema-design.
[3] https://www.facebook.com.

in *Metadata Service*. Leaf nodes in Fig. 2 are small sets, the remaining nodes are meta data. *IDGenerator* in Fig. 3 is used to generate Small Set ID, Meta data ID to make keys in *Small Set Service* and *Metadata Service* auto incremental. These services are co-ordinated by *BigSet Service*. That data organization of big set is similar to B+Tree with some modifications as follow:

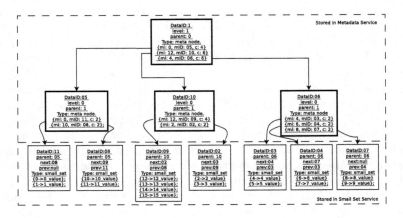

Fig. 2. Big-set sample data for key-value items, each node has at most 4 children

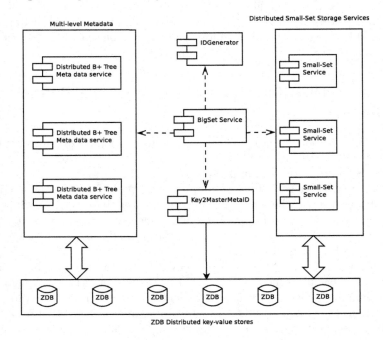

Fig. 3. Big-set components overview

– Tolerate constraints in the number of children in B+Tree nodes. It can be greater than B+Tree defined maximum number for a time before asynchronous splitting operation is completed. And it can be smaller than minimum defined number in B+tree when items are removed from big set.
– Maximum number of items in leaf-node can be different from maximum number of children in non-leaf node.

Each big-set is presented as a Distributed Modified B+Tree where B+Tree nodes are distributed in key-value stores. *BigSet Service* can manage large number of big-sets and called "Forest of Distributed B+Tree".

3.1   Small Set Service

This service is a key-value store where the key is set id and the value is a set of items. It uses architecture described in Subsect. 1.1. This service supports atomic functions for manipulating small sets. Each set has a limited size and information about adjacent set in the big set. It contains a range of sorted items in the big set. Listing 1.1 below is a sample Thrift type definition in the interface of a Small Set Service.

Listing 1.1. Type definition in a Small Set Service with key-value item

```
struct TItem{
    1:  binary key,
    2:  binary value,
}
typedef binary TItemKey
typedef list <TItem> TItemList
typedef list <TItemKey> TItemKeySet
typedef i64 TSmallSetIDKey
```

The structure of a small set information:

– entries: TItemList - items in the set, this list has a limited size.
– prevSetID: TKey - id of the previous small set in the Big Set. If this field equal to zero, it means that this small set is the first chunk of the Big Set.
– nextSetID: TKey - id of the next small set in the Big Set. If this field equal to zero, it means that this small set is the last chunk of the Big Set.

Small Set Service has functions for query, atomically manipulate data and split when a small set reaches limited size. It consists of these important functions:

– getSlice(TSmallSetIDKey smallSetID, i32 startIndex, i32 number): TItemSet - get a range of *number* items from $startIndex^{th}$ item in small set by *smallSetID*
– getSliceFromItem(TSmallSetIDKey smallSetID, TItemKey startItem, i32 number) - get up to *number* items in small set, from item *startItem*.

- addItem(TSmallSetIDKey smallSetID, TItem item): TAddItemResult - add *item* into small set associate with *smallSetID*. *TAddItemResult* is a return type with following information: error code of operation (EAdded, ENoChange, EReplaced); current size of list after operation; old item if this operation overwrites it by *item* (in case of a complex TItem type).
- splitSet(TSmallSetIDKey setID, TSmallSetIDKey newSetID, i32 currentSize, TItemKey newSetHead): TSplitResult - split a small set into two smaller sets.

This service stores frequently accessed small sets in main memory and flushes them into persistent key-value when a set is modified. In the production environment, we deploy multiple instances of ZDBService in many servers for persisting data of small sets. This design of *Small Set Service* is similar to the design of leaf nodes in B+Tree. Each set can be linked to a next small set and a previous small set. Moreover, they are not resided in the only process space, they can be distributed in multiple key-value stores. It is able to travel backward and forward the big set from a specific item or a position in a small set. Figure 2 demonstrates small sets in leaf nodes. In this figure, *DataID* field is the key of key-value store of Small Set Service, *mi*, *mID*, *c* stand for fields: *minItem*, *metaID* and *count* in TMetaItem described below.

3.2   Metadata Service

Metadata Service is the most important service in our Big-Set architecture. It stores information about every big-set, and shows how it is organized as multiple small-sets stored in distributed key-value store. This service has algorithms to manipulate metadata and to locate big-set's items. Each Big-Set has a root metadata item, each metadata item contains information about its children and so on till the leaf which stores information about small sets.

Metadata Information.   Meta data of a big set is organized as a multi-level structure. Each element is a structure called *TMetaItem* and consists of: *minItem*: lower-bound item of the current branch of B+Tree; *metaID*: identity of meta data (key) in the Metadata Service; *count*: number of item in a branch of B+Tree identified by *metaID*.

Each set's meta-data is described in *TSetMeta* structure. It consists of multiple items of *TMetaItem* type sorted ascending by *minItem* field. *TMetaItem* is the information used to lookup lower level *TSetMeta* or a Small Set stored in Small-Set-Service. *minItem* field is the lower-bound item key of total items managed by the current meta data node and its children. When *level* of a *TSetMeta* is greater than zero, its *children* describe lower level *TSetMeta*, otherwise, they describe small sets. Meta-data Service mapping from *metaID* to *TSetMeta* structure. Each Big-Set is mapped with a root metaID. From this root metaID, we can travel through whole big-set metadata. The size of the *children* array is limited by a constant called KMAX_META_SIZE. Figure 2 demonstrates metadata in non-leaf nodes.

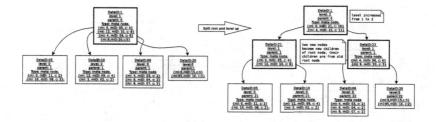

Fig. 4. Split root node and level up

TSetMetaPath is a structure in Metadata Service. It is used to describe a path from a root meta node to a leaf node. It is used in algorithms for querying and manipulating meta data and big-sets as well. It consists of a split info and multiple children called *TSetMetaPathItem*

In TSetMetaPath structure, elements of *metaPath* list has decreasing level, the ith element is parent of $i + 1^{th}$ element. The final element in *metaPath* is the parent of *smallSetInfo*.

Metadata Service Algorithms. There are important algorithms in Metadata Service. They are parts of big set operations. The *Traversal* algorithm is used to locate the path from root node to small set associate with an item for adding, modifying or reading. Splitting Algorithms are used to split full Meta-Data node when it reaches the limited number of children. These algorithms are implemented in *Service Model* layer of the framework described Subsect. 1.1.

Traversal Algorithm: To insert an item into a big set, we have to find the small set which will store that item. In this algorithm, a path from root node to leaf node (small set) which may contain a searched itemKey is constructed. At each meta data node, it decides which children node to travel in next step by using binary search in *minItem* field of TMetaItem objects in children of current TSetMeta. And a node with largest number of children will also be updated in *splitInfo* of the meta path. The found small set information will be updated into *smallSetInfo* of TSetMetaPath. Big Set is organized as a distributed B+Tree and the number of node this algorithm will visit is proportional to the height of the B+Tree, so the complexity of this algorithm is $O(log(N))$, while N is number of items in the Big Set.

Split Root Metadata Node: When a root meta node reaches the limited maximum size, it will be split into two smaller nodes. Two new smaller metadata nodes will be created and the original root node will have two meta items and its level will be increased by 1. Figure 4 demonstrates an example of splitting root meta data node with KMAX_META_SIZE=3.

Split Middle Meta-data Node: This algorithm is used when children number of a non-root meta-data node is greater than limited maximum size caused by

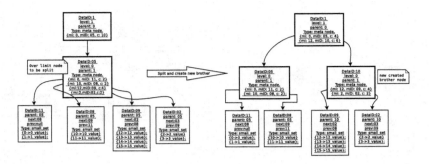

Fig. 5. Split and create new brother meta data node

smaller level node splitting or small set splitting. The algorithm will create a new same-level metadata node (new brother node), moving second-half children into new node and add new meta item that described new node into the parent meta node. This algorithm is demonstrated in Fig. 5 with KMAX_META_SIZE=3.

3.3 BigSet Service

This service is a server application and cares about big set functions for clients. It coordinates other services and uses algorithms from *Small Set Service*, *Metadata Service*. All bigset operations are implemented in this services. We use ZooKeeper [6] to co-ordinate configuration of *Small Set Service* and *Metadata Service* as long as their persistent key-value stores. Each Big Set is organized as a distributed B+Tree. Total Big-Sets managed by *BigSet Service* can be viewed as a "Forest of Distributed B+Tree".

Add an Item to a Big Set. To add item into a big set, it uses the *Traversal* Algorithm of **Metadata Service** to fill data to an object of class TSetMetaPath. The result describes a path from root meta data node to associated small set that will contain adding item.

Item will be added to the found small set from the path. If this is a new item in that small set, the *count* field in associated children of meta data node in the path will be increased by one (or decreased by one in *Remove* function). After adding an item into the small set, it may reach the limited size l_{max} or a full meta data node may be found. If the small set reaches the limited size l_{max}, it will be split into two smaller set in function *checkSplit*. A new small set is created with second-half items from original small set. New *small set id* or key in Small Set Service is generated using *IDGenerator* service. This make Small Sets have auto increasing integer key, that can take advantage of ZDB [15] efficiently when persistently storing data. The Fig. 6 shows an example of splitting small set. This operation always ensures that items in next small set is greater than items in current small set, so we can easily travel backward and forward through big set from a specific position in a small set. This is a characteristic of B+Tree and we can do it on key-value store easily.

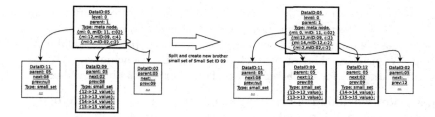

Fig. 6. Split small set example

In case of full a metadata node found in the insertion path, that node in *Metadata Service* will be split using *Split Root Metadata node and Level Up* Algorithm or Algorithm presented in Fig. 5 depend on if that metadata node is root or not. ID of new metadata node or key in *Metadata service* is also generated using *IDGenerator*. It makes the keys of *Metadata serivce* auto increasing integer and can also take advantage of ZDB efficiently like *Small Set Service*.

The algorithm for *Remove* function of Big Set service is similar to the algorithm for adding an item into big set. However, it eliminates the *checkSplit* step.

Query Items from Big Set Algorithms. Big Set Service provides several type of query in big set:

– **GetSlice**: Query n items from a specific position s.
– **GetSliceFromItem**: Query n_q items from an item key.
– **RangeQuery**: Query items in a range between two item keys such as [startItemKey, endItemKey).

We developed two approaches for solving these query types. The first approach is suitable for sequential processing in Big Set. While all items in big set are sorted by item keys and according to the design of *Metadata Service* and *Small Set Service* presented above, we can easily do these query types by specifying the starting small set ID of the result. The first small set ID is got by Traversal Algorithm for *GetSliceFromItem* query and *RangeQuery*.

The second approach is suitable for parallel processing in Big Set. In each query, we firstly estimate all small set IDs using Metadata Service for the query. Each meta data element describes a branch of B+Tree. Because meta data elements in children of $TSetMeta$ have fields $minItem$ - the lower bound item key it stores and $count$ - total number of items it stores. These fields are used to retrieve total small set IDs for processing the query efficiently. Then it is able to query these small set data in parallel in *Small Set Service*.

4 Evaluation

In this section, we present workload design, configuration we used to benchmark and compare the proposed big set with existing popular solutions: Redis [20]

and Cassandra [9]. The main purpose of benchmarks is to verify the efficiency
of proposed architecture when facing big set storage problems.

We measured number of operations per second (ops), the scalability of the
system. We used method of Yahoo! Cloud Serving Benchmark (YCSB) [4] to
generate workloads for the benchmarks. Four workloads are described below:

- **Insertion workload**: all clients generate and add new items into big sets.
- **Read Only workload**: all clients check the existence of items in the big set.
- **Mixed workload R**: 90 % client operations read items from big set, 10 %
  client operations update new value to big set.
- **Mixed workload W**: 10 % client operations read items from big set, 90 %
  client operations update new value to big set.

In these benchmark workloads, we used 8 storage servers and 4 machines to run
clients with configuration below: Ubuntu Server 64 bits, CPU Intel Xeon Quad
core, 64 GB DDRam, 600 GB SSD ext4 filesystem, 1 Gbps Network.

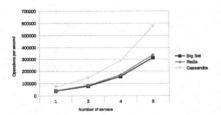

Fig. 7. Write only throughput

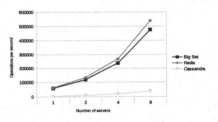

Fig. 8. Read only throughput

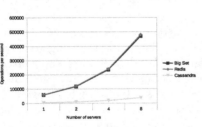

Fig. 9. Mixed Workload R

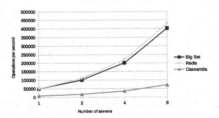

Fig. 10. Mixed Workload W

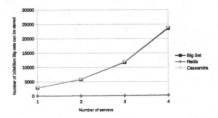

Fig. 11. Capacity comparison

Items in both Cassandra and proposed Big Set for this benchmark are key-value pairs (A simple column in Cassandra is a key-value pair), each big set in Cassandra is a wide row. Items in Redis is a string with the same size to key-value pair in Cassandra.

In client machines, we run 16 processes for each workload. Figure 7 shows the result for write benchmark throughput. Cassandra has the best performance in Write Only Workloads. Proposed Big Set has a bit lower performance compared to Redis Ordered Sets. All these storage system has linear scalability when number of servers increasing.

Figure 8 shows the result of read only benchmark throughput. In spite of having a good write performance, number of read operations per second of Cassandra is much lower than Redis and Proposed Big Set. The same result in Mixed workload is showed in Figs. 9 and 10. Proposed Big Set and Redis out perform Cassandra in reading and mixed workloads. Figure 11 shows the capacity comparison.

5 Conclusion

The proposed Big-Set architecture can be viewed as a *Forest of Distributed B+Tree* with many advantages to build various data storage services. Every Big Set as a value in a key-value store is split into multiple small sets. Small Sets are managed in storage backend service using *Memory Cache* for fast accessing and distributed key-value store to persist data. A *Metadata Service* is designed based on key-value store to manage the big set structure as a distributed B+Tree. It supports building scalable data storage system for big data structure. While other works often build key-value store using B+Tree, this research build distributed B+Tree using key-value store from scratch. This work can easily distribute big value such as big set, big dictionary into multiple servers using advantages of distributed key-value store while others such as Cassandra and Redis have not supported this feature yet. Proposed Distributed Modified B+Tree has a relative high performance in both read and write operations and capability to distributively store big data structures. The proposed work is used to store data for friend suggestion in Zing Me Social Network and to store time series of user action for game data anomaly detection system for games of VNG.

Acknowledgment. This research is funded by Research and Development Department of VNG.

References

1. Aguilera, M.K., Golab, W., Shah, M.A.: A practical scalable distributed B-tree. Proc. VLDB Endowment **1**(1), 598–609 (2008)
2. Burrows, M.: The chubby lock service for loosely-coupled distributed systems. In: Proceedings of the 7th Symposium on Operating Systems Design and Implementation, pp. 335–350. USENIX Association (2006)

3. Chang, F., Dean, J., Ghemawat, S., Hsieh, W.C., Wallach, D.A., Burrows, M., Chandra, T., Fikes, A., Gruber, R.E.: Bigtable: a distributed storage system for structured data. ACM Trans. Comput. Syst. (TOCS) **26**(2), 4 (2008)
4. Cooper, B.F., Silberstein, A., Tam, E., Ramakrishnan, R., Sears, R.: Benchmarking cloud serving systems with YCSB. In: Proceedings of the 1st ACM Symposium on Cloud Computing, pp. 143–154. ACM (2010)
5. Ghemawat, S., Gobioff, H., Leung, S.-T.: The Google file system. ACM SIGOPS Oper. Syst. Rev. **37**, 29–43 (2003). ACM
6. Hunt, P., Konar, M., Junqueira, F.P., Reed, B.: ZooKeeper: wait-free coordination for internet-scale systems. In: Proceedings of the 2010 USENIX Conference on USENIX Annual Technical Conference, vol. 8, p. 11 (2010)
7. Google Inc.: LevelDB - A fast and lightweight key/value database library by Google (2013). http://code.google.com/p/leveldb. Accessed on 23 July 2013
8. FAL Labs: Kyoto Cabinet: a straightforward implementation of DBM (2013). http://fallabs.com/kyotocabinet. Accessed on 1 May 2013
9. Lakshman, A., Malik, P.: Cassandra: a decentralized structured storage system. ACM SIGOPS Oper. Syst. Rev. **44**(2), 35–40 (2010)
10. Lim, H., Fan, B., Andersen, D.G., Kaminsky, M.: SILT: a memory-efficient, high-performance key-value store. In: Proceedings of the Twenty-Third ACM Symposium on Operating Systems Principles, pp. 1–13. ACM (2011)
11. Litwin, W., Neimat, M.-A., Schneider, D.: RP*: a family of order preserving scalable distributed data structures. VLDB **94**, 12–15 (1994)
12. Mao, Y., Kohler, E., Morris, R.T.: Cache craftiness for fast multicore key-value storage. In: Proceedings of the 7th ACM European Conference on Computer Systems, pp. 183–196. ACM (2012)
13. Megiddo, N., Modha, D.S.: ARC: a self-tuning, low overhead replacement cache. In: FAST, vol. 3, pp. 115–130 (2003)
14. Megiddo, N., Modha, D.S.: Outperforming LRU with an adaptive replacement cache algorithm. Computer **37**(4), 58–65 (2004)
15. Nguyen, T., Nguyen, M.: Zing Database: high-performance key-value store for large-scale storage service. Vietnam J. Comput. Sci. **2**(1), 13–23 (2015)
16. Nguyen, T.T., Nguyen, A.T., Nguyen, T.A.H., Vu, L.T., Nguyen, Q.U., Hai, L.D.: Unsupervised anomaly detection in online game. In: Proceedings of the Sixth International Symposium on Information and Communication Technology, SoICT 2015, pp. 4–10. ACM, New York (2015)
17. O'neil, E.J., O'neil, P.E., Weikum, G.: The LRU-K page replacement algorithm for database disk buffering. ACM SIGMOD Rec. **22**(2), 297–306 (1993)
18. O'neil, E.J., O'Neil, P.E., Weikum, G.: An optimality proof of the LRU-K page replacement algorithm. J. ACM (JACM) **46**(1), 92–112 (1999)
19. Oracle: Oracle Berkeley DB 12c: Persistent key value store (2013). http://www.oracle.com/technetwork/products/berkeleydb
20. Sanfilippo, S., Noordhuis, P.: Redis. http://redis.io. Accessed on 07 June 2013
21. Sowell, B., Golab, W., Shah, M.A.: Minuet: a scalable distributed multiversion B-tree. Proc. VLDB Endowment **5**(9), 884–895 (2012)
22. Zhang, K., Wang, K., Yuan, Y., Guo, L., Lee, R., Zhang, X.: Mega-KV: a case for GPUs to maximize the throughput of in-memory key-value stores. Proc. VLDB Endowment **8**(11), 1226–1237 (2015)

BDQM 2016

An Efficient Schema Matching Approach Using Previous Mapping Result Set

Hongjie Fan[1], Junfei Liu[2(✉)], Wenfeng Luo[1], and Kejun Deng[1]

[1] School of Electronics Engineering and Computer Science, Peking University,
Beijing, China
{hjfan,1201214087,kejund}@pku.edu.cn
[2] National Engineering Research Center for Software Engineering,
Peking University, Beijing, China
liujunfei@pku.edu.cn

Abstract. The widespread adoption of eXtensible Markup Language pushed a growing number of researchers to design XML specific Schema Matching approaches, aiming at finding the semantic correspondence of concepts between different data sources. In the latest years, there has been a growing need for developing high performance matching systems in order to identify and discover such semantic correspondence across XML data. XML schema matching methods face several challenges in the form of definition, utilization, and combination of element similarity measures. In this paper, we propose the XML schema matching framework based on previous mapping result set (*PMRS*). We first parse XML schemas as schema trees and extract schema feature. Then we construct *PMRS* as the auxiliary information and conduct the retrieving algorithm based on *PMRS*. To cope with complex matching discovery, we compute the similarity among XML schemas semantic information carried by XML data. Our experimental results demonstrate the performance benefits of the schema matching framework using *PMRS*.

Keywords: Schema matching · XML · Previous mapping result set

1 Introduction

eXtensible Markup Language, because of the flexibility of self-description, has become a standard information representation and exchange of data in a wide range of scenarios [1,2]. XML has been widely used in many domains, such as biology [3], business [4], chemistry [5], and geography/geology [6], to name a few. To make data exchange easier, organizations like the World Wide Web Consortium (W3C) are increasingly committed to define an advanced languages to describe the structure and content of XML data source, such as DTD/XSD. Despite the presence of powerful languages, the achievement of the full interoperability among applications based on XML data is often illusory. One of the biggest obstacles to the development of this technology is how to effectively

© Springer International Publishing Switzerland 2016
H. Gao et al. (Eds.): DASFAA 2016 Workshops, LNCS 9645, pp. 285–293, 2016.
DOI: 10.1007/978-3-319-32055-7_23

identify and correspondence between the semantic nodes, called Schema Matching [7]. One of the most important steps of schema matching between source and target data is to select an appropriate measure which can best calculate an amount of similarity between documents based on their representation, but these measures are time consuming.

A promising approach to improve both the effectiveness and efficiency of schema matching is reusing of previous match results [7]. Exploiting the reuse potential requires a comprehensive repository to maintain previously determined correspondences and match results. Schema matching tools such as COMA [8] and its successor COMA++ [9] apply a so-called *MatchCompose* operator for a join-like combination of two match mappings to indirectly match schemas. [10] is the corpus-based match approach uses a domain specific corpus of schemas and focuses on the reuse of element correspondences. They augment schema elements with matching elements from the corpus and assume that two schema elements match if they match with the same corpus element(s), and use a machine learning approach to find matches between schema and corpus elements. The OpenII project is developing an infrastructure, *Harmony*, for information integration of schemas to permit their reuse [11]. [12] describes an approach called *schema covering* to partition the input schemas such that the partitions can be matched to schema fragments in the repository. Such techniques are not yet common in current match systems, and more research is needed in reuse of previous determined matching result.

In this paper, we develop and implement a schema matching framework based on previous mapping result set, *PMRS*. This matching framework consists of three phases. (1)*Parse Schema and Extract Feature*. During this step, we pre-process the XML data and extract the specific features, such as name, attribute, and comment. (2)*Construct the Matching Framework*. Similarity among XML schemas are determined by exploiting semantic information. In this step, we develop the schema matching framework. We construct *PMRS* as the auxiliary information, and conduct the retrieving algorithm based on *PMRS*. To cope with complex matching discovery, we need to compute the similarity among XML schemas. (3)*Experiment Demonstration*. We carried out a set of experiments to evaluate the proposed framework. Our experiment results show that the proposed framework is useful and efficient in heterogeneous XML data matching issue, especially for the situation of reusing previous match results.

2   Framework

This section gives an overview of our proposed method and our processing can be divided into two major steps: XML Preprocess including feature extraction and Retrieve Process.

Figure 1 depicts the overall framework of our method. During the matching, we present the previous result as the auxiliary information. Data entity e from source schema X_s firstly retrieve from the *PRMS*. If we find the semantic correspondence in this step, then output the mapping result directly, and delete e

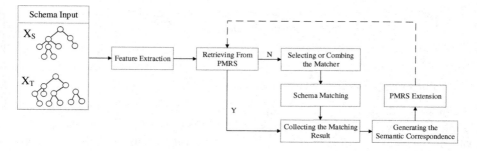

Fig. 1. The framework of our method

from *Entity Set* consequently. This action would increasingly reduce the number of matching entity candidates and boost the matching speed with extension of *PMRS*. If there is no semantic correspondence in this step, we switch into the normal matching workflow. As we known, after collecting the matching result, we generate the semantic correspondence in succession, and can put these collections as *PRMS*. With the process of matching, the *PRMS* will be expanded and enriched. Finally we can achieve the well performance of matching.

The *PMRS* is composed of two parts:

(1) the *Entity Set(ES)* of reference schema and source schema. Entity is the basic matching element. We construct the $ES(E_i)$ for storing all the matching entities E_i from reference schema.
    The structure of (ES) is represented as $ES(E_i)=\{E_i,1,e_1,S_{e_1},e_2,S_{e_2},...,e_n,S_{e_n}\}$. In this structure, e_i means the entity match E_i. S_{e_i} means the original matching similarity, and calculated as $S_{e_i}=Sim(e_i,E_i)$. Because entity E_i has three types: *concept, attribute,* and *individual,* all these types storing in *PMRS*, the entity belongs to specific type could not match with entity from other types.
(2) the *Comment Key Words Library (CKWL)*. We need to maintain the *(CKWL)* storing e_i, keywords after splitting *comment*, and the frequency of these keywords. we need to point out if the frequency exceed the threshold, this keyword is called *stopping word* and will be blocked.

3   Quick Retrieving Algorithm Construction

Each XML schema may contain a large scale of computational attributes. Considering some attributes information would be meaningless, in this paper, we use three typical attributes: *name, label,* and *comment.* We pick up these features in data preprocess, and construct them as $e_i= < name, label, comment >$. During the quick retrieving algorithm, we calculate the similarity using *e.name, e.label,* and *e.comment* with *E.name, E.label,* and *E.comment,* e represented as source entity and E represented as result entity. The similarity calculation between e and E as:

$$\begin{cases} Sim(e, E) = \lambda_1 \, Sim(e.name, E.name) + \lambda_2 \, Sim(e.label, E.label) + \\ \qquad\qquad \lambda_3 \, Sim(e.comm, E.comm) \\ \\ \sum_{i=1}^{3} \lambda_i = 1 \end{cases} \qquad (1)$$

(1) *Name/Label Simiarity Computation*

We present $Sim(e.name, E.name)$ and $Sim(e.label, E.label)$ to caculate the entity similarity of name/label between source schema and reference schema. The values of *e.name* and *E.name* are often the words, but may contain some special symbol such as "-". We process all these issues including format the letters in lowercase, get rid of special symbol. Then we use *Levenshtein Distance Algorithm* [13] to calculate the similarity these strings. It is the basic programming algorithm for computing the edit distance. Several variants of the edit distance have been proposed, such as the normalized edit distance [14]. There are many methods to compare strings depending on the way the string is coded (as exact sequence of characters, an erroneous sequence of characters, a set of characters, etc.) [15–17].

(2) *Comment Computation*

We present $Sim(e.comment, E.comment)$ to calculate the similarity of comment between source schema and reference schema. Considering *comment* always contain phrase or sentence, we need to split *comment* into keywords set. During this step, another issue, blocking the *stopping words* which exceed threshold, should be pay attention to. The *stopping words* are composed of two parts: *Static Stopping Words* and *Dynamic Stopping Words*. *CKWL* store the frequency of keyword(kw), represented as f_{kw}. The frequency is calculate as $p(kw)=f_{kw}+N_g$, while N_g is the number of entities in reference schema. If $p(kw)$ is large than τ (τ is the threshold we set), we judge kw as stopping word. After that, Then we use the classic *Cosine Method* to calculate the similarity. The cosine similarity is a well known measure from information retrieval. It computes the cosine of the angle between the two d-dimensional vectors $\overrightarrow{\sigma_1}$ and $\overrightarrow{\sigma_2}$ of the two string σ_1 and σ_2. The d dimensions of these vectors correspond to all d distinct tokens that appear in any string in a given finite domain. For example, we assume that σ_1 and σ_2 originate from the same attribute A. The Cosine similarity is calculated as $Cosine(\sigma_1, \sigma_2) = \frac{\overrightarrow{\sigma_1} \cdot \overrightarrow{\sigma_2}}{||\overrightarrow{\sigma_1}|| \cdot ||\overrightarrow{\sigma_2}||}$. Several variants of the Cosine similarity have been proposed for comment computation, such as the *TF-IDF* [18] *Soft-TFIDF* [16].

(3) *Quick Rretrieving Algorithm*

Based on above algorithm, we calculate the transfer similarity for all same type entities between e from source schema and E_i from *PMRS*. We pick up the highest similarity, and compare it with τ_{PMRS}. If it is larger than τ_{PMRS}, we output it as the final matching result. The transfer similarity is represented as $Sim_{trans}(e, e_i) = Sim(e, e_i)^* S_{e_i}$, while $Sim_{trans}(e, e_i)$ is caculated from the context similarity, S_{e_i} is the original similarity which denote the similarity between e and E.

Algorithm 1. Fast Retrieving Algorithm

Require: Input: Entity e.
    Output: $< e, E_i, Sim_{trans} >$.
 1: **while** $ES(E_i)$ in all ES **do**
 2:    **while** e_i in all $ES(E_i)$ **do**
 3:       $Sim(e,e_i)$.
 4:       $Sim_{trans}(e,e_i){=}Sim(e,e_i)^{*}S_{e_i}$.
 5: **if** $MAX(Sim_{trans}(e,e_i)){>}\tau_{PMRS}$ **then**
 6:    Return $< e, E_i, Sim_{trans} >$.

There are two points need to pay attention to in the algorithm implementation process: (1) Only retrieve the same type ES. *Concept, attribute* and *individual* are the different types of entity, which cannot match with each other. We need to determine e type with ES. (2) Entity e firstly need to match entities from reference schema, so we calculate the transfer similarity rather than the similarity between entity e and e_i. The existing result utilized in quick retrieval is an intermediary, so under this circumstance the transfer similarity is used as confidence of similarity equivalence.

(4) PMRS *Extension*

We set the similarity between e and E_i as original similarity, put (e,S) in $ES(E)$, and add $e.comment$ into $CKWL$. Since the error information of $PMRS$ will effect the final match result, we need to set τ_{PMRS} and τ_{ePMRS} precisely in order to ensure the accuracy of the existing result. We set an expansion threshold τ_{ePMRS}, if similarity is bigger than τ_{ePMRS} after matching between source schema X_S and reference schema X_R, we determine it as the matching pair (e,E_i).

There is no previous matching result in $PMRS$ at the time $PMRS$ constructed initially. The algorithm is facing the "*cold start*" problem. In order to properly use $PMRS$, we put the entities of reference schema into $PMRS$, and set the original similarity manually. So the quick retrieval algorithm would be operated normally without "*cold start*".

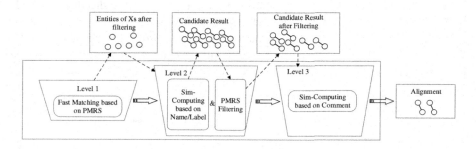

Fig. 2. Matching Workflow by 3-level Filter

Figure 2 depicts the overall matching workflow by 3-level filter. In *level-1*, the entity e firstly match with entity E from reference schema using *Quick Retrieving Algorithm*. If the threshold exceeds τ_{PMRS}, then we output the mapping result directly, and delete data entity e from *Entity Set* consequently. The rest entities take participate in the following matching. In *level-2* and *level-3*, we compute the similarity among XML schemas using similarity measures based on *name*, *label*, and *comment*, respectively. The main reason of schema matching in this way is to reduce the computation cost. During such three level filter, we can get the optimistic matching result in well time cost.

4   Experiment Demostration

In this section, we present the experimental results to demonstrate efficiency and effectiveness of our method. *F-measure* combined precision and recall to present the ratio of error match and missing correct match to evaluate the matching result comprehensively.

4.1   Datasets & Setup

We use OAEI benchmark as our experiment data set to evaluate our schema matching algorithm with other algorithms. All experiments are implemented in Java. Our method is based on disk, and our experiment is conduct on a machine with Intel Corei7 CPU processor, 8G RAM memory and running Ubuntu 14.04 LTS (64bits). Here, we provide more details and statistics about the datasets.

OAEI[1] Since 2004, OAEI organizes evaluation campaigns aiming at evaluating ontology matching technologies, and benchmark is an important part oriented to specific domains. The datasets include 51 schemas come from reference bibliography fields. #101 is the example schema; #101 and #1XX are substantially the same; #2XX lacks some essential elements; #3XX comes from real existing schemas. Figure 3 gives the suitable range of each schema.

Testing purpose	Number	Explanation
Easy test	#103-104	Similar with #101
Comprehensive test	#201-266	To test the robustness and availability of the algorithm. Some elements from the schema (*name, attributes, constraints*) are excluded.
Application test	#301-304	Real existing schemas. The purpose is to test the performance and availability of the algorithm in practical application.

Fig. 3. Testing Purpose using OAEI Dataset

---

[1] http://oaei.ontologymatching.org/.

4.2   Computational Result and Efficiency

We designed a comparative experiment using *PMRS* and without using *PMRS* to calculate the similarity between #101 schema and #103–304 schemas. Figure 4 presents results for our two quality metrics, the *F-measure* and *time-cost*, respectively.

(1) Similarity Measure Quality

They show that schema matching using *PMRS* in most cases is better than normal matching method. To get better matching quality, different similarity measures have been used, such as similarity measure based on *comment*. Compared to the studied approaches, it improves *F-measure* by +7.3% and +14.6% for #251–260, #261–266, respectively. For the schemas #254–257 and #261–266, the original *F-measure* is barely null, but in our experiment *F-measure* is 0.12 and 0.14. It proves the schema matching method using *PMRS* can improve the similarity quality.

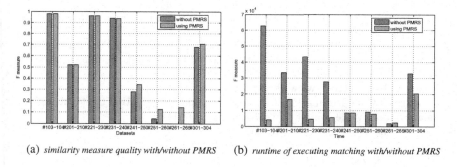

(a) *similarity measure quality with/without PMRS*   (b) *runtime of executing matching with/without PMRS*

Fig. 4. Similarity measure quality and efficiency with/without PMRS

(2) Effectiveness of *PMRS*

We measure the runtime of computing and executing the schema matching. The results for different datasets are shown in Fig. 4(b). They show the schema matching algorithm using *PMRS* achieves stable running time and reduce the running time efficiently. The average saving runtime achieves at 18 s. Compared to the studied approaches, it reduce runtime by 58 s at most for #103–104. For the complex schema such as #301-#304, the improvement achieves at 12 s (reduce 38% runtime). Our experimental results demonstrate the effectiveness of the schema matching framework using *PMRS*.

5   Conclusions

In this paper, we propose the XML schema matching framework based on previous mapping result set (*PMRS*). We construct *PMRS* as the auxiliary information and conduct the retrieving algorithm based on *PMRS*. To cope with complex matching discovery, we compute the similarity among XML schemas semantic information. Our experimental results demonstrate the performance benefits of the schema matching framework using *PMRS*. Future research is geared towards efficiently generating candidate queries for similarity evaluations.

Acknowledgements. This research is supported by The National Natural Science Foundation of China under Grant No. 61272159 and No. 61402125. All opinions, findings, conclusions and recommendations in this paper are those of the authors and do not necessarily reflect the views of the funding agencies.

References

1. XML schema part 1: Structures. http://www.w3.org/TR/xmlschema-1
2. De Meo, P., Quattrone, G., Terracina, G., Ursino, D.: Integration of XML schemas at various "severity" levels. Inf. Syst. **31**(6), 397–434 (2006)
3. Hucka, M., et al.: The systems biology markup language (SBML): a medium for representation and exchange of biochemical network models. Bioinformatics **19**(4), 524–531 (2003)
4. Ebxml website. http://www.ebxml.org
5. Murray, P.: Chemical markup language: a simple introduction to structured documents. World Wide Web J. **2**(4), 135–147 (1997)
6. Gml website. http://www.opengis.net/gml/
7. Rahm, E., Bernstein, P.A.: A survey of approaches to automatic schema matching. VLDB J. **10**(4), 334–350 (2001)
8. Do, H.H., Rahm, E.: COMA - a system for flexible combination of schema matching approaches. In: Proceedings of 28th International Conference on Very Large Data Bases. VLDB 2002, 20–23 August 2002, Hong Kong, China, pp. 610–621 (2002)
9. Aumueller, D., Do, H.H., Massmann, S., Rahm, E.: Schema and ontology matching with COMA++. In: Proceedings of the ACM SIGMOD International Conference on Management of Data, Baltimore, Maryland, USA, 14–16 June 2005, pp. 906–908 (2005)
10. Aberer, K., Franklin, M.J., Nishio, S. (eds.). Proceedings of the 21st International Conference on Data Engineering, ICDE 2005, 5–8 April 2005, Tokyo, Japan. IEEE Computer Society (2005)
11. Seligman, L., Mork, P., Halevy, A.Y., Smith, K.P., Carey, M.J., Chen, K., Wolf, C., Madhavan, J., Kannan, A., Burdick, D.: Openii: an open source information integration toolkit. In: Proceedings of the ACM SIGMOD International Conference on Management of Data. SIGMOD 2010, Indianapolis, Indiana, USA, 6–10 June 2010, pp. 1057–1060 (2010)
12. Saha, B., Stanoi, I., Clarkson, K.L.: Schema covering: a step towards enabling reuse in information integration. In: Proceedings of the 26th International Conference on Data Engineering. ICDE 2010, 1–6 March 2010, Long Beach, California, USA, pp. 285–296 (2010)

13. Levenshtein, V.: Binary codes capable of correcting deletions, insertions, and reversals. In: Soviet physics doklady, pp. 707–710 (1966)
14. Marzal, A., Vidal, E.: Computation of normalized edit distance and applications. IEEE Trans. Pattern Anal. Mach. Intell. **15**(9), 926–932 (1993)
15. Navarro, G.: A guided tour to approximate string matching. ACM Comput. Surv. **33**(1), 31–88 (2001)
16. Cohen, W.W., Ravikumar, P.D., Fienberg, S.E.: A comparison of string distance metrics for name-matching tasks. In: Proceedings of IJCAI-03 Workshop on Information Integration on the Web (IIWeb 2003), 9–10 August 2003, Acapulco, Mexico, pp. 73–78 (2003)
17. Formica, A.: Similarity of XML-schema elements: a structural and information content approach. Comput. J. **51**(2), 240–254 (2008)
18. Joachims, T.: A probabilistic analysis of the rocchio algorithm with TFIDF for text categorization. In: Proceedings of the Fourteenth International Conference on Machine Learning (ICML 1997), Nashville, Tennessee, USA, 8–12 July 1997, pp. 143–151 (1997)

A Distributed Load Balance Algorithm of MapReduce for Data Quality Detection

Yitong Gao$^{(\boxtimes)}$, Yan Zhang, Hongzhi Wang, Jianzhong Li, and Hong Gao

School of Computer Science and Technology,
Harbin Institute of Technology, Harbin, China
gaoyitonghit@163.com, {zhangy,wangzh,lijzh,honggao}@hit.edu.cn

Abstract. Big data quality detection is a valuable problem in data quality field. MapReduce is an important distributed data processing model mainly for big data processing. Load balance is a key factor that influences the property of MapReduce. In this paper, we propose a distributed greedy approximation algorithm for load balance problem in MapReduce for data quality detection. There are three key challenges: (a) reduce the problem to NP-complete and prove a considerable approximation ratio of the proposed algorithm, (b) just impose one more round of MapReduce than conventional processing and occupy minimal time in the total process, (c) be simple and convenient feasible. Experimental results on real-life and synthetic data demonstrate that the proposed algorithm in this paper is effective for load balance.

Keywords: Load balance · Mapreduce · Data quality detection · Distributed approximation greedy algorithm

1 Introduction

In the contemporary era, a flood of data strongly flows out in various ways every day. The analysis and processing of big data in all trades is a particularly significant problem. Data quality detection plays an important role in big data analysis and processing and faces challenges. MapReduce is an important distributed programing model which is transparent for programmers to create user-defined map and reduce operations. Big data quality detection based on MapReduce is an effectively considerable method. The input data to be detected quality on MapReduce has an admirable characteristic. This kind of data can be conveniently split to some independent task process units and the computing load of each task process unit is foreseeable. For example, the number of each entitys tuples is foreseeable in the entity currency and consistency problems [16,17,19,20] which are two of data quality detection problems and the process of every entity is independent. MapReduce is expert in dealing with this kind of data with the above characteristic. However, load balance is a key factor that influences the property of MapReduce when dealing with big data quality detection problem. Load unbalance mainly includes two types: map phase

© Springer International Publishing Switzerland 2016
H. Gao et al. (Eds.): DASFAA 2016 Workshops, LNCS 9645, pp. 294–306, 2016.
DOI: 10.1007/978-3-319-32055-7_24

unbalance and reduce phase unbalance. Map phase unbalance mainly results from the uneven user data, which means some parts of the datas processing time are far more than others. This nodes heavy workload slows down entire processing time in the cluster. Reduce phase unbalance mainly results from the improper partition methods, which distribute the middle results unevenly to the reduce nodes. The reduce phase load unbalance problem reflects obviously in data quality detection of MapReduce methods.

Example 1. Consider the workload distribution on different reduce task nodes shown in Fig. 1. It shows the case that when we detect the currency [16,17] of input data on MapReduce. We can see that Recude1 and Reduce4 deal with the great mass of workload: 5059 and 6000 tuples respectively; however, Reduce2 and Reduce3 are relatively idle: 530 and 710 tuples respectively. The entire processing time of MapReduce is determined by Reduce1:6000 tuples workload processing time. If the workload of reduce task nodes is distributed evenly, that is to say, each reduce task node has the homogeneous workload partition: $(5059 + 530 + 710 + 6000)/4 = 3075$, the total processing time is 3075 tuples workload processing time. It doubles the speed of load unbalance process.

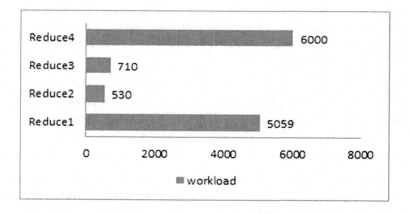

Fig. 1. Reduce phase unbalance

Example 1 illustrates the problem. Load unbalanced cases can delay seriously the entire running time of MapReduce, but load balanced cases dramatically improve the performance of MapReduce. In this paper, we focus on the load balance problem of the reduce phase in MapReduce for the data to be detected quality or the data with similar characteristics. For instance, in the data of massive matrix computations, each matrix calculation is independent and the calculation is predictable. MapReduce includes two main partition functions which are HashPartitioner and TotalOrderPartitioner. TotalOrderPartitioner majors in the total sort, HashPartitioner is widely used in balancing workload, but it works well only if the key is linear. However, most data are not even-distributed, which results in data skew by using HashPartitioner. Many researchers try to ameliorate the Partitioner or even to modify the kernel of MapReduce for fitting every kind of data distribution, but sometimes it is not a relatively convenient approach to solve the load balance problem in reduce phase. In this paper, we propose a distributed load balance algorithm by fitting the default HashPartitioner. The algorithm only appends a round of MapReduce to preprocess the user data depending on the workload of each node. In the second round of MapReduce, we apply our algorithm to redistribute the user data to reducers evenly and the reducers can process almost the same workload. It is just one more round of MapReduce than the conventional processing.

1.1   Challenges

Note that the problem which redistributes the input data and assign them to reducers evenly is not solvable in polynomial time, it is NP-complete (as proven are in Sect. 2). In light of this, we have to relax the condition which means that the difference between the workload of each reducer minimize to the extent that we can accept, and design a simple but feasible approximation algorithm with a relatively high approximation ratio. If the algorithm is too complicated to perform, the cost of achieving load balance will be unacceptable. Moveover, it is important to understand the advantages and disadvantages about the HashPartitioner if we expect to design an approach of redistributing the data to adapt for the HashPartitioner. From (1) in Sect. 2.1, we can know that only if k2 is linear-continuous, HashPartitioner will more evenly assign the input data to different reducers. If not, HashPartitioner may not work well. We redistribute the data and generate linear-continuous keys in the second round of MapReduce to make HashPartitioner work well.

1.2   Our Contributions

We study that our problem which distributes the uneven user data to reducers evenly can be reduced to K-Partition Problem [4] which is NP-complete. We propose a distributed greedy approximation algorithm with assurance for load balance in MapReduce by means of extending the greedy approximation algorithm of Partition Problem [3]. It firstly sorts the input data totally in descending order according to the every independent workload in the first round of MapReduce, and then puts every task in the input data to the current reducer with minimum workload in turn in the second round of MapReduce. In our experimental study, we have shown that the balance results are superior to the unbalance results, and the balance MapReduce is more effective than the unbalance MapReduce. Furthermore, we prove our idea has simplicity and feasibility.

1.3   Paper Outline

The remaining parts of this paper are organized as follows. We introduce MapReduce fundamental and our problem definition in Sect. 2. In Sect. 3, we propose the distributed load balance algorithm and provide the approximation ratio bound of the algorithm. We conduct the experimental evaluation in Sect. 4. Related works are in Sect. 5. Conclusions are in Sect. 6.

2   Background

2.1   MapReduce Fundamental

MapReduce is a distributed programming model. It mainly includes two phases, which are mapper and reducer. Mapper executes map function and reducer with reduce function separately. In the map and reduce function, programmers can create user-defined operations to process input data. The data in map function and reduce function is expressed through key-value pairs.

$$Map : \langle k_1, v_1 \rangle \rightarrow list \langle k_2, v_2 \rangle$$

$$HashPartitioner : (k_2.hashCode() \& Integer.MAX\_VALUE)) \% numReduceTasks \tag{1}$$

$$Reduce : \langle k_2, list(v_1) \rangle \rightarrow list \langle k_3, v_3 \rangle$$

Figure 2 simply shows how MapReduce works. The input data are divided into several splits and each input split corresponds to one map task. Each map task generates a map output which will be divides into several partitions by default HashPartitioner. Since output data will be assigned to reduce tasks as uniformly as possible, the number of partitions is equal to the number of reducers. The data in each partition is sorted by key, and the data with the same key will be transferred into the same reduce task processing. The middle process shuffle is the heart of MapReduce where the miracle happens. In the shuffle process, reduce tasks get the map outputs through HTTP. The map tasks may finish at different times, so the reduce task starts copying their outputs as soon as each completes. [1] During this time until all the map outputs have been copied,

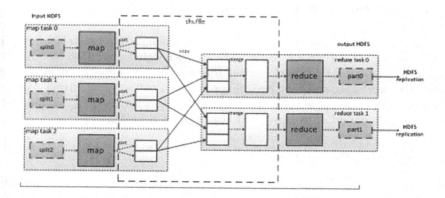

Fig. 2. How MapReduce works [1]

reduce task sorts the data by key circularly and merges it with the merge factor. After that, the sorted and merged data will be processed by reduce function who generates the final results.

2.2   Problem Definition

Before we describe our problem, we define our problem and explain several related content. We preprocess the data which will be processed by our algorithm. Now we define the concepts and the problems studied in this paper.

Definition 1. (Input Data) *The data have been preprocessed before performing our algorithm. It is arranged to many task sets with calculable workload which can be absolutely processed in parallel, every task includes task ID, task workload and task content.*

Definition 2. (Independent Tasks) *The tasks sets with calculable workload in the Input Data. They can be independently processed in parallel.*

Definition 3. (Redistributing Input Data Problem (RIDP)) *To redistribute the input data to make Independent Tasks can be distributed evenly to r reducers (r is the number of the reducers) for fitting HashPartitioner.*

Definition 4. (K-Partition Problem (KPP) [4]) *The enhanced version of Partition Problem (PP), and the simplest PP is 2-Partition Problem [3].*

Note that the problem is that assigning the independent tasks with different work-load to finite reducers evenly, the essence of RIDP is to distribute a big set with a huge of different values to several finite small sets satisfying the condition that the sum of every set is almost equal. Clearly, it can be reduced to the K-Partition Problem. In [2–4], we know that it is NP-complete. Thus we attempt to solve this problem by the approximation algorithm.

3   The Distributed Load Balance Algorithm

In this section, we propose the Distributed Load Balance Algorithm (DLBA) for RIDP. The algorithm extends the greedy approximation algorithm of PP. The mean idea of DLBA is that firstly sorting the tasks in descending order according to the tasks workload, and then assigning the sorted tasks to the current reducer with minimum workload in turn. We describe the algorithm in Sect. 3.1 and then prove the ratio bound of the proposed algorithm in Sect. 3.2.

3.1   Algorithm Description

The distributed load balance algorithm mainly includes two rounds of MapReduce. Algorithm 1 explains the first round of MapReduce. In the first round of MapReduce, we preprocess the input data, which is sorting the independent tasks

---

Algorithm 1. Algorithm to compute the tasks workload in the input data and totally sort them in descending order

---

Mapper: Input: $\langle key_1, value_1 \rangle$ $=<$ $rowpartialorder, taskcontent(taskID +$ $taskWorkload + taskContent) >$

Output: $list \langle key_2, value_2 \rangle = list < -taskWorkload, taskID + taskWorkload +$ $taskContent >$

Reducer:

Input: $list \langle key_2, value_2 \rangle = list < -taskWorkload, list(taskID + taskWorkload +$ $taskContent) >$

Output: $list \langle key_3, value_3 \rangle = list < taskWorkload, taskID + taskWorkload +$ $taskContent >$

1: //Mapper:
2: Obtain each taskWorkload in $value_1$;
3: Print$\langle key_2, value_2 \rangle = list < -taskWorkload, value_1 >$
4: //Reducer:
5: **for** $\forall p \in (-taskWorkload)$ **do**
6:   **if** list(taskID+taskWorkload+taskContent).hasNext() **then**
7:     print$< taskWorkload, list(taskID+taskWorkload+taskContent).Next() >$;

---

in the input data totally on the basis of the each task workload in descending order. It aims to optimize the assignment of the second round of MapReduce. Besides, the technique of total sort which sorts the whole user data can be seen in [1].

Since the sort function is an inherent character of MapReduce, map and reduce functions have no need to sort, and we set TotalOrderPartitioner to achieve total sort. Map function only needs to obtain every workload and make the negative value of workload be the output key (Line 2, 3), because we expect to sort the tasks in descending order. The keys of reduce inputs are sorted negative taskWorkloads, so reduce function is just required to print every taskWorkload and corresponding list of task information in turn, and the results are the taskWorkloads with descending order (Line 5, 6, 7). Algorithm 2 describes the second round of MapReduce. The second round of MapReduce mainly redistributes the sorted independent tasks to the current reducer with minimum workload for balancing the workload in reducers. As we know, the tasks with the same key will be assigned to the same reducer, so we design some partitionIDs which can identify each reducer simply, the number of partitionIDs is equal to the number of reducers. In light of this, we abstract the reducers to simple and unique partitionIDs, and we just need to assign the current task to reducer partitionID with the minimun workload.

In Algorithm 2, every mapper in the second round of MapReduce maintains a min-heap H. Map function gets the reducer partitionID with minimun workload in H.top() every time (Line 2), and assigns the current task to this reducer partitionID (Line 3). Since H has been modified, it destroys the character of the min-heap, and H.meapify function recovers the character to make sure of its next use (Line 4). The last step of map is to print key-value pairs formating that key

Algorithm 2. Algorithm to redistribute the tasks distribution to balance the reducers workload

Mapper: Input: $\langle key_1, value_1 \rangle$ $=<$ $taskWorkload, taskID + taskWorkload +$ $taskContent >$
Output: $list \langle key_2, value_2 \rangle$ $=$ $list$ $<$ $partitionID, taskID + taskWorkload +$ $taskContent >$
Reducer: Input: $list \langle key_2, value_2 \rangle$ $=$ $list$ $<$ $partitionID, list(taskID +$ $taskWorkload + taskContent) >$
Output: $list \langle key_3, value_3 \rangle$ $=$ $list$ $<$ $partitionID, partitionWorkload$ $>$

```
 1: //Mapper:
 2: partitionID = H.top(); //H is a min heap
 3: H.get(partitionID) += taskWorkload;
 4: H.meapify(H, partitionID);
 5: Print¡partitionID, taskID+taskWorkload+taskContent¿
 6: //Ruducer:
 7: totalPartitionLoad = 0;
 8: for list(taskID+ taskWorkload+taskContent).hasNext() do
 9:    totalPartitionLoad += taskWorkload;
10:    Implement the task and save task execution results;
11: Print< partitionID, totalPartitionLoad >;
```

is the reducer partitionID and value is the task which is assigned to the reducer corresponding to this partitionID (Line 5). After the map function, the main work of the reducer is to collect the sum of workload in each partitionID (Line 7, 8, 9), and implement every task (Line 10). Finally reducer print key-value pairs which are partitionID and the load of the partitionID (Line 11).

Example 2. Suppose there are 7 matrix multiplication tasks with taskIDs 1, 2, 3, 4, 5, 6, 7 in the input data, the matrix dimensions which determine processing times are 2, 14, 4, 16, 5, 3, 6 separately. We process them with 3 reducers. DLBA performs the first round of MapReduce, we will get the results representing as key-values pairs like that:16, 4 + 16 + matrix multiplication 4; 14, 2 + 14 + matrix multiplication 2; 6, 7 + 6 + matrix multiplication 7; 5, 5 + 5 + matrix multiplication 5; 4, 3 + 4 + matrix multiplication 3; 3, 6 + 3 + matrix multiplication 6; 2, 1 + 2 + matrix multiplication 1. Figure 3 shows the results of the second round of MapReduce.

3.2   Approximation Ratio Bound

In this section, we provide the proof of DLBA approximation ratio bound. In Table 1, we declare some symbols in our proof.

In order to facilitate our proof, we firstly illustrate the approximation ratio of List Scheduling Algorithm (LST) is 2 [5]. LST assigns tasks in the input data to the current reducer with minimum workload in turn but without sorting the

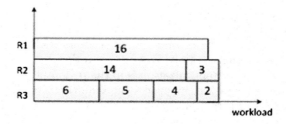

Fig. 3. The second round of MapReduce in DLBA

Table 1. Symbol table

Symbols	Explanation
n	The number of tasks in the input data
r	The number of reducers
p_j	The processing requirement time (the workload) of task j
C_{max}	If the task j complete at the time C_j, $C_{max} = max_{j=1,2...,n} C_j$
C_{max}^*	The optimal processing requirement time of all the tasks in the input data
S_l	The start processing time of the last task l, i.e. $C_l - p_l$

tasks in descending order. The proof of LST is shown in [5] Theorem 2.7. Note that the approximation ratio of LST is not the tight one. We know that different task sequences will generate different schedules, there must exists the best one in all of the schedules, we try to the better one (maybe it is not the best one). Thus we make a fine-tuning about LST with adding an auxiliary condition which is sorting the tasks in the light of tasks workload in descending order. This process is called longest processing time rule (LPT) in [5] in Sect. 2.3, it provides the approximation ratio of LPT is $\frac{3}{4}$. We know that in our algorithm, a reducer will process a lot of tasks, we suppose the reducer with the lightest workload processes k tasks. Theorem 1 demonstrates the approximation ratio bound of our algorithm.

Theorem 1. *if a reducer processes k tasks at least, the approximation ratio of DLBA is* $\frac{k+1}{k}$.

Proof. The proof of the Theorem 1 is similar to the proof of Theorem 2.7 in [5]. Without loss of generality, suppose that $p_1 \geq p_2 \geq \cdots \geq p_n$. Suppose that the theorem is not true, i.e. there exists a counter-example that the tasks list is not in the descending order about the tasks workload but the schedule of the tasks list is the better one, that is to say, the approximation ratio of the schedule and the optimization is in the region $\left(1, \frac{k+1}{k}\right]$. Suppose that the last processed task is task l in the counter-example, its workload (processing requiring time) is not the smallest. Now we firstly take out tasks $l + 1, ..., n$ in the counter-example, then we get a schedule of the shorter tasks list corresponding to a completed time C'_{max}, which is definitely smaller than C_{max} (the completed time of LTP schedule). Here we put the sorted task $l + 1, ..., n$ into the taken-out schedule by LST for getting a new schedule, corresponding to the completed time C''_{max} with a fact $C''_{max} \leq C_{max}$. Consequently, there is no such a counter-example that tasks list without the workload descending order, but the approximation ratio of the schedule and the optimization is in the region $\left(1, \frac{k+1}{k}\right]$. Suppose that the last processed task is task n, its processing requiring time is $p_n (= p_l)$. Since a reduce node processes k tasks at least, so $p_l \leq C^*_{max}/k$ (if $p_l > C^*_{max}/k$, the optimally completed time will go beyond C^*_{max}). Thus:

$$C_{max} = S_l + p_l \leq \sum_{j \neq l} p_j/r + p_l = \left(1 - \frac{1}{r}\right) p_l + \sum_{j=1}^{n} p_j/r \leq \left(1 - \frac{1}{r}\right) \quad (2)$$

So, we get the following formula:

$$\frac{C^*_{max}}{k} + C_{max} = \left(\frac{k+1}{k} - \frac{1}{kr}\right) C^*_{max} \leq \frac{k+1}{k} C^*_{max} \quad (3)$$

Based on the inequality $(2, 3)$, we know the completed time of tasks by the DLBA schedule is at most $\frac{k+1}{k} C^*_{max}$. Thus the approximation ratio of DLBA is $\frac{k+1}{k}$. $\square$

Example 3. If the tasks in Example 3.1 are processed by LST (a counter-example schedule), we can get the results shown in Fig. 4. $C_{max-ce} = 2 + 16 = 4 + 5 + 3 + 6 = 18$ (C_{max-ce} is the completed time of LST schedule). The last task is 6 task, so taking out 5,4,3,2 tasks. The remaining tasks painting white are shown in Fig. 4.

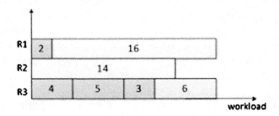

Fig. 4. Scheduling by LST

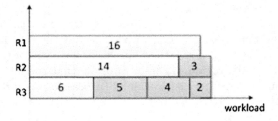

Fig. 5. The new schedule

According to Fig. 5 about the tasks painting white, we draw the conclusion that $C'_{max} = 16$, $C''_{max} = 17$. If we process the tasks using DLBA, we will get the schedule about all the tasks shown in Fig. 5. We get $C_{max} = 17$. Thus $C'_{max} \leq C_{max}$, $C''_{max} \leq C_{max}$.

4   Experimental Evaluation

We use real-life data to evaluate the effectiveness of the balance reducer workload in comparison with the unbalance results. Because we cannot get many real-life data sets with regular size changes, so we generate some synthetic data sets with monotone increasing data sizes to evaluate the advantage of load balance MapReduce in time.

4.1   Experiment Environment

We perform the experiments on a 64-bit Linux System DELL computer, the configuration of the machine is as follows: CPU: Inter(R) Core(TM)2 Quad CPU Q8300 @ 2.50 GHz 2.49 GHz; RAM: 4.00 GB. The machine has been installed the Ubuntu 14.04 and Hadoop-1.2.1 for a pseudo-distributed configuration.

4.2   Experiments on Real Data

We download the experimental real data train_v2.txt (4.1 GB) from the URL http://www.kaggle.com/c/billion-word-imputation/data. We preprocess the data for each line with the pattern of (lineID, wordcount+words). We want to search the most frequent word on each line. The tasks of each line are independent with each other, and the workloads of each line are foreseeable. We do it by DLBA, and the comparison of the balance results using DLBA in parallel and the unbalance results without DLBA is shown in Fig. 6. We can see that the balance results are better than the un-balance results.

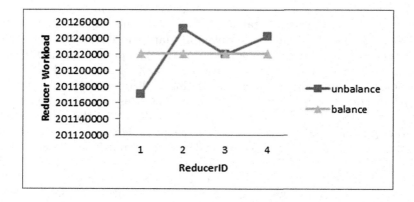

Fig. 6. The comparison of reducer-workloads on the real data

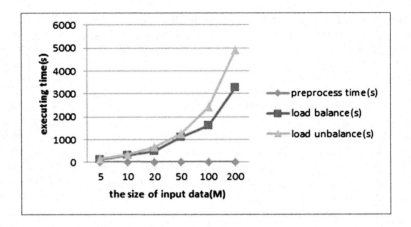

Fig. 7. The efficiency comparison on the synthetic data

4.3   Experiment on Synthetic Data

In this part, we use the synthetic data sets with different sizes to compare
the running time of balance MapReduce and unbalance MapReduce. The synthetic data sets are to complete massive matrix calculations. The data pattern
is (lineID, matrix dimension+two matrixs), and many matrix dimensions differs
widely. The efficiency comparison is shown in Fig. 7, we get that the first round
of MapReduce just occupies a little time in the total process of the balance
MapReduce. Moreover, the size of the data sets grows at twice the last data
set, and the advantage of balance MapReduce in time is more obvious than the
unbalance MapReduce.

5   Related Works

Load balance is an important problem in Distributed System such as distributed
caching, storage, computing, database and so on. In this paper, we concentrate
on the reduce phase load balance problem in MapReduce. Now the dominating
research achievements include initiative prevention and passive solving. Initiative
prevention essentially falls into two categories containing static load balance distributing and dynamic load balance distributing, utilizing users future knowledge
sometimes. The main thought of static load balance is adapting the partition
interface in Hadoop according to the future knowledge about the current job, the
input data property and resources of each node. [6] designs a particular key-value
pairs that map outputs generates a block distribution matrix for BlockSplit, but
it aims at the data of Entity Resolution with no catholicness. In [9], OS4M
(Operation Scheduling for MapReduce) collects statistics of all Map operations,
optimizes distribution of keys, adapts partitioner, and utilizes reduce pipelining
for designing an operation scheduling, but it is not parallel. [12] meliorates the

partitioner to improve the load balance in MapReduce. [13] proposes a hierarchical MapReduce with an idea that dividing heavy tasks into subtasks to balance reduce task nodes according to the cost function of every task, but cost function is defined by the user, the workload for users is tremendous with more and more data. Moreover, some papers involve the sampling idea in TotalOrderPartitioner. More specifically, firstly they sample and analyse the input data, estimate the cost of every block data, and define the stragglers which are blocks beyond the average, then, they transfer the stragglers to the free blocks which are under the average. [8, 11] take advantage of sampling thought. Static load balance distributing is capable of the data skew, but unable to deal with computing skew. [14, 15] discuss dynamic load balance distributing. Dynamic load balance regards the uncompleted running tasks with the existence of free slots (unit of computing resources corresponding to the cpu) as stragglers. It migrates the uncompleted running tasks to free slots by direct method or iterative method. [14] proposes a 2-dimensional hashing to get a better partition. It is mainly for clusterjoin data. [15] focus on the load balance of Hadoop different racks by analyzing its log files. In paper [18], authors implement an approach as a drop-in replacement for an existing MapReduce implementation: SkewTune, it can mitigate map and reduce phases unbalance, and impose minimal overhead. But the idea is too complex for the kind of the data referring in this paper. In this paper, we just use one more round of MapReduce to balance the workload of reduce phase, it is more simple and convenient.

6   Conclusion

We have proposed a distributed greedy approximation algorithm of K-Partition Problem. It is a general load balance approach for the independent tasks data processed on MapReduce. Our experimental evaluation has verified that our approach works well on balancing the reducers workload. To better verify the efficiency of balance results, we compare the running time of balance MapReduce with unbalance MapReduce, and the experiment effects are considerable. In the further work, we are supposed to study a more perfect treatment for the situation that some tasks workload are so heavy than all others workload that the heaviest tasks become a bottleneck of the load balance. Meanwhile, we will improve our approach for dynamic load balance of MapReduce.

References

1. Hadoop, W.T.: The Definitive Guide. O'Reilly Media Inc., Sebastopol (2012)
2. Michael, R.G., David, S.J.: Computers and Intractability: A Guide to the Theory of NP-Completeness. W.H. Freeman and Co., San Francisco (1979)
3. http://en.wikipedia.org/wiki/Partition_problem
4. http://en.wikipedia.org/wiki/3-partition_problem
5. Williamson, D.P., Shmoys, D.B.: The Design of Approximation Algorithms. Cambridge University Press, Cambridge (2011)

6. Kolb, L., Thor, A., Rahm, E.: Block-based load balancing for entity resolution with MapReduce. In: Proceedings of the 20th ACM International Conference on Information, Knowledge Management, pp. 2397–2400. ACM (2011)

7. Kolb, L., Thor, A., Rahm, E.: Load balancing for mapreduce-based entity resolution. In: 2012 IEEE 28th International Conference on Data Engineering (ICDE), pp. 618–629. IEEE (2012)

8. Ramakrishnan, S.R., Swart, G., Urmanov, A.: Balancing reducer skew in MapReduce workloads using progressive sampling. In: Proceedings of the Third ACM Symposium on Cloud Computing, p. 16. ACM (2012)

9. Fan, L., Gao, B., Zhang, F., et al.: OS4M: achieving global load balance of MapReduce workload by scheduling at the operation level (2014). arXiv preprint arXiv:1406.3901

10. Fan, L., Gao, B., Sun, X., et al.: Improving the load balance of mapreduce operations based on the key distribution of pairs (2014). arXiv preprint arXiv:1401.0355

11. Xu, Y., Zou, P., Qu, W., et al.: Sampling-based partitioning in MapReduce for skewed data. In: ChinaGrid Annual Conference (ChinaGrid), 2012 Seventh, pp. 1–8. IEEE (2012)

12. Fan, Y., Wu, W., Cao, H., et al.: LBVP: a load balance algorithm based on virtual partition in Hadoop cluster. In: Cloud Computing Congress (APCloudCC), 2012 IEEE Asia Pacific, pp. 37–41. IEEE (2012)

13. Martha, V.S., Zhao, W., Xu, X.: h-MapReduce: a framework for workload balancing in MapReduce. In: 2013 IEEE 27th International Conference on Advanced Information Networking and Applications (AINA), pp. 637–644. IEEE (2013)

14. Sarma, A.D., He, Y., Chaudhuri, S.: Clusterjoin: a similarity joins framework using map-reduce. Proc. VLDB Endowment **7**(12), 1059–1070 (2014)

15. Hou, X., Thomas, J.P., Varadharajan V.: Dynamic workload balancing for Hadoop MapReduce. In: 2014 IEEE Fourth International Conference on Big Data and Cloud Computing (BdCloud), pp. 56–62. IEEE (2014)

16. Fan, W., Geerts, F., Wijsen, J.: Determining the currency of data. ACM Trans. Database Syst. (TODS) **37**(4), 25 (2012)

17. Cao, Y., Fan, W., Yu, W.: Determining the relative accuracy of attributes. In: Proceedings of the 2013 ACM SIGMOD International Conference on Management of Data, pp. 565–576. ACM (2013)

18. Kwon, Y.C., Balazinska, M., Howe, B., et al.: Skewtune: mitigating skew in mapreduce applications. In: Proceedings of the 2012 ACM SIGMOD International Conference on Management of Data, pp. 25–36. ACM (2012)

19. Fan, W., Geerts, F., Tang, N., et al.: Inferring data currency and consistency for conictresolution. In: 2013 IEEE 29th International Conference on Data Engineering (ICDE), pp. 470–481. IEEE (2013)

20. Fan, W., Geerts, F., Tang, N., et al.: Conflict resolution with data currency and consistency. J. Data Inf. Qual. (JDIQ) **5**(1–2), 6 (2014)

A Formal Taxonomy to Improve Data Defect Description

João Marcelo Borovina Josko[1]([✉]), Marcio Katsumi Oikawa[2],
and João Eduardo Ferreira[1]

[1] Institute of Mathematics and Statistics, University of São Paulo,
Sao Paulo, Brazil
{jmbj,jef}@ime.usp.br
[2] Center of Mathematics, Computing and Cognition, Federal University of ABC,
Santo Andre, Brazil
marcio.oikawa@ufabc.edu.br

Abstract. Data quality assessment outcomes are essential for analytical processes, especially for big data environment. Its efficiency and efficacy depends on automated solutions, which are determined by understanding the problem associated with each data defect. Despite the considerable number of works that describe data defects regarding to accuracy, completeness and consistency, there is a significant heterogeneity of terminology, nomenclature, description depth and number of examined defects. To cover this gap, this work reports a taxonomy that organizes data defects according to a three-step methodology. The proposed taxonomy enhances the descriptions and coverage of defects with regard to the related works, and also supports certain requirements of data quality assessment, including the design of semi-supervised solutions to data defect detection.

Keywords: Data defects · Dirty data · Formal taxonomy · Data quality assessment · Relational database · Big data

1   Introduction

The effects of poor data quality on the reliability of the outcomes of analytical processes are notorious, especially for big data environment. Improving data quality requires alternatives that combine procedures, methods and techniques. The Data Quality Assessment process (DQAp) provides practical inputs for choosing the most suitable alternative through its mapping of data defects. To provide a reliable outcome, this process requires *know about* data defect structures to *know how* to assess them.

Data defects descriptions provide structural understanding of the problem associated with each defect. Descriptions are relevant for different DQAp issues such as the rule definition of certain computational approaches of data assessment and the establishment of measurement-based data quality program.

© Springer International Publishing Switzerland 2016
H. Gao et al. (Eds.): DASFAA 2016 Workshops, LNCS 9645, pp. 307–320, 2016.
DOI: 10.1007/978-3-319-32055-7_25

Much literature has described data defects through hierarchical [14], formal [13] or formal-textual-example [5] models. However, the analysis of this literature shows remarkable differences in terminology, nomenclature, coverage, granularity of description, and the description model used (as also mentioned by [8]). This poor organization and description cause uncertainties on which data defects should be assessed, which their structures are, and it also hampers the ability to determine the corresponding detection approaches.

To address this situation, this work reports a taxonomy of data defects related to the *accuracy, completeness* and *consistency* quality dimensions. The taxonomy is characterized and categorized according to a three-step methodology and its main contribution is a major coverage of data defects and enhanced descriptions in terms of terminology, examples and mathematical formalism.

The work reported here is organized as follows: Sect. 2 reviews all related works. Section 3 presents the methodology applied to the development of the proposed taxonomy, while Sect. 4 describes the taxonomy and its basic concepts. Section 5 presents the conclusions of this work.

2   Related Works

In literature, works that describe data defects are common in certain research areas, including Information and Data Quality Management, Data Mining and Statistics. Here, these works are analysed according to how the following questions are answered: *What is the representative set of defects related to the quality dimensions of accuracy, completeness and consistency? What is the problem structure behind each defect?*

Certain works describe data defects as a complementary topic for their main subject of interest. This approach by [3,4,15] leads to ambiguous structure descriptions of the data defects. Moreover, the data defects representativeness is assigned by common sense within a context.

In contrast, the data defects issues are relevant for Data Profiling [12], Statistics [16] and Data Cleaning [5] works. Such works expose the data defect structure through the combination of textual, instance-based examples and formal resources. However, they cover few defects.

Lastly, taxonomies intend to provide reasonable descriptions aligned to a broad coverage and classification of data defects. However, a review of state-of-art taxonomies [1,2,7,9,10,13,14] reveals heterogeneous descriptions and coverage. This situation is caused by the *degree of accuracy afforded by the defect definition, terminology* and *absence of theoretical support on defect selection*.

The *defect description model* determines the degree of accuracy afforded by the defect definition. Except by [13], the taxonomies use an informal or example-driven description model which require considerable interpretations when considered from a technical perspective.

Regarding *terminology* and *nomenclature*, the taxonomies use distinct terms to well known database jargon defects. For instance, "Domain format errors" by [10] and "Wrong data type" by [7,9] are the different terms applied to the same database jargon of "Domain Constraint Violation".

The *shared absence of theoretical support* to identify the set of data defects and the lack of concern with extending the descriptions contribute to an incomplete coverage of data defects. For instance, despite the sequence of citation between [13,14] and [9], data defects as "Embedded values" by [14] do not appear in [9,13]. Moreover, defects regarding data modelling rules (e.g., Cardinality Ratio) and data life cycle failures (e.g., Missing References, False Tuple) are not addressed by any taxonomy. Further examples of data defect heterogeneities are also mentioned by [8].

3 Methodological Approach to Organize the Taxonomy

The taxonomy proposed to address the limitations and the questions in Sect. 2 resulted from applying three steps in sequence. The first step *re-examined a broad set of topics related to relational theory*. Among the topics, but no restricted to them, it can be mentioned conceptual data modelling, transformation decisions between conceptual and logical models, and constraints [5,6,11].

These topics revealed a broad rule set that may be applied on a relational schema to represent properties and behavior of the Universe of Discourse (UoD). Each rule was basis to identify one or more violations (data defects) that leads data to defective states. Furthermore, the review also determined the terminology, nomenclature and the description model.

The second step *classified the data defects in layers according to their shared properties*. The first layer determined whether or not data defects violate rules about the UoD, named respectively as "Data Constraint" and "Fact Representation Deviation". "Data Constraint" gathers data defects that violate static or dynamic rules. The former denotes explicit rules or inherent characteristics of relational model (implicit rules) that a valid state of a relation must satisfy, including data domains, integrity and participation in data relationships. In contrast, the latter comprises rules applied during state transitions of a relation. "Fact Representation Deviation" denotes defects related to differences between data representation and the corresponding fact about the objects of the UoD, including meaning, content and element of representation.

The third step *classified each data defect based on its place or granularity of occurrence*, which are attribute value (V), single attribute (A), single tuple (T), single relation (R) or interrelation (IR), which may involve one or more database instances. The outcome of this three-step methodology is observed in Fig. 1. This figure provides an effective arrangement to identify data defects and comprehend their properties (denoted by the class hierarchy) and interrelationships. Moreover, this arrangement is basis to incorporate additional data defects to the taxonomy, such as the time-related ones.

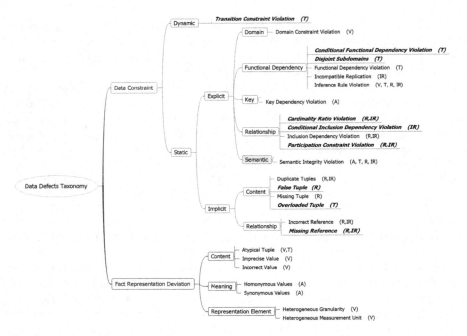

Fig. 1. Taxonomy of data defects (*Granularity:* V-Attribute Value, A-Single Attribute, T-Single Tuple, R-Single Relation, IR-Interrelations. *Note:* Data Defects in *italic*, **bold** and <u>underline</u> were not addressed by any of the state-of-art taxonomies.)

4 Data Defect Taxonomy

4.1 Structural Background

This work applies a formalization language well known by database community based on [5,6,11], which the main elements are observed in Table 1. It is beyond the scope of this work to evaluate the most proper language for this goal.

Moreover, each defect is illustrated by examples selected from a simple financial domain, as shown in the logical model below.

> *Customer* (<u>CID</u>, Name, Job, Salary, State, City, Zip, Age, Ms, SpouID)
> SpouID references Customer
> *CreditCardAccount* (<u>CAccID</u>, ActivationDate, UsageTime, IsFreeOffer)
> *CustomerCreditAccount* (<u>CCAID</u>, CAccID, CID, Score, IsHolder, State)
> CAccID references CreditCardAccount
> CID references Customer

In this model, *Customer* has certain properties of Natural People and Legal Entities in regard to owners of credit cards. *CreditCardAccount* denotes the properties of acquired credit cards. *CustomerCreditAccount* represents all of the relationship roles (holder or joint holder) between the customers and the credit cards. An instance I_0 of each logical relation is observed in Tables 2, 3 and 4.

Table 1. Main elements of the formalization language

Relational	Relational Database Schema	Set of relations schemas $BD = \{R_1, R_2, R_3..., R_m\}$, $m \geq 1$
	Relation Schema	Set of attributes $A = \{a_1,..., a_k\}$ denoted by $R(A)$, $R \in DB$
	Attribute	Each a_j, $j = [1, k]$, is regulated by a domain D_j, given as $dom(a_j)$
	Subset of a Relation	Set of attributes $X, Y \subset R(A)$, where $R \in DB$ and $X \cap Y = \varnothing$
	State of Relation	Set of n tuples, $r = \{t_1, t_2, t_3..., t_n\}$, denoted by $r(R)$
	Tuple	Each tuple t_p, $p \in [1, n]$, is a list of q values $t_p = \{v_1, v_2, ..., v_q\}$
	Tuple Value	Each value v_s, $s \in [1, q]$, is a domain element of the corresponding attribute a_s, denoted as $t[a_s]$
	Relationship	Referential integrity rule between relations W (refer to) and U (referred), denoted by $Rel : R_W \rightarrow R_U$
	Universal Thesaurus	Lexical definitions, relationships and similarity degrees of terms in common usage, denote by LEX.
Operational	Value Predicate Symbols	$\ominus = \{<, \leq, =, \neq, \geq, >\}$
	Set Elements and Operators	$Q = \{\in, \notin, \subseteq, \subset, \cup, \cap\}$
	Logical Connectives	$\{\wedge, \vee\}$ of type $Boolean \times Boolean \rightarrow Boolean$
	Unary Connective	$\{\neg\}$ of type $Boolean \rightarrow Boolean$
	Quantifiers	$\{\forall, \exists\}$ are the universal and existential quantifiers

Table 2. An instance of customer

	CID	Name	Job	Salary	State	City	Zip	Age	Ms	SpouID
c1:	1	John Taylor	Bassist	20k	SP	SP	08000	52	E	19
c2:	3	Joan Ripley	Tapster	320k	BHZ	BHZ	03000	20	M	40
c3:	8	John T	Bartender	20k	MG	BHZ	08200	52	W	NULL
c4:	13	Ann P. Taylor	Barkeeper	NULL	MG	BHZ	31000	44	U	1
c5:	19	Chris Taylor	NULL	8k	SP	SP	08100	39	J	28
c6:	28	Carl de la Poll	Student	21k	SP	SP	08400	34	M	13
c7:	29	James Bond	Bassist	22k	SP	SJC	08000	53	W	NULL
c8:	40	Alice Bond	Principle Manager	1k	SP	SP	08501	53	E	49
c9:	41	John N. T	Principal Manager	40k	MG	BHZ	03099	17	Y	NULL
c10:	3	Ann P. Taylor	Writer	38k	MG	BHZ	03100	44	J	1
c11:	52	Jean P. Jones	Student	33k	SP	SJC	08400	15	S	NULL
c12:	53	Dick Rhodes	Writer	35k	SP	SJC	12200	45	W	NULL

Table 3. An instance of CreditCardAccount

	CAccID	ActivationDate	UsageTime	IsFreeOffer
cr1:	100	07/30/2001	13	No
cr2:	155	01/19/2004	10	No
cr3:	199	05/12/2005	9	Yes
cr4:	200	01/19/2004	1	No
cr5:	201	04/11/2013	1	No

Table 4. An instance of CustomerCreditAccount

	CCAID	CAccID	CID	Score	IsHolder	State
cc1:	120	100	1	2.12307	Yes	SP
cc2:	312	100	13	3.00999	No	MG
cc3:	138	100	19	1.80500	No	SP
cc4:	813	100	3	3.10999	Yes	MG
cc5:	883	155	28	2.11001	Yes	SP
cc6:	901	199	44	3.89099	Yes	SP
cc7:	902	200	40	2.12320	Yes	MG
cc8:	903	201	52	1.83449	Yes	MG
cc9:	909	201	41	1.80011	No	MG
cc10:	911	100	3	19.13329	No	SP

4.2 Taxonomy Description

This section formally describes the higher granularity of each data defect, as observed in Fig. 1. For each definition, there is an example based on the financial logical model aforementioned. The taxonomy has a subset of defect that requires deep knowledge of data context to determine its occurrence, i.e., requires high human curation. Therefore, this subset's formalization applies elements that denote a specialist knowledge.

Definition 1 (Atypical Tuple). Let $outl : R(A) \rightarrow \{true, false\}$ be a function that maps an attribute from relation R to a statistical result of outlier detection methods. An atypical value occurs *iff* $\exists t \in R, \exists a \in R(A)$ such that $outl(t[a])$ is *true*.

An atypical tuple deviates from the common behavior of most tuples within a relation. The unusual composition of attributes values or the unusual value of an isolated attribute are instances of atypical tuple.

Example: A Customer tuple $c2$ reveals an uncommon situation due to the composition of its Salary, Job and Age values.

Definition 2 (Cardinality Ratio Violation). Let $cr : R_1 \times R_2 \rightarrow \mathbb{N}$ be a function that maps the cardinality association of $Rel : R_1 \rightarrow R_2$. Let $s : r(R_1) \times r(R_2) \rightarrow \mathbb{N}$ be a function defined as follows:

$$s(a, b) = \begin{cases} 1 & \text{if } a = b, \\ 0 & \text{if } a \neq b. \end{cases}$$

The cardinality violation occurs *iff* $\exists t_i \in R_1, t_j \in R_2$ such that $\sum s(t_i[W], t_j[U]) > cr(R_1, R_2)$.

The cardinality ratio establishes the maximum number of relationship instances so that each tuple from a relation can participate within a binary relationship. A violation occurs when any tuple does not comply with the maximum as the *referred* or *refer to* role.

Example: In regard to the self-relationship of the "marriage" role (SpouID), each customer must be referred to at most once. However, the customer on tuple $c1$ is referred twice (tuples $c4, c10$).

Definition 3 (Conditional Functional Dependency Violation). Let Tp be a pattern tableau with attributes in X_1 and Y_1, where for each attribute $a \in X_1 \cup Y_1$ and for each pattern tuple $tp \in Tp$, $tp[a]$ is either a particular value in dom(a) or a "wildcard" symbol "_" that draws values from dom(a). Let Y_1 be conditionally dependent on X_1 defined as $(X_1 \to Y_1, Tp)$ on relation R. This conditional dependency is violated *iff* $\exists t_i, t_j \in r(R)$, $i \neq j$, such that $t_i[X_1] = t_j[X_1] = tp[X_1]$ and $t_i[Y_1] = t_j[Y_1] \neq tp[Y_1]$ or $t_i[Y_1] \neq t_j[Y_1]$.

A conditional functional dependency $(A \to B, Tp)$ on a relation R denotes that B values depend functionally on A values only for the tuples of R that satisfies a data pattern specified at Tp. A violation arises when this constraint is not obeyed by a some B value.

Example: The Customer relation has a conditional functional dependency between State, City and Zipcode, denoted by $[State, City \to ZipCode, Tp]$. Pattern Tableau Tp specifies that state "SP" and city "SP" uses the zipcode range between 08000 and 08499, while the same state and city "SJC" has a zipcode range from 12200 to 12248. However, the tuples $c7$ and $c11$ violate these patterns.

Definition 4 (Conditional Inclusion Dependency Violation). Let Tp be a pattern tableau with attributes in X_p and Y_p, where for each attribute $a \in X_p \cup Y_p$ and for each pattern tuple $tp \in Tp$, $tp[a]$ is a particular value in dom(a). Let R_1 be conditionally dependent on R_2, as represented by $(R_1[X; X_p] \subseteq R_2[Y; Y_p], Tp)$. This dependency is violated *iff* $\exists t_i \in r(R_1), \exists t_j \in r(R_2)$ such that $t_i[X_p] = tp[X_p]$ and $t_j[Y_p] = tp[Y_p]$ and $t_i[X] \neq t_j[Y]$.

The OR exclusive constraint specifies that relationships set to a root relation must be disjunct. A violation of this constraint arises when there are root relation tuples that participate in two or more mutually exclusive relationships.

Example: An disjoint constraint prohibits that holder and joint holder roles be exerted by the same customer. However, there are a customer $(c2)$ with both roles on the same card $(cc4$ and $cc10)$.

Definition 5 (Disjoint Subdomains). The problem of disjoint subdomains holds when exists subdomains $S_1, S_2, ..., S_n$ for an attribute a_j such that $dom(a_j) = \bigcup_{i=1}^{n} S_i$, and exists a function $f : a_i \to \{S_1, S_2, \ldots, S_n\}$ that maps a_i to one subdomain of a_j, $i \neq j$ and $a_i, a_j \in R(A)$. It establishes that values of a_j depend on values of a_i.

An attribute has disjoint subdomains (or multiple uses) when its values represent different facts about the objects of the UoD, according to some assignment predicate.

Example: The Salary attribute may represent an adjusted remuneration for providing services (e.g., tuple $c1$) or an estimated family income (e.g., tuples $c6, c11$). The latter occurs when the customer does not have an income source, i.e., Job equal to "Student" or Age lower than sixteen years old.

Definition 6 (Domain Constraint Violation). The domain violation occurs *iff* $\exists t \in R, \exists a \in R(A)$ such that $t[a] \notin dom(a)$.

A domain constraint regulates the allowed values for an attribute domain. In this work, a domain constraint denotes a set of values (such as enumerations), a interval or semi-interval constraint, a mandatory (Not-Null) constraint or a regular expression. A domain constraint violation arises when a value does not match the permissible values of the attribute.

Example 1: The Ms attribute (abbreviation of "Marital Status") of the customer relation has a domain defined as: "M", "E", "J", "U", "D", "W", or "S". However, the tuple $c9$ contains "Y" for this attribute.

Example 2: The Score attribute of the CustomerCreditAccount relation has a domain constraint between -2.99999 and 9.9999. However, the tuple $cc10$ has a score of 19.13329.

Definition 7 (Duplicate Tuples). Let X_1 and X_2 be attribute subsets, where $X_1 \subset R_1$ and $X_2 \subset R_2$. Let X_1 and X_2 be pairwise compatible, where for all $a_1^i \in X_1$ and $a_2^i \in X_2$, $i \in [1, k]$, $k \geq 1$, $dom(a_1^i)$ and $dom(a_2^i)$ are identical. Let $\simeq_i$ be the record matching similarity predicate on attributes $a_1^1 \simeq_1 a_2^1 \wedge ... \wedge a_1^k \simeq_k a_2^k$, denote as $X_1 \simeq X_2$. There are duplicate tuples *iff* $\exists t_1 \in r(R_1), \exists t_2 \in r(R_2)$ such that $t_1[X_1] \simeq t_2[X_2]$.

This defect denotes multiple tuples from one or more relations that refer to the same object in the UoD. The content of these tuples may have identical values, have a certain similarity degree or be mostly divergent.

Example: The Customer tuples $c4$ and $c10$ represent the same object in the UoD with equal values in almost all the attributes.

Definition 8 (False Tuple). Let $ftup : r(R) \rightarrow \{true, false\}$ be a function which returns if a tuple from R complies with the rules that define its usefulness for the UoD. A tuple is false *iff* $\exists t \in r(R)$ such that $ftup(t)$ is *false*.

A database schema must represent only the required objects of an UoD. A tuple is named false when it represents an object *beyond* the UoD interest.

Example: The tuple $c7$ represent a customer who have never had a credit card.

Definition 9 (Functional Dependency Violation). Let Y be functionally dependent on X, denoted by $X \rightarrow Y$, $X, Y \subseteq R(A)$. This dependency is violated *iff* $\exists t_i, t_j \in R$, $i \neq j$, such that $t_i[X] = t_j[X]$ and $t_i[Y] \neq t_j[Y]$.

A functional dependency $A \rightarrow B$ denotes that each B values is associated with precisely one A value. A violation arises when this constraint is not obeyed by some B value.

Example: The Customer relation has the functional dependency $State \rightarrow City$. However, this dependency is violated by the tuple $c2$.

Definition 10 (Heterogeneous Granularity). Let G be the granularity defined for the attribute a from relation R, $a \in R(A)$. Let $grain : a \to \{true, false\}$ be a function which returns if an value of attribute a complies with the granularity G. This defect occurs *iff* $\exists t_i \in r(R)$ such that $grain(t_i[a])$ is *false*.

Granularity denotes the abstraction level of value representation. There is a heterogeneous granularity attribute when some of its values represent facts about objects of the UoD using different abstraction levels. These abstraction levels may have distinct degrees of disparity and also expose a random or attribute-driven pattern.

Example: Salary attribute of Customer relation must represent monthly pay. However, the tuple $c2$ has the annual payment.

Definition 11 (Heterogeneous Measurement Unit). Let Ω be an equivalence relation on the values of attribute a, $a \in R(A)$, such that $v_i \Omega v_j$ iff v_i and v_j have the same measurement unity. Let MU_Ω be an equivalence class on Ω which contains all the values of attribute a which has the measurement unity required by the UoD. Heterogeneous measurement unit holds *iff* $\exists t_i \in r(R)$ such that $t_i[a] \notin MU_\Omega$.

A measurement unit denotes the magnitude of a given physical quantity, which provides useful basis for comparison. This defect occurs when certain values of an attribute represent facts about objects of the UoD using different measurement units. These units denote dissimilar magnitudes and also expose a random or attribute-driven pattern.

Example: Customer relation must represent salaries in American dollars. However, the customers whose job is "Writer" (tuples $c10, c12$) have their salaries represented in Euros.

Definition 12 (Homonymous Values). Let $sp : r(R) \times r(R) \to \{true, false\}$ be a function which returns if the graphy and pronunciation of attributes values are equal, according to LEX. Let $me : r(R) \times r(R) \to \{true, false\}$ be a function which returns if the meaning of attributes values are equal or nearly the same, according to LEX. An attribute has homonymous values *iff* $\exists a \in R(A), \exists t_i, t_j \in r(R), i \neq j$, such that $sp(t_i[a], t_j[a])$ is *true* and $me(t_i[a], t_j[a])$ is *false*.

The homonym denotes terms pronounced in the same way but with distinct meanings. This defect arises when homonymous terms are applied interchangeably and indicate the same fact about objects of the UoD.

Example: The customer tuples $c8$ and $c9$ has different jobs ("Main Manager" and "Principal", respectively) represented with homonymous terms.

Definition 13 (Imprecise Value). Let σ be the imprecise degree allowed to each value v of the attribute a, $a \in R(A)$, such that $v - \sigma \leqslant v \leqslant v + \sigma$. Let $rv : a \to dom(a)$ be a function on relation R which returns the value of attribute a in the UoD. Imprecise value occurs *iff* $\exists t \in r(R)$ such that $|t[a] - rv(a)| > \sigma$.

The term "imprecise" denotes a value that is close to the fact about of an object of the UoD. This closeness denotes an certain accuracy level determined by a range of values within which the accurate value is asserted to be.

Example: The Score attribute of CustomerCreditAccount allows a imprecision degree between $\pm$ 0.00049. The tuple $cc3$ has the score of 1.80500 for an accurate value of 1.80538.

Definition 14 (Inclusion Dependency Violation). Let $Rel : R_1 \rightarrow R_2$ be a relationship set between relations R_1, R_2. There is an inclusion dependency violation *iff* $\exists t_i \in r(R_1)$, $\forall t_j \in r(R_2)$ such that $t_i[W] \neq t_j[U]$.

Inclusion dependency imposes acceptance conditions on actions over instances of relationships to ensure referential integrity consistency. A violation is created when a tuple t_1 refers to tuple t_2, which is not available for the referred relation.

Example: The CustomerCreditAccount tuple $cc6$ refers to a customer who is absent within the Customer relation.

Definition 15 (Incompatible Replication). Let $X \subset R_1(A)$, $Y \subset R_2(A)$ be two subsets of relations R_1, R_2. A replication $X \rightrightarrows Y$ occurs if $\forall t_i \in R_1, t_j \in R_2$, $t_i[W] = t_j[U] \Rightarrow t_i[X] = t_j[Y]$. A replication defect occurs *iff* $\exists t_i \in r(R_1)$, $t_j \in r(R_2)$, such that $t_i[W] = t_j[U]$ and $t_i[X] \neq t_j[Y]$.

Due to certain reasons (including performance and poor data modelling), a base attribute may have its content replicated into multiple attribute copies. There is a contradictory situation when these attributes have different values.

Example: The State attribute of the Customer relation should have been replicated to the State attribute of the CustomerCreditAccount relation. However, the two hold different values for the same client (tuples $cc4, cc7, cc8$).

Definition 16 (Incorrect Reference). Let $Rel : R_1 \rightarrow R_2$ be a relationship set between relations R_1, R_2. Let $rrel : Rel \rightarrow \{true, false\}$ be a function which returns if an instance of relationship Rel holds in the UoD. Incorrect reference occurs *iff* $\exists (t_i, t_j) \in Rel$ such that $rrel(t_i, t_j)$ is *false*.

This occasion refers to a relationship instance that does not represent a fact about an object of the UoD, although it obeys all of the other rules.

Example: Customer tuple $c8$ is owner of credit card number 199 (tuple $cc6$). However, this customer is related to the credit card number 200 (tuple $cc7$).

Definition 17 (Incorrect Value). Let $rv : a \rightarrow dom(a)$ be a function on relation R which returns the value of attribute a in the UoD. Incorrect values occurs *iff* $\exists t \in r(R)$ such that $t[a] \neq rv(a)$.

An incorrect value is an unfaithful or contradictory representation of a fact about an object of the UoD. In other words, such a defect denotes a large discrepancy between the represented value and the real value of the object.

Example: A customer's salary is 73.4k, but it was represented as 8k on tuple $c5$.

Definition 18 (Inference Rule Violation). Let $Rel : R_1 \rightarrow R_2$ be a relationship set between relations R_1, R_2. Let $ir : Rel \rightarrow \{true, false\}$ be a function which returns if an instance of relationship Rel complies with its inference rule. There is an inference rule violation iff $\exists(t_i, t_j) \in Rel$ such that $ir(t_i, t_j)$ is $false$.

An inference rule is a procedure that generates new facts based on the ones available in a database. A violation arises when an inferred attribute value, tuple or relationship instance is not represented, or when it is different from the one that was determined by the rule.

Example: The Score attribute of the CustomerCreditAccount relation is inferred by a complex analysis of credit card usage for the last six months. This relation has a tuple ($cc8$) where Score is 1.83449 instead of 1.01553, which was inferred.

Definition 19 (Key Dependency Violation). Let X be an attribute subset of relation R, $X \subseteq R(A)$. Let $R(A)$ be key dependent on X, as represented by $X \rightarrow R(A)$. This dependency is violated iff $\exists t_i, t_j \in r(R)$, $i \neq j$, such that $t_i[X] = t_j[X]$.

The purpose of the identifier attribute subset is to uniquely identify all relation tuples to enable data relationship. This situation is violated when two or more tuples share the same value for their identifiers' attributes.

Example: The tuples $c2$ and $c10$ share their CID, but they are distinct customers.

Definition 20 (Missing Reference). Let $Rel : R_1 \rightarrow R_2$ be a relationship set between relations R_1, R_2. Let $rrel : Rel \rightarrow \{true, false\}$ be a function which returns if an instance of relationship Rel holds in the UoD. A reference is absent iff $\exists t_i \in r(R_1), \exists t_j \in r(R_2)$ such that $(t_i, t_j) \notin Rel$ and $rrel(t_i, t_j)$ is $true$.

Relationship instances represent facts about objects of the UoD. The missing reference defect arises when a required relationship instance is not represented.

Example: The Customers tuples $c11$ and $c12$ are married. However, this marriage relationship has not been represented.

Definition 21 (Missing Tuple). Let $mist : DB \rightarrow \{true, false\}$ be a function which returns if a relation from DB represents all of the required objects of the UoD. A tuple is absent iff $\exists R_i \in DB$ such that $mist(R_i)$ is $false$.

A database schema must represent only the required objects of the UoD and their properties. The missing tuple defect denotes the lack of representation of certain important objects of the UoD.

Example: Certain joint holders of credit card accounts (tuples $cc5$ to $cc8$) are not represented in the Customer relation.

Definition 22 (Overloaded Tuple). Let $overt : r(R) \rightarrow \{true, false\}$ be a function which returns if a tuple from R represents a single object of the UoD. A tuple is overloaded iff $\exists t_i \in r(R)$ such that $overt(t_i)$ is $false$.

A single tuple represents facts about a single object of an UoD. An overload denotes an excessive representation (more than one) of objects by one tuple.

Example: "Dick Rhodes" and "Dick Rhodes" are two distinct people of the UoD. Nonetheless, only a single customer tuple $c12$ represents both.

Definition 23 (Participation Constraint Violation). Let $pc : R_1 \times R_2 \to \mathbb{N}$ be a function that maps the minimal participation of tuples from R_1 to R_2. Let $s : r(R_1) \times r(R_2) \to \mathbb{N}$ be a function defined as follows:

$$s(a, b) = \begin{cases} 1 & \text{if } a = b, \\ 0 & \text{if } a \neq b. \end{cases}$$

A minimum participation violation occurs *iff* $\exists t_i \in R_1$, $t_j \in R_2$ such that $\sum s(t_i[W], t_j[U]) < pc(R_1, R_2)$.

The participation constraint determines the minimum number of relationship instances in which each tuple from a relation must participate in a binary relationship. The violation occurs when a tuple does not comply with the minimum *referred* or *refer to* role.

Example: A credit card must be associated with at least two customers. However, there are credit cards in the CustomerCreditAccount relation with only one customer (tuples $cc5$ to $cc7$).

Definition 24 (Semantic Integrity Violation). Let $Rel : R_1 \to R_2$ be a relationship set between relations R_1 and R_2. Let $rule : Rel \to \{true, false\}$ be a function which returns if an instance of relationship Rel complies with its semantic integrity rule. Let RU^{Rel} be the set of semantic rules on relationship Rel, denoted as $RU^{Rel} = \{rule_1, ..., rule_z\}$, $z \geqslant 1$. There is a semantic integrity violation *iff* $\exists (t_i, t_j) \in Rel, \exists rule_h \in RU^{REL}$ such that $rule_h(t_i, t_j)$ is *false*.

The semantic integrity comprises a set of complex rules for an UoD that guarantees a state of data consistency. A violation arises when one of these rules is disobeyed.

Example: A Customer relation rule determines that only 9–17 years old people living at "SP" state can possess a credit card as joint holder. For the remaining states this relationship is forbidden. However, such a rule is disobeyed by a certain customer (tuple $c9$) of "MG" state that has a credit card (tuple $cc9$).

Definition 25 (Synonymous Values). Let $sp : r(R) \times r(R) \to \{true, false\}$ be a function which returns if the graphy and pronunciation of attributes values are equal, according to LEX. Let $me : r(R) \times r(R) \to \{true, false\}$ be a function which returns if the meaning of attributes values are equal or nearly the same, according to LEX. An attribute has synonymous values *iff* $\exists a \in R(A), \exists t_i, t_j \in r(R), i \neq j$, such that $sp(t_i[a], t_j[a])$ is *false* and $me(t_i[a], t_j[a])$ is *true*.

Synonyms denote distinct terms in writing that share the same or similar meanings. Such terms can be expressed as vernacular words, acronyms, abbreviations or symbols. This defect arises when synonymous terms are used interchangeably to indicate the same fact about objects of the UoD.

Example 1: Customer relation tuples $c1, c4, c5, c8, c10$ have marital statuses such as "E" (espoused), "J" (joined), "U" (united) that designate married ("M") in each case.

Example 2: Customer relation tuples $c2$ and $c4$ have job titles as "Tapster" and "Barkeeper" that designate "Bartender" in each case.

Definition 26 (Transition Constraint Violation). Let $tran : R \to R$ be a transitional function that leads the original state of R to another state R', according to a inference system $R \to^{tran} R'$. A transition violation occurs *iff* $\exists t \in R$ such that $t[R'(A)] \neq t[tran(R(A))]$.

The transition or dynamic constraints represent a set of rules that enforces the valid state transitions of data. These constraints are evaluated on a pair of successive pre and post-transaction states of a database relation. A violation arises when a tuple possesses an invalid post-transaction state.

Example: There is a rule that regulates the transitions between the valid states of MS ("MaritalStatus") attribute. The tuple $c12$ violates this rule because his marital status has changed from "S" (Single) to "W" (Widower).

5   Conclusions

This work reports a taxonomy that organized a detailed description of data defects regarding to the quality dimensions of accuracy, completeness and consistency. The taxonomy applied a three-step methodology to address all the issues discussed in Sect. 2: the theoretical review enabled the systematic and broad coverage of data defects (nine defects were not addressed by the state-of-art taxonomies, as highlighted in Fig. 1), and improved the data defect descriptions; the classification steps organized data defects according to their properties and granularity of occurrence.

The taxonomy structure can support relevant issues in data quality assessment, including the training of data quality appraisers, the establishment of measurement-based process and guiding the design decisions in regard to semi-supervised approaches for detecting data defects. Nevertheless, the taxonomy does not address time-related data defects, as well as it offers high level formal descriptions of some defects since they involve human curation or complex and broad rules. In future works, this taxonomy will be the basis for classifying data defects according to time-related data and designing a supervised approach for data defect detection.

Acknowledgments. This work has been supported by CNPq (Brazilian National Research Council) grant number 141647/2011-6 and FAPESP (Sao Paulo State Research Foundation) grant number 2015/01587-0.

References

1. Almutiry, O., Wills, G., Crowder, R.: A dimension-oriented taxonomy of data quality problems in electronic health records. In: 13th IADIS International Conference on e-Society, pp. 98–114. IADIS, Portugal (2015)
2. Barateiro, J., Galhardas, H.: A survey of data quality tools. Datenbank-Spektrum **14**, 15–21 (2005)

3. Borek, A., Woodall, P., Oberhofer, M., Parlikad, A.K.: A classification of data quality assessment methods. In: 16th International Conference on Information Quality, pp. 189–203. IEEE Press, New York (2011)
4. English, L.P.: Improving Data Warehouse and Business Information Quality: Methods for Reducing Costs and Increasing Profits. Wiley, New York (1999)
5. Fan, W., Geerts, F.: Foundations of Data Quality Management. Morgan & Claypool Publishers, San Rafael (2012)
6. Grefen, P.: Combining theory and practice in integrity control: a declarative approach to the specification of a transaction modification subsystem. In: 19th International Conference on Very Large Data Bases, pp. 581–591. Morgan Kaufmann Publishers Inc., Dublin, Ireland (1993)
7. Kim, W., Choi, B.-J., Hong, E.-K., Kim, S.-K., Lee, D.: A taxonomy of dirty data. Data Min. Knowl. Discov. **7**, 81–99 (2003)
8. Laranjeiro, N., Soydemir, S.N., Bernardino, J.: A survey on data quality: classifying poor data. In: 21st Pacific Rim International Symposium on Dependable Computing, pp. 179–188. IEEE Press, Zhangjiajie, China (2015)
9. Li, L., Peng, T., Kennedy, J.: A rule based taxonomy of dirty data. GSTF Int. J. Comput. **1**, 140–148 (2011)
10. Müller, H., Freytag, J.C.: Problems, methods, and challenges in comprehensive data cleansing. Technical report, Humboldt University Berlin (2005)
11. Maier, D.: The Theory of Relational Databases. Computer Science Press, Rockville (1983)
12. Naumann, F.: Data profiling revisited. ACM SIGMOD Rec. **42**, 40–49 (2014)
13. Oliveira, P., Rodrigues, F., Henriques, P.: A formal definition of data quality problems. In: International Conference on Information Quality, pp. 181–184. IEEE Press, New York (2005)
14. Rahm, E., Do, H.H.: Data cleaning: problems and current approaches. IEEE Bull. Tech. Comm. Data Eng. **23**, 3–13 (2000)
15. Schmid, J.: The main steps to data quality. In: Perner, P. (ed.) ICDM 2004. LNCS (LNAI), vol. 3275, pp. 69–77. Springer, Heidelberg (2004)
16. Winkler, W.E.: Methods for evaluating and creating data quality. Inf. Syst. **29**, 531–550 (2004)

ISSA: Efficient Skyline Computation for Incomplete Data

Kaiqi Zhang$^{(\boxtimes)}$, Hong Gao, Hongzhi Wang, and Jianzhong Li

School of Computer Science and Technology, Harbin Institute of Technology,
Harbin, China
{zhangkaiqi,honggao,wangzh,lijzh}@hit.edu.cn

Abstract. Over the past years, the skyline query has already caused wide attention in database community. For the skyline computation over incomplete data, the existing algorithms focus mainly on reducing the dominance tests among these points with the same bitmap representation by exploiting *Bucket* technique. While, the issue of exhaustive comparisons among those points in different buckets remains unsolved, which is the major cost. In this paper, we present a general framework COBO for skyline computation over incomplete data. And based on COBO, we develop an efficient algorithm ISSA in two phases: *pruning compared list* and *reducing expected comparison times*. We construct a compared list order according to ACD to diminish significantly the total comparisons among the points in different buckets. The experimental evaluation on synthetic and real data sets indicates that our algorithm outperforms existing state-of-the-art algorithm 1 to 2 orders of magnitude in comparisons.

1 Introduction

The skyline query coined out by Börzsönyi et al. in [1] is an important query in database community. Over the years, it has caused wide attention to researchers, especially in multi-criteria decision making field. For a data set S with multiple dimensions, the skyline query retrieves out the interesting points which are not dominated by others. Any two points p and q in S, p dominates q if p is smaller or equal to q in every dimension and smaller than q in at least one dimension.

The skyline query can diminish the amount of the concern points down to the skyline set and non-skyline points will never take up users' decision time. In the light of the importance of the skyline query, a series of efficient algorithms [1,3,9] for the skyline computation have been proposed. They are devised based on the *completeness assumption*, *i.e.*, all dimensions are available for every data point. While there are many real life scenarios do not hold this assumption. For instance, there is a movie rating data set (MovieLens [5]). Users usually rate a few of all the movies, which results in many *null* value for the movie that a user does not rate. In this incomplete data set, [6] gives the new definition about skyline, which makes all aforementioned traditional algorithms invalid. Also, proposes a *ISkyline* algorithm. It reduces the dominance tests among these points

© Springer International Publishing Switzerland 2016
H. Gao et al. (Eds.): DASFAA 2016 Workshops, LNCS 9645, pp. 321–328, 2016.
DOI: 10.1007/978-3-319-32055-7_26

with the same bitmap representation by exploiting *Bucket* technique. While, the issue of exhaustive comparisons among those points in different buckets remains unsolved, which is the major cost for computing skyline set over incomplete data.

In this paper, we first illustrate the challenge brought by *non-transitive* dominance relation for incomplete data set. Then, a general framework for skyline computation over incomplete data, named COBO, is presented to accommodate the relation. Later, based on COBO, we propose an efficient algorithm ISSA in two phases: *pruning compared list* and *reducing expected comparison times*. *Pruning compared list* could delete safely these points which are eliminated that have no effect on the result of skyline computation. And *reducing expected comparison times* constructs a compared list order according to ACD to strive to diminish the total expected comparisons among points in different buckets. We summarize our contributions as follows:

- We generate a general framework named COBO for skyline computation over incomplete data.
- We analyze that there are only two techniques to improve the general framework: *pruning compared list* and *reducing expected comparison times*.
- We propose a compared list order according to ACD to strive to diminish the total expected comparisons.
- We evaluate our presented algorithm with state-of-the-art algorithm on synthetic and real data sets.

2   Related Work

The skyline query is first pioneered in [1] in the database community, and afterward, many skyline algorithms have been presented, such as SFS [3], OSPS [9] and so on. In addition, many efforts have been paid to the variants of traditional skyline, for instance k-dominant skyline [2], probabilistic skyline [7], reverse skyline [4], and skyline cube [8].

However, for incomplete data set, there are quite few of literatures up to now. The issue is first introduced in [6] which gives the new definition about skyline in incomplete data set. Unfortunately, the transitive dominance relation cannot remain hold, *i.e*, dominance relation become *non-transitive*, which leads to that all the traditional skyline algorithms are not applicable. In the light of this, [6] presents an algorithm ISkyline for skyline computation over incomplete data. ISkyline stores automatically the points with the same bitmap representation into the identical bucket. It reduces the dominance tests among the points in the same bucket, in which the transitive dominance relation become valid. While, the issue of exhaustive comparisons among those points in different buckets remains unsolved, which is the major cost for computing skyline set over incomplete data.

3   Preliminaries

This section first introduces the description about incomplete data set. And then the definitions of dominance and skyline are both given referring to the prior work [6].

Up to now, we have referred to many times of incomplete data, so what is it and how to represent it? In incomplete data set, there maybe exist missing value in any dimension for every data point. For each data point, we call the dimension with missing value as *incomplete* dimension, otherwise the dimension is *complete* dimension. For example, $p(1, -, 4)$ is an incomplete point in 3-dimensional data space, where '–' denotes the missing value. Also, the ith dimensional value of p is represented by p^i, we know that $p^3 = 4$. Here, we first explain some symbols used through the paper from now on. Incomplete data set S has d dimensions whose value is positive number. Specially, any data point in S has 1 to $d - 1$ *incomplete* dimensions. Now, we introduce formally the definitions about skyline over incomplete data in the following:

Definition 1 (Dominance for Incomplete Data). *Given any two points p, $q \in S$, it is said that p dominates q, denoted by $p \succ q$, if and only if the following conditions both satisfy:*

(1) $\exists i \in [1, d]$, p^i and q^i both exist and $p^i < q^i$.
(2) $\forall i \in [1, d]$, p^i is missing or q^i is missing or $p^i \leq q^i$.

If $p \succ q$, we call that p is the *dominating* point of q and q is the *dominated* point of p.

Definition 2 (Skyline for Incomplete Data). *Using $SKY(S)$ represents the skyline set of S, where $SKY(S) = \{\forall p \in S \mid \nexists\, q \in S, q \succ p\}$.*

4   A General Framework for Skyline Computation over Incomplete Data

In this section, we first present a general framework named Compared One by One (COBO) for computing skyline result over incomplete data. Based on it, we propose an efficient algorithm ISSA in two phases: *pruning compared list* and *reducing expected comparison times*. And then we illuminate them in detail.

4.1   A General Framework

For the skyline computation over complete data, points hold the transitive dominance relation. For instance, there are three points $p1 = (2, 3, 2)$, $p2 = (3, 3, 4)$ and $p3 = (4, 5, 4)$. $p1$ dominates $p2$ and $p2$ dominates $p3$, therefore $p1$ must dominate $p3$. All existing approaches are based on transitive dominance relation. While, the dominance relation is *non-transitive* [6] among incomplete points according to the Definition 1. An example as follows: $p1 = (2, 3, -)$, $p2 = (4, -, 1)$, $p3 = (-, -, 2)$. Apparently, $p1$ dominates $p2$ and $p2$ dominates $p3$, while $p1$ cannot dominate $p3$. Non-transitive makes it extremely difficult that the skyline computation over incomplete data. Not like computing skyline result in complete data set, once one point is checked out to be non-skyline, it can be removed immediately. Because all points it dominates must be dominated by the points which dominate it. So traditional skyline algorithms exploiting the property can only maintain a skyline window organized as sorted list [3] or *skyline* tree [9] to justify the subsequent points. Unfortunately, for incomplete data set, it is not safe to discard the points which, even though, are determined to be non-skyline. Any point, whether it is skyline or not, must be contained in the window to check other points.

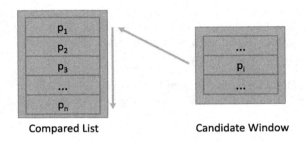

Compared List   **Candidate Window**

Fig. 1. A general framework: COBO

Based on above analysis, we present a general framework named Compared One by One (COBO) for computing skyline result over incomplete data as shown in Fig. 1. The framework is made of two parts, compared list (CL) and candidate window (CW). Compared list is mainly used to justify whether a point is skyline or not, and it generally contains all points of the data set because of the *non-transitive* dominance relation. For candidate window, it stores these points which maybe belong to the skyline result. Simply initialize candidate window by all points since any point has the possibility to be skyline. Now we will illustrate the process of skyline computation in the framework. After initialization of two parts, every point in candidate window do this operation: For any being checked point p_i in candidate window, generating a compared list order according to p_i. Then compare the points in this order, one by one, with p_i. Once one point dominating p_i occurs, eliminate p_i from candidate window. Finally, the remaining points in candidate window are the skyline result of the incomplete data set.

The basic process of the framework COBO is shown in Algorithm 1. While, its cost is significantly expensive. Now, based on COBO, we propose an efficient algorithm ISSA in two phases: *pruning compared list* and *reducing expected comparison times*.

4.2   Phase I: Pruning Compared List

The first phase is pruning compared list, *i.e.*, remove all these points which are eliminated that have no effect on the result of skyline computation. While, what the points can be removed safely in the compared list? This issue has already been figured out in [6] by introducing the *bucket* technique. These points with the same bitmap representation are stored automatically into the identical bucket. The transitivity of dominance relation becomes valid among the points in the

---

Algorithm 1. COBO(S)

Input: A d-dimensional positive numerical incomplete data set S
Output: The skyline result of data set S

1: Candidate Window $candidate = S$
2: Compared List $CL = S$
3: **for** $\forall p \in candidate$ **do**
4:    determine the order of CL as CL_p according to p.
5:    **for** $\forall q \in CL_p$ **do**
6:        **if** q dominates p **then**
7:            delete p from $candidate$;
8:            break;
9: return $candidate$

---

same bucket. Therefore we can straightforwardly discard the dominated points in each bucket.

For example, $p1 = (2, 3, -)$, $p2 = (3, 4, -)$, $p3 = (-, 5, 1)$. The points $p1$ and $p2$ are in the same bucket 110 and $p1$ dominates $p2$. Then we can remove $p2$ safely, the remaining point $p3$ dominated by $p2$ must be dominated by $p1$.

Theorem 1. *Bucket technique can maximize to delete these points which are eliminated that have no effect on the result of skyline computation.*

Proof. Suppose that there is one point p which can be discarded safely and it does not belong to these points that are eliminated by bucket technique. It is obvious to infer that there must exist a point q which satisfy following conditions if p can be safely removed: q is not worse than p in every dimension and q and p have the same representation. Then q dominates p and they are in the identical bucket. So point p must be pruned by bucket technique. This contradicts with origin assumptions.   □

Bucket technique prunes as many safe points as possible. Then, use the remaining points to initialize CL and CW as shown in Algorithm 1 in line 1–2. Although the cost is still extremely expensive, there is no better way since the dominance relation is *non-transitive* in incomplete data.

4.3   Phase II: Reducing Expected Comparison Times

Only adopting phase I makes it still inefficient, which is mainly due to the point in candidate list stopping being checked until find its dominating point in compared list. So we should strive to find out the dominating point of the being checked candidate point as soon as possible. As the skyline points, they must be compared with all the points in compare list. While for the other points, the ideal way is to only need one comparison, *i.e.*, compared with one of their dominating points. Unfortunately, if we know which points are their dominating points, the skyline result will be given immediately.

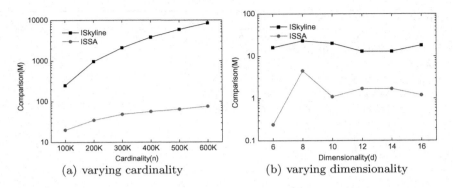

Fig. 2. Performance on synthetic data set

Now, the most important task is to pay low cost to construct a compared list order, in which all points can quickly find their dominating points. So as to achieve the goal of reducing expected comparison times. Under UI condition [3], especially in incomplete data set, for those points with the same size of *complete* dimensions, we observe that the point with lower sum have greater probability of being not dominated by any point than that with higher sum. To extend it to general incomplete situation, we sort all points in compared list according to their average value in all *complete* dimensions (ACD).

For instance, $p1 = (2, 7, -)$, $p2 = (4, -, 3)$ and $p3 = (-, 3, 2)$ in compared list. The order is $p1$, $p2$, $p3$ before sorting. Now we sort the list according to ACD, and result in the sorted list $p3$, $p2$, $p1$. We call the list sorted compared list. Then, all the points in candidate window are checked by the order of sorted compared list.

5   Experimental Evaluation

In this section, we evaluate our algorithm by experiments. We compare our algorithm ISSA with state-of-the-art approach ISkyline [6] in dimensionality and cardinality in synthetic and real data sets. All algorithms are performed on a Microsoft Windows 7 computer with an Intel Core i7-4790 CPU at 3.6 GHz and 8 GB memory.

5.1   The Performance on Synthetic Data Sets

This section describes the performance of our algorithm ISSA and state-of-the-art approach ISkyline [6] on synthetic data sets. We generate several incomplete synthetic data sets with dimensions and incompleteness ratio ranging from 6 to 16 and 25 % to 50 %, respectively. First, we produce the complete data set under the UI condition in [3], and the domain of complete dimensions is [0,1000]. Then, randomly remove some dimensions to make them incompleteness according to

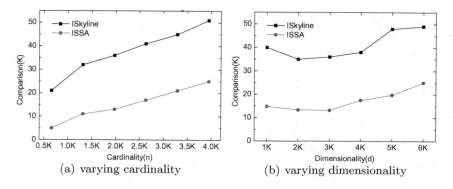

(a) varying cardinality   (b) varying dimensionality

Fig. 3. Performance on real data set

identical ratio for every data point. These synthetic incomplete data sets are generated as above methods.

Figure 2(a) shows the total comparisons of them over cardinality variation from 100K to 600K, where the dimensionality is 12 all the time and the incompleteness ratio is 25 %. Apparently, ISSA outperforms ISkyline 1 to 2 orders of magnitude in comparisons. Figure 2(b) describes the performance of algorithms by varying dimensionality from 6 to 16. The cardinality and incompleteness ratio of the data sets are all 200K and 50 %, respectively. As Fig. 2(b) reports, ISSA is always faster than ISkyline.

5.2 The Performance on Real Data Sets

This section describes the performance of ISSA and ISkyline on real data set MovieLens [5]. It is one of data sets in [5] and contains 1 million ratings. Actually, 95.75 % is incomplete for the data set which includes 3952 points(movies) and each of them has 6040 dimensions (users). We conduct experiments by varying cardinality and dimensionality. In Fig. 3(a), the cardinality is ranging from 658 to 3952 where the dimension is 6040. In Fig. 3(b), the dimension is ranging from 1 K to 6 K where the cardinality is 3952. It is clear that ISSA is always several times faster than ISkyline as shown in Fig. 3.

6 Conclusions

In this paper, we presented a general framework COBO for skyline computation over incomplete data. And based on COBO, we developed an efficient algorithm ISSA in two phases: *pruning compared list* and *reducing expected comparison times*. *Pruning compared list* could delete safely these points which are eliminated that have no effect on the result of skyline computation. And *reducing expected comparison times* constructs a compared list order according to ACD to strive to diminish the total expected comparisons among points in different buckets. Experimental results on synthetic and real data sets indicated that our algorithm outperforms existing state-of-the-art algorithm 1 to 2 orders of magnitude in comparisons.

References

1. Börzsönyi, S., Kossmann, D., Stocker, K.: The skyline operator. In: Proceedings of the 17th International Conference on Data Engineering, Heidelberg, Germany, pp. 421–430, 2–6 April 2001
2. Chan, C.Y., Jagadish, H.V., Tan, K., Tung, A.K.H., Zhang, Z.: Finding k-dominant skylines in high dimensional space. In: Proceedings of the ACM SIGMOD International Conference on Management of Data, Chicago, Illinois, USA, pp. 503–514, 27–29 June 2006
3. Chomicki, J., Godfrey, P., Gryz, J., Liang, D.: Skyline with presorting. In: Proceedings of the 19th International Conference on Data Engineering, Bangalore, India, pp. 717–719, 5–8 March 2003
4. Dellis, E., Seeger, B.: Efficient computation of reverse skyline queries. In: Proceedings of the 33rd International Conference on Very Large Data Bases, University of Vienna, Austria, pp. 291–302, 23–27 September 2007
5. http://movielens.umn.edu
6. Khalefa, M.E., Mokbel, M.F., Levandoski, J.J.: Skyline query processing for incomplete data. In: Proceedings of the 24th International Conference on Data Engineering, ICDE 2008, Cancún, México, pp. 556–565, 7–12 April 2008
7. Pei, J., Jiang, B., Lin, X., Yuan, Y.: Probabilistic skylines on uncertain data. In: Proceedings of the 33rd International Conference on Very Large Data Bases, University of Vienna, Austria, pp. 15–26, 23–27 September 2007
8. Yuan, Y., Lin, X., Liu, Q., Wang, W., Yu, J.X., Zhang, Q.: Efficient computation of the skyline cube. In: Proceedings of the 31st International Conference on Very Large Data Bases, Trondheim, Norway, pp. 241–252, 30 August–2 September 2005
9. Zhang, S., Mamoulis, N., Cheung, D.W.: Scalable skyline computation using object-based space partitioning. In: Proceedings of the ACM SIGMOD International Conference on Management of Data, SIGMOD, Providence, Rhode Island, USA, pp. 483–494, 29 June–2 July 2009

Join Query Processing in Data Quality Management

Mingliang Yue$^{(\boxtimes)}$, Hong Gao, Shengfei Shi, and Hongzhi Wang

School of Computer Science and Technology,
Harbin Institute of Technology, Harbin, China
{ml_yue, honggao, wangzh}@hit.edu.cn

Abstract. Data quality management is the essential problem for information systems. As a basic operation of Data quality management, joins on large-scale data play an important role in document clustering. MapReduce is a programming model which is usually applied to process large-scale data. Many tasks can be implemented under the framework, such as data processing of search engines and machine learning. However, there is no efficient support for join operation in current implementations of MapReduce. In this paper, we present a strategies to build the extend bloom filter for the large dataset using MapReduce. We use the extend bloom filter to improve the performance of two-way and multi-way joins.

Keywords: Data quality management · MapReduce · Bloom filter · Join

1 Introduction

In recent year, with the wide popularity of Internet technology, along with the rapid development of cloud computing technology, the data in the Internet is growing at an unprecedented speed and accumulation. Data quality management [1] is the essential problem for information systems. As a basic operation of Data quality management, joins on large-scale data play an important role in document clustering.

Hadoop [2] provides a default join mechanism for relational data, the MapReduce programming model is widely applied to massive data based processing because of its good scalability, fault tolerance and usability. However because of its own limitation, the performance of MapReduce [3] is slow when it is adopted to perform complex data analysis tasks that require the joining of data sets in order to compute certain aggregates. Therefore, it is necessary to design an improved method for Join operation under MapReduce framework.

Aiming at the shortage for processing join operations based on MapReduce. In this paper we presents a join algorithm based on extend Bloom Filter, whose core idea is to use Bloom Filter to decrease the network overhead between the map and reduce phases so as to improve the efficiency. First of all, an efficient algorithm building a Bloom Filter for a dataset is proposed; secondly, join algorithms based on Bloom Filter axe proposed, including two-way and multi-way. In this paper, we study relational data join within MapReduce, and make the following contributions:

© Springer International Publishing Switzerland 2016
H. Gao et al. (Eds.): DASFAA 2016 Workshops, LNCS 9645, pp. 329–342, 2016.
DOI: 10.1007/978-3-319-32055-7_27

1. We present and compare an extended bloom filter [4] for a large dataset using MapReduce.
2. We consider the optimization of joins using mutual filtering policy based on extended Bloom Filter and conduct an extensive experimental evaluation.

The rest of this paper is organized as follows. We present related work in Sect. 2. In Sect. 3, we compute the Bloom Filter Using MapReduce. In Sect. 4, we present our approach for join operator. We will study how use the bloom filter to improve the efficiency of the join algorithm. In Sect. 5, experimental results demonstrating the performance of proposed join implementation are presented. We conclude the whole work in Sect. 6.

2   Related Work

In this section, we first review the Hadoop and join processing techniques in MapReduce. Then, we describe the Bloom filter.

2.1   Hadoop

Hadoop, the open source project of Apache, is a distributed parallel computing platform for large-scale data, including Hadoop Distributed File System (HDFS) and MapReduce.

MapReduce programs run on HDFS, which is the primary storage system used by Hadoop applications and provides high throughput access to application data. Data on HDFS is usually divided into many small blocks (splits). HDFS creates multiple replicas for each data block and distributes them on computing nodes throughout a cluster. One replica of a block would be processed by a Mapper locally on the node where it is distributed. This mechanism enables reliable and extremely rapid computations.

2.2   Joins in MapReduce

Join algorithms in MapReduce [12, 14] are roughly classified into two categories: map-side joins and reduce-side joins [5]. Map-side join algorithms are more efficient than reduce-side joins, because they only produce the final result of the join in map phase. However, they can be used only in particular circumstances. For Map-Merge join [6, 13], two input datasets should be partitioned and sorted on the join keys in advance, or an additional MapReduce job is required to meet the condition. Broadcast join [6] is effective when the size of one dataset is small.

Map-Reduce-Merge [7] adds merge phase after the reduce phase to support operations with multiple heterogeneous datasets, but it has the same drawback as reduce-side join algorithms. There are some attempts to optimize multiway joins in MapReduce. They discuss the same idea to minimize the size of the records replicated to reduce processes. In this paper, we address only two-way joins. However, our approach can be extended to multi-way joins by combining these work.

2.3   Bloom Filter

Bloom filter consists of an array of m bits and a set of k hash functions, which hash the element of the dataset to an integer in the range of [1, m]. The example for a bloom filter is shown in Fig. 1. All bits of the array are initialized to zero. Each hash function maps an element to some bits of the filter. In order to check the membership of an element, we must look at k positions. We answer positively only if all k bits are set to 1. The bloom filter allows false positives, but never false negatives.

Bloom join [4, 11] is a join algorithm which uses the Bloom filter to filter out tuples not matched in a join. Suppose relations $R(a, b)$ and $S(a, c)$ that reside in site 1 and site 2 respectively. In order to join these two relations, Bloom join algorithm generates a Bloom filter with the join key a of a relation, say R. Then, it sends the filter to site 2 where R resides. At the site 2, the algorithm scans R and sends the only tuples which are set in the received Bloom filter to site 1. Finally, a join of the filtered R and S is performed at site 1.

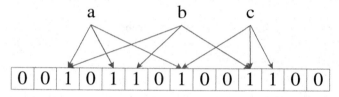

Fig. 1. Example for a bloom filter

Give a set $R(x)$, m is the size of the bit array, k is the number of hash functions, p is a false positive, we denote the bloom filter for the relation R on the attribute x by $BFR(x)$.

$$m_p = \frac{-\ln p}{(\ln 2)^2}$$

The total size of the bloom filter for the whole set $R(x)$ is

$$m = m_p * |R| = \frac{-|R| * \ln p}{(\ln 2)^2}$$

3   Computing Bloom Filter Using MapReduce

A Bloom filter is a probabilistic data structure used to test whether an element is a member of a set. It consists of an array of m bits and k independent hash functions. All bits in the array are initially set to 0. When an element is inserted into the array, the element is hashed k times with k hash functions, and the positions in the array corresponding to the hash values are set to 1. To test membership of an element, if all bits of its k hash positions of the array are 1, we can conclude that the element is in the set. Bloom filter may yield false positives, but false negatives are not generated.

The false positive of standard bloom filter:

$$p = \left(1 - \left(1 - \frac{1}{m}\right)^{kn}\right)^k \approx \left(1 - e^{\frac{kn}{m}}\right)^k$$

Because of the false positive of standard bloom filter, many attributes that are not matched with the join condition have been transferred to the reduce phase. That will bring some of the network transmission overhead, *I/O* overhead. Therefore if we can further reduce the false positive of the bloom filter, so we can use bloom filter to decrease the network overhead between the map and reduce phases so as to improve the efficiency.

On the basis of the data structure, we define a new bit array of m. All bits in the array are initially set to 0. The element is hashed k times with k hash functions, we performed *XOR* operations on the results. The position in the array corresponding to the result is set to 1.

In the extend bloom filter, in this array, the probability of assigning a value to 1 is 1/m. When the hash address of each element of the XOR operation is mapped to the array, One bit is the probability of 0:

$$p' = \left(1 - \frac{1}{m}\right)^n \approx e^{-\frac{n}{m}}$$

So the false positive of the extend bloom filter:

$$p = \left(1 - e^{-\frac{kn}{m}}\right)^k * \left(1 - e^{-\frac{n}{m}}\right)$$

The map phase: Each map function build a bloom filter for its local data of its own partition named *BFR(x)*. $|BFR(x)| = m_p * |R|$. And the map function also make the XOR bloom filter named *BFK(x)*. The intermediate results of the map function output will be sent to a single reducer.

The reduce phase: The reduce function unions the intermediate results by a bit-wise OR operation.

The example of the strategy is shown in Fig. 2. There are two relation *R(A, B)* and *S (B, C)*. First, the two tables of all the local filter files for "OR" operation. Such as the relation R has m slices. We build *BFR(B_1)* and *BFK(B_1)*, *BFR(B_2)* and *BFK(B_2)*... *BFR (B_m)* and *BFK(B_m)* in the map phase. We unions the result by OR operation in the reduce phase. We build *BFR(B)* and *BFK(B)* for the relation *R(A,B)* and use *BFR(B)* and *BFK(B)* to filter the relation *S(B, C)*. We do the same to the relation *S*.

4   Extend Bloom Join Using MapReduce

In this section, we will study how to use the bloom filter to improve the efficiency of the join algorithm. The concept of using the bloom filter to improve the efficiency is based on the semi-join technique.

4.1   Two-Way Joins Using MapReduce

Aiming at the shortage for processing join operations based on MapReduce, we proposed the optimized strategy. When processing two tables join based on MapReduce model, we use mutual filtering policy based on extended bloom filter. After the extended bloom filter process, the two tables join attribute values are extracted respectively, and form the file. Then the files are used to filter two tables do not meet the join condition. The optimized method of join, can be achieved to extend bloom filter and reduce the false positive rate, reducing shuffle phase time, improving the execution efficiency of the system.

We present a algorithms using the extend bloom filter to compute $R(A, B)$ and $S(B, C)$, each record of R and S has two attributes. This algorithm has two phases, each corresponding to a separate.

MapReduce job. In the first MapReduce job, we build a extend bloom filter $EBF(R)$ on the attribute B for the relation R and S. The extend bloom filter $EBFR(R)$ consists of $BFR(B)$ and $BFK1(B)$ for the relation R. The relation S includes $BFS(B)$ and $BFK2(B)$. $BFK(B)$ is the array which the hash address of each element of the XOR operation is mapped to. $BFK(B)$ can reduce the false positive of the bloom filter.

In the second MapReduce job, Map function reads the two relations R and S, while reading $EBFR(B)$ and $EBFS(B)$. For the elements in the relation S. The map function uses $EBFR(B)$ to filter the relation S. The detection of the attribute is true means that the values of the join attributes need to send out. In the map phase, to express the natural join $R(A, B)$ and $S(B, C)$, the input to the map function is key-value pairs (r, t), where r is the relation name (either R or S), and t is a tuple of the relation named by r. the output is a key-value pair (r, t). All these output records form the intermediate result.

The key is a composite of two elements. The first element is the value of the join attribute B, the second element is a flag bit, indicates that this property is derived from the relation S. The value contains an element, its value is the value of the attribute B. We do the same to the relation R. The map function uses $EBFS(B)$ to filter the relation R for the elements in the relation R.

In the suffer phase, because the key is a composite value, If the entire key value is used as a partition element, It cannot guarantee that the same value is sent to the same reduce node and satisfy the equijoin condition. So we define a Partition function in advance, the first value of the key is used as the partition element. In the reduce phase, although the same key value is assigned to the same reduce, But extend bloom filter has the possibility of a miscarriage of justice too. So we need to add a step in the reduce function to check if the join property is equal. The final results are computed in the reduce function just like the improve repartition join described in [9].

The algorithm of Two-Way Joins Using MapReduce is as follows.

Algorithm 1 Two-Way Joins Using MapReduce

Input: the relation R *R(A, B)*,the relation S *S(B, C)*,the extend bloom filter *EBFR(B)*,the extend bloom filter *EBFS(B)*

Output: the result of joins r $R \bowtie S$

1	map(key = relationName, value = tuple)
2	joinAttrs = tuple.getValues(*B*);
3	**if** relationName == *R* **then**
4	**if**(EBFS(joinAttrs)==true) **then**
5	nonJoinAttrs = tuple.getValues(*A*);
6	output(joinAttrs+ relationName, nonJoinAttrs+relationName);
7	**end if**
8	**end if**
9	**if** relationName = *S* **then**
10	**if**(EBFR(joinAttrs)==true) **then**
11	nonJoinAttrs = tuple.getValues(*C*);
12	output(joinAttrs+ relationName, nonJoinAttrs+relationName);
13	**end if**
14	**end if**
15	partition(pair *p*, *values*)
16	hashcode = hashfunc(*p*.firstKey)
17	return hashcode % reducersNumber
18	reduce(key=joinAttrs+relationName, value=list(nonJoinAttrs+relationName))
19	rList = new List();
20	tag=key.getSecond();
21	**for each** (nonJoinAttrs+relationName) in value **then**
22	if(relationName == tag) **then**
23	rList.add(nonJoinAttrs);
24	**end if**
25	**else then**
26	**for** each a in rList **then**
27	output(a + key.getfirst() + nonJoinAttrs);
28	**end for**
29	**end else**
30	**end for**

HDFS   Map   Reduce   HDFS

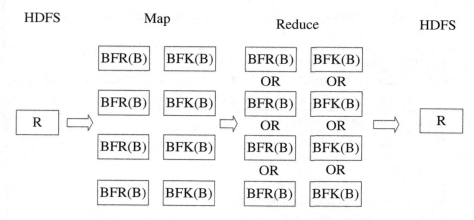

Fig. 2. Example for a extend bloom (*EBFR(B)*) filter generating

Mutual filtering strategy by improved Filter Bloom is mainly in order to filter out the two table does not meet the join conditions of the tuple, reduce the output of the map phase, greatly reduce the shuffle phase of the time, as well as the I/O overhead of the data. The efficiency of the system to perform the join task will be greatly improved.

Figure 3 is an example of the second phase of the algorithm. Relation R and S are stored in HDFS. The first phase of the MapReduce program for relation R generates *EBFR(B)*. and it do the same thing for the relation S. The second stage is shown in Fig. 3. Each map function reads the block of relation R and S. The map function reads the *EBFR(B)* and *EBFS(B)*. The map function uses it *EBFR(B)* to filter relation S. Therefore, the values of join attribute B *(6, 7, 8, 9)* are filtered out without passing through the network transmission in the relation S, just send a value of 1 and 2 to the reduce phase.

4.2 Multi-way Joins Using MapReduce

In this section, we solve the problem how to use bloom filters to a multi-way joins. Let us consider the case of a 3-way joins: $R(A, B) \bowtie S(B, C) \bowtie T(C, D)$. We can implement this join by a sequence of two two-way joins, choosing either to join R and S first, and then join T with the result. In [10], Afrati proposed another algorithm to deal with this join using only one MapReduce job. However, there are still a lot of tuples to be copied in this process. Naturally, we can use extend bloom filters to filter useless data to improve the efficiency of the multi-way joins.

We introduce an algorithm called multi-way-bf join having two MapReduce phases. In the first MapReduce phase, we build *EBFS(B)* and *EBFS(C)* for the relation S on the attribute B and C. In the second MapReduce phase, we use the *EBFS(B)* and *EBFS (C)* and adopt the algorithm [10] to get the final results.

The algorithm of the Replicated join with extend bloom filter as follows.

Algorithm 2 Multi-way Joins Using MapReduce

Input: the relation R $R(A, B)$, the relation S $S(B, C)$, the relation T $T(C, D)$, the extend bloom
filter $EBFS(B)$, the extend bloom filter $EBFS(C)$

Output: the result of (key, value)

1	map(key = relationName, value = tuple)
2	joinAttrs = tuple.getValues(B);
3	**if**(relationName == R) **then**
4	**if**(EBFS(joinAttrs)==true) **then**
5	nonJoinAttrs = tuple.getValues(A);
6	output(joinAttrs+ relationName, value);
7	**end if**
8	**end if**
9	**else if**(relationName = T) **then**
10	**if**(EBFR(joinAttrs)==true) **then**
11	nonJoinAttrs = tuple.getValues(C);
12	output(joinAttrs+ relationName, value);
13	**end if**
14	**end if**
15	**else then**
16	nonJoinAttrs = tuple.getValues(D);
17	output(joinAttrs+ relationName, value);
18	**end else**

The algorithm of the Replicated join with extend bloom filter as follows.

Algorithm 3

Input: the (key, value), the number of partitions numPartitions
Output: the array of (key ,value)

1	Partition(pair p, value, int numPartitions)
2	We are connected to the attribute column B, C set a hash bucket value $n1, n2$
3	n1*n2=numPartition;
4	sList=new List();
5	**if**(p.secondKey $\in R$) **then**
6	$H(B)$ = hash(value.B) % n_1;
7	**for** i = 0 in n_2 - 1 **then**
8	sList[i] = $H(B)$ + $i * n_2$;
9	**end for**
10	**end if**
11	**else if**(p.secondKey $\in R$) **then**
12	$H(B)$ = hash(value.B) % n_1;
13	H(C) = hash(value.C) % n_2;
14	List[0] = $H(B)$ + $H(C)$ * n_2;
15	**end else if**
16	**else then**
17	$H(C)$ = hash(value.C) % n_2;
18	for i = 0 in n_1 – 1 then
19	sList[i] = $H(C)$ + $i * n_1$;
20	**end for**
21	**end else**
22	return List;

The algorithm of the Replicated join with extend bloom filter as follows.

	Algorithm 4 The reduce of Multi-way Joins Using MapReduce
	Input: (key ,value)
	Output: the result of joins $R \bowtie S \bowtie T$
1	reduce(key=joinAttrs+relationName, values=list(value))
2	rList = new List();
3	sList = new List();
4	**for each** value in values **then**
5	**if**(value.tag == R) **then**
6	rList.add(value);
7	**end if**
8	**else if**(value.tag==S) **then**
9	sList.add(value);
10	**end else if**
11	**else then**
12	**for each** a in rList **then**
13	**for each** b in sList **then**
14	**if**(a.B==b.B&&b.C==value.C) **then**
15	output(a + b + value);
16	**end if**
17	**end for**
18	**end for**
19	**end for**

The general process of the algorithm is as follows, Let H be a hash function. The hash range is from 1 to n. Use *(i, j)* label each reduce, the range of values of I and j is 1 to n. Each element *S(B, C)* is sent to the label *(h(B), h(C))* for the reduce, Each element R(A, B) is sent to the label *(h(B), *)* for the reduce, * represents any value, Each element *T(C, D)* is sent to the label *(*, h(C))* for the reduce. In the end, the equivalent connection operation is done in each reduce.

5 Experiments

Our experiments run on a cluster consisting of 1 master node (served as both Namenode and Jobtracker) and 3 slave nodes. The release of Hadoop is 2.6.0. The size of each data block, which has 3 replicas on HDFS, is no more than 64 MB. In this

HDFS   Map   Reduce   HDFS

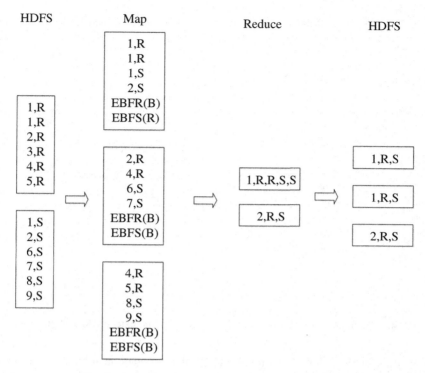

Fig. 3. Example of the second phase of the algorithm

paper, the data used in the experiments are derived from TPC-H. We compare our algorithm with the algorithm called EBF-M and EBF-RSJ.

We used the default Hadoop join RSJ and the EBF-RSJ to join these two relations on the cluster respectively. In the experiment of two table joins, we use relation ORDERS and CUSTOMER to do the join experiment in the TPC-H. Six sets of data of different sizes are selected in the experiment. The different datas are show as follows. Figure 4 shows the performance of the two joins (Table 1).

Table 1. Six groups of data for two relations

Data	1	2	3	4	5	6
CUSTOMER	160M	225M	525M	750M	900M	1.1G
ORDERS	175M	275M	400M	760M	940M	1.2G

From Fig. 4, when the size of the relation is small, RSJ is as the same as EBF-RSJ, because they add the additional MapReduce rounds to waste time. When the size of the relation is bigger, EBF-RSJ is more efficient than RSJ, because this filters a lot of useless data to save network overhead and processing overhead.

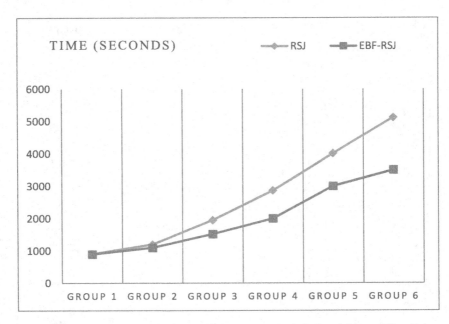

Fig. 4. MapReduce time for the default Hadoop join and extend Bloom Filter Join

For joins of three (or more) relations, the default Hadoop join has to join two relations and then join the intermediate result with the third relation. Figure 5 shows the performance comparison of Hadoop join and EBF-M. The different datas are show as follows (Table 2).

Table 2. Six groups of data for three relations

DATA	1	2	3	4	5	6
CUSTOMER	65M	125M	200M	400M	640M	780M
ORDERS	70M	100M	240M	390M	520M	680M
LINEITEM	81M	110M	234M	350M	510M	620M

We compare our algorithm with the algorithm called EBF-M and the default Hadoop join. The result is shown in Fig. 5. From Fig. 5, we can observe that our method is as efficient as the default Hadoop join when the relations are small, because our method adds an additional MapReduce phase to build the bloom filters. While the size of relations become large, our method is more efficient, as our method uses the extend bloom filter to filter a lot of useless data to save the network overhead and processing overhead.

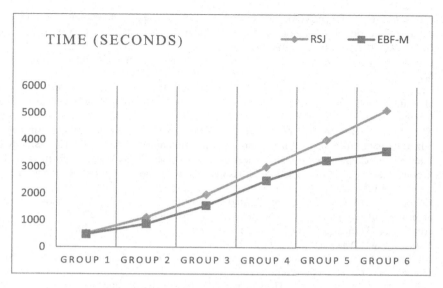

Fig. 5. MapReduce time for multi-way joins

6   Conclusion

In this paper, We present and compare an extended bloom filter for a large dataset using MapReduce, and we present a way to join relations within MapReduce framework by extend bloom filter. For joins of two relations, our approach avoids sorting large data records and reduces network overhead. It proves more efficient than the default join algorithm provided by Hadoop. For multi-joins, which the default join in Hadoop does not support directly, our method can improve the performance of multi-way joins.

In the future, we are planning to develop a dynamic cost analyzer. This will help us to implement a best MapReduce approach to any multi-way joins problems. We are also planning to investigate techniques of incorporating Hadoop parameters into the cost model to improve the join efficiency.

Acknowledgements. This paper was partially supported by National Sci-Tech Support Plan 2015BAH10F01 and NSFC grant U1509216, 61472099, 61133002.

References

1. Lueebber D, Grimmer U.: Systematic development of data mining based data quality tools. In: 29th VLDB (2003)
2. Dean, J., Ghemawat, S.: MapReduce: simplified data processing on large clusters. In: OSDI, pp. 137–150 (2004)
3. Apache Software Foundation. Hadoop, April 2010. http://hadoop.apache.org

4. Mackert, L.F., Lohman, G.M.: R* optimizer validation and performance evaluation for distributed queries. In: Proceedings of the 12th International Conference on Very Large Data Bases (VLDB), pp. 149–159 (1986)

5. Lee, K.-H., Lee, Y.-J., Choi, H., Chung, Y.D., Moon, B.: Parallel data processing with MapReduce: a survey. ACM SIGMOD Rec. **40**(4), 11–20 (2011)

6. Blanas, S., Patel, J.M., Ercegovac, V., Rao, J., Shekita, E.J., Tian, Y.: A comparison of join algorithms for log processing. In: Proceedings of the 2010 ACM SIGMOD International Conference on Management of Data (SIGMOD 2010), pp. 975–986 (2010)

7. Yang, H.-C., Dasdan, A., Hsiao, R.-L., Parker, D.S.: Map-reduce-merge: simplified relational data processing on large clusters. In: Proceedings of the 2007 ACM SIGMOD International Conference on Management of Data (SIGMOD 2007), pp. 1029–1040 (2007)

8. Bloom, B.H.: Space/time trade-offs in hash coding with allowable errors. Commun. ACM (CACM) **13**(7), 422–426 (1970)

9. Blanas, S., Patel, J.M., Ercegovac, V., Rao, J., Shekita, E.J., Tian, Y.: A comparison of join algorithms for log processing in MapReduce. In: SIGMOD, pp. 975–986 (2010)

10. Afrati, F.N., Ullman, J.D.: Optimizing multiway joins in a map-reduce environment. IEEE Trans. Knowl. Data Eng. **23**(9), 1282–1297 (2011)

11. Broder, A., Mitzenmacher, M.: Network applications of bloom filters: a survey. In: Internet Mathematics, pp. 636–646 (2002)

12. Lee, K.-H., Lee, Y.-J., Choi, H., Chung, Y.D., Moon, B.: Parallel data processing with MapReduce: a survey. In: SIGMOD, pp. 11–20 (2011)

13. Yang, H.C., Dasdan, A., Hsiao, R.-L., Parker, D.S.: Map-reduce-merge: simplified relational data processing on large clusters. In: SIGMOD 2007, pp. 1029–1040 (2007)

14. Friedman, E., Pawlowski, P., Cieslewicz, J.: SQL/MapReduce: a practical approach to self-describing, polymorphic, and parallelizable user-defined functions. In: Proceedings of VLDB (2009)

Similarity Search on Massive Data Based on FPGA

Yanzheng Wang, Hong Gao, Shengfei Shi, and Hongzhi Wang[✉]

School of Computer Science and Technology,
Harbin Institute of Technology, Harbin, China
{yz_wang, honggao, shengfei, wangzh}@hit.edu.cn

Abstract. Data quality is a very important question in massive data process. When we want to distill valuable knowledge from a mass set of data, the key point is to know whether the dataset is clean. So before we extract useful massage from the dataset we'd better do some data clean job. Similarity search is a very important method in data clean. MapReduce will be used to do similarity search in our data clean system. But the efficiency is very low. We found that when we process the massive data stored in HDFS with MapReduce programing model every part of the dataset will be scanned and this is very time-consuming especially for large scale dataset. In this paper we will do filter operation on original data with hardware before we use similarity search to do data clean.

Keywords: Data clean · FPGA · Similarity search · MapReduce

1 Introduction

There is growing enthusiasm for the notion of "Big Data". More and more people want to find treasure from "Big Data". However data quality issues will result in lethal effects of big data applications. Therefore clean the massive data with the problem of quality is very important. Real treasure will be found only the data quality issue is taken seriously [1].

In traditional relation database, multi-tuples representing the same entity is the most common type of poor-quality data. Organizing the multi-tuples which represents the same entity is an effective method of management of poor-quality data. A Similarity search [2–4] problem is that given a query, one can get a list of results and each pair of them meets the similarity threshold. Similarity search is a very important technique in massive data clean.

In order to clean large amounts of data, we use MapReduce [5] to do similarity search on massive data stored in HDFS [6]. MapReduce is a part of Hadoop environment and it is a programming model for processing large data sets with a parallel, distributed algorithm on a cluster.

Although the performance is very good when we do similarity search with MapReduce on massive data, it becomes slow as data size stored in HDFS grows fast. MapReduce will scan every part of the table stored in HDFS and this is very

© Springer International Publishing Switzerland 2016
H. Gao et al. (Eds.): DASFAA 2016 Workshops, LNCS 9645, pp. 343–352, 2016.
DOI: 10.1007/978-3-319-32055-7_28

time-consuming. In order to fix this problem we will use FPGA [7] to filter the original data. And FPGA does better in this job.

The main contributions of the paper are summarized as follows:

1. Put FPGA into hadoop environment and call FPGA to do filtration job. In order to use FPGA in hadoop, we changed Mapreduce programing model.
2. Two algorithm was proposed and implemented. A lot of experiments was performed to test our system.

The rest of the paper is organized as follows: Sect. 2 describes the background of our work. We present two algorithms in Sect. 3 and we did some explain for each of them. In Sect. 4, we give the results of our experiment evaluation. Finally, Sect. 5 concludes this paper.

2   Background

2.1   Similarity Search

Due to data reported many times or other human factor, it's quite normal for data repeat in real work environment. Field similarity was used to judge repeated data. The similarity factor S $(0 < S < 1)$ between two fields represent the level of similarity. It is calculated according to the content of the fields. The smaller of S, the similarity between the fields. S = 0 means the two fields completely same. The method to calculate S is different according to the field type.

For bool type, if the two fields are equal, S is zero; otherwise, S is one.

For numeric field, we use relative difference to get the similarity factor. It can be represent by

$$S(S\_1, S\_2) = |S\_1 - S\_2| / \max(S\_1, S\_2) \tag{1}$$

For character type, there is a relatively easy method to calculate the similarity factor. Divide the number of matching character by the average number of the two character string.

$$S(S\_1, S\_2) = |K| / ((|S\_1| + |S\_2|)/2) \tag{2}$$

In this formula, K is matching character of the two character string.

Set the threshold and discover similar objects with similarity search. Then we can do delete operation or other data clean operations.

2.2   MapReduce Programing Model

In recent years Hadoop was used to solve massive data problem due to its distributed file system and MapReduce programing model. MapReduce is a software framework capable of processing large amounts of data-sets in parallel across a distributed cluster of processors or stand-alone computers. A MapReduce program is composed of two procedures:

- **Map()** procedure performs filtering and sorting
- **Reduce()** procedure performs a summary operation

2.3   System Architecture

Our data clean system is based on Hadoop environment, data-sets stored in HDFS (Hadoop Distributed File System) and processed with MapReduce. In order to speed up similarity search module, we use FPGA to do filter operation instead of CPU. We can see FPGA does better than CPU in this job in many previous papers. As we write before, Map procedure performs filtering and sorting. So we will add FPGA into Map procedure to do filter job (Fig. 1).

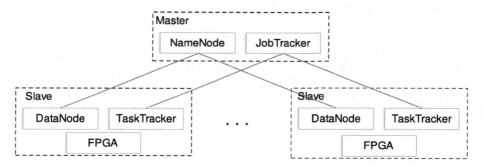

Fig. 1. Hadoop system with FPGA. As we can see from this architecture, FPGA will insert into each slave.

When we use MapReduce to do data clean job, we will change that job into another form which can be done with FPGA.

2.4   File Format

ORCFile [8, 9] was introduced in Hive 0.11 and each ORCFile is composed of one or several stripes. The default size of stripe is 250 MB. Stripes have three sections: a set of indexes for the rows within the stripe, the data itself, and a stripe footer.

This file format will be used as default file format in our data clean system because of its excellent compression. This file format is convenient for hardware to process (Fig. 2).

The stripe footer contains the encoding of each column and the directory of the streams including their location. In row data each column is stored separately. Index data includes min and max values for each column and the row positions within each column. We will use the statistic information of each column stored in index data to do coarse filtration.

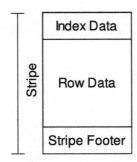

Fig. 2. Stripe's structure, ORCFile is composed of stripes.

3 Filter Operation

ORCFile was used as our default file format. Our dataset was stored in HDFS in this format. Each file stored in HDFS will be sliced into several ORCFiles and every ORCFile composed by some stripes. Stripe will be processed as a whole, it won't be split at here. This operation was added into map function. Map function was changed by us to process with FPGA (Fig. 3).

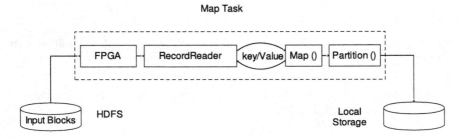

Fig. 3. Map task's execution with FPGA

3.1 Coarse Filtration

If we just use a part of data of one file and the rest of the file has nothing to do with the result, we do not want to scan all of the file. If we can read the data related the result only we can avoid many I/O time-consuming.

Stripe has some index data, we use this information to do coarse filtration. When we store data file in HDFS, program will calculate statistical information for each column. Each stripe has statistical information for its row data. This information will be used for coarse filtration.

Algorithm 1. Coarse Filtration

Input: *sinfo* – Stripe Information
Input: *ficon* – filter conditions
Output: *true* iff this stripe can be ignored without read
 1: **Function** coarseFil(*sinfo, ficon*)
 2: **if** we want to use FPGA to do filter job **then**
 3:     **for** each filter condition *c* ∈ *ficon* **do**
 4:         **if** the variable in condition *c* can be used to do filtration **then**
 5:             get *s* ∈ *sinfo*
 6:             result ← compare(*c, s*);
 7: update the ficon with result;
 8: **for** each operator *op* ∈ *ficon* **do**
 9:     **if** *op* is 'a|b' **then**
10:         return a&b;
11:     **if** *op* is 'a&b' **then**
12:         return a|b;

For every stripe, we can get its statistic information by its id. Then the information will be used by function coarseFil. In function coarseFil we can decide whether this stripe needs to be read from disk through compare stripe's statistic information with filter conditions. If one stripe needs not to be read from disk, we can avoid I/O operations on the stripe. This can save our time.

We analyze the filter condition and compare the data user defined with every stripes statistic information in Line 2–6. After we will replace the filter condition with the compare result. And the final result will be calculated in Line 8–12.

3.2   Add Hardware into Software

In this paper, we use FPGA to do filtration job. We expect FPGA will give us a good performance in this kind of job. In order to use FPGA in hadoop environment we will change original system.

We pass the whole stripe and some useful information to FPGA. FPGA will do filter operations with the information on that stripe. Result will be returned to software after filtration. Software will use the result to do similarity search. Because the data used to do similarity search was filtered by FPGA. Unnecessary data won't be used at this time.

For each stripe, it will be read from disk and passed to FPGA through interface. Along with the stripe, some useful information will be used by FPGA as parameter. We get the filter results from FPGA and put it into memory, it will used to do other things.

Algorithm 2. Interface to access Hardware

Input: *finfo* – ORCFile which we are processing
Input: *ficon* – filter conditions
Input: *rdata* – raw data will be used by FPGA
 1: **Function** visit_FPGA(*finfo, ficon, rdata*)
 2: **if** we want to use FPGA to do filter job **then**
 3:    **for** each stripe *s* ∈ *finfo* **do**
 4:        *t* ← coarseFil(*ficon, s*);
 5:        **if** *t* is true **then**
 6:            get *sinfo* from stripe;
 7:            get *raw_data* from stripe;
 8:            set_para_FPGA(*sinfo*);
 9:            get_result(*raw_data*);
10: put each result for stripe into memory pool;
11: **for** each key/value in memory pool **do**
12: pass down the data to do similarity search;

We get each stripe information from metadata (Line 6). Then the raw data waiting to be processed will be read from disk (Line 7). Function set_para_FPGA will be used to set parameters for FPGA (Line 8). We can get the final result from FPGA through function get_result (Line 9).

3.3 Mechanism of FPGA

In recent years, many researches use FPGA (field-programing gate arrays) to process high-volume data, e.g., data mining [10, 11], image processing [12–14], or other high-throughput applications [15, 16]. It seems that we can make use of FPGA in the field of data clean.

The core component of FPGA is a series of processing unit. Each unit can process two kinds of judge sentences.

1. Sentence like *column θ constants*. In this sentence θ is compare symbol, it include =, <>, >, <, >=, <=. *Column* is the column used to do compare. *Constant* represent a constant or string, it was used to be compared.
2. Sentence like column θ column. In this sentence θ is compare symbol, it include =, <>, >, <, >=, <=. Two columns will be used to do compare.

When a query comes, CPU will analyze that query and pass the parameter to FPGA. Processing unit of FPGA can calculate whether the data meet the condition based on parameter immediately. It will send signal "1" when the condition are met. Otherwise, it will send signal "0".

4 Experiments

4.1 Framework

The algorithms introduced in Sect. 4 have been implemented in Hadoop-2.6.0. The experiments were performed on Hadoop system with one namenode and three datanode. Each node running Ubuntu 14.04 equipped with one Intel Core i5-2400 3.10 GHz quad-core processors and 8 GB DRAM.

The dataset we used in experiments is generated by TPC-H. We will use different size to test our system because of we assumed that the lager the dataset the better the performance of this system. And our FPGA will just do filter job this time, so we will just use one table in TPC-H.

The experiments will tested with and without coarse filtration in a series of size of dataset. In this way we can see the performance of coarse filtration and FPGA alone.

4.2 With FPGA Alone

Set the MapReduce run with FPGA and run test query in this model. Compare the result with the original Hadoop. Do this in different size of dataset.

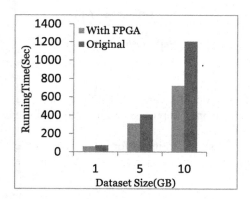

Fig. 4. Running time on three kind of dataset size   **Fig. 5.** Performance improved ratio

In Fig. 4 we can see two system's time cost on three different size of dataset. It shows that we can improve the performance with FPGA. Time cost by system with FPGA less than original Hadoop. Figure 5 shows the ratio between our system and original one. With the increase of the amount of data, the ratio will increase. It means our system perform better on high-volume dataset.

4.3   Add Coarse Filtration

We know how coarse filtration works from algorithm 1. It performs filtration based on statistic information of the whole stripe. So it only works on stripe rather than rows. This means the performance relate with the query sentence. And if one column of the file was ordered, the performance will be better.

In this section query sentence include a range query on the ordered column. Therefore, the coarse filtration will be used. Otherwise, we cannot test the performance of it (Figs. 6 and 7).

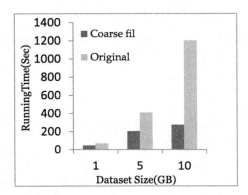
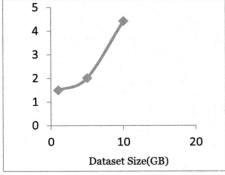

Fig. 6. Running time on three kind of dataset size

Fig. 7. Performance improved ratio

We can imagine that if the range query on ordered column is fixed, the large the dataset the more data will be filtered. The performance improved significantly because of the filtered data won't be read from disk.

However, we can't say this System is the best due to its performance rely on the query sentence. If we can't filter data from it, its performance is not better than the system with FPGA alone.

5   Conclusion and Outlook

In this paper, we proposed filtration based on FPGA to improve the performance of data clean system based on Hadoop. We want to reduce the running time of the most time-consuming part. We use FPGA to do filtration job to reduce I/O time and ease CPU's pressure. The experiment results show that our system performs better than the original one.

Coarse filtration performs better during the query sentence include range query on the ordered column. We can filter large amount of data through it when the dataset is ordered. If the dataset is disorganized, coarse filtration will not bring us benefits. In order to fix this problem we can chose use it or not based on the query sentence.

We can do many things with FPGA [17–19] because of its inherent advantages in data process. We want to implement join, group by and other operations with FPGA. We hope FPGA and other hardware play a huge role in massive data process in the future.

Acknowledgements. This paper was partially supported by National Sci-Tech Support Plan 2015BAH10F01 and NSFC grant U1509216, 61472099, 61133002.

References

1. Rahm, E., Do, H.: Data cleaning: problems and current approaches. IEEE Data Eng. Bull. **23**(4), 3–13 (2000)
2. Morales, G.D.F., Lucchese, C., Baraglia, R.: Scaling out all pairs similarity search with mapreduce. In: 8th Workshop on LargeScale Distributed System for Information Retrieval (2010)
3. Bayardo, R.J., Ma, Y., Srikant, R.: Scaling uop all pairs similarity search. In: Proceeding of WWW (2007)
4. Awekar, A., Samatova, N.F.: Fast matching for all pairs similarity search. In: Intelligent Agent Technology Workshop (2009)
5. Dean, J., Ghemawat, S.: MapReduce: simplified data processing on large clusters. In: Proceedings of the 6th OSDI, vol. 51, no. 1, pp. 107–113 (2004)
6. HDFS (Hadoop Distributed File System) Architecture. http://hadoop.apache.org/core/docs/current/hdfs_design.html
7. Sukhwani, B., Hong, M., Thoennes, M., Dube, P., lyer, B.: Database analytics acceleration using FPGAs. In: International Conference on Parallel Architectures and Compilation Techniques, pp. 411–420 (2012)
8. http://docs.hortonworks.com/HDPDocuments/HDP2/HDP-2.0.0.2/ds_Hive/orcfile.html
9. https://cwiki.apache.org/confluence/display/Hive/LanguageManual+ORC
10. Woods, L., Teubner, J., Alonso, G.: Real-time pattern matching with FPGAs. In: IEEE International Conference on Data Engineering, pp. 1292–1295 (2011)
11. Teubner, J., Muller, R., Alonso, G.: Frequent item computation on a chip. IEEE Trans. Knowl. Data Eng. **23**(8), 1169–1181 (2011)
12. Zarifi, T., Malek, M.: FPGA implementation of image processing technique for blood samples characterization. Comput. Electr. Eng. **40**(5), 1750–1757 (2014)
13. Brost, V., Yang, F., Meunier, C.: Flexible VLIW processor based on FPGA for efficient embedded real-time image processing. J. Real-Time Image Process. **9**(1), 47–59 (2014)
14. Chenini, H., Dérutin, J.P., Aufrère, R., Chapuis, R.: Parallel embedded processor architecture for FPGA-based image processing using parallel software skeletons. J. Adv. Sig. Process. **2013**(1), 1–23 (2013)
15. Choi, Y.M., So, K.H.: Map-reduce processing of K-means algorithm with FPGA-accelerated computer cluster. In: IEEE International Conference on Application-specific System, Architectures and Processors, pp. 9–16 (2014)
16. Belean, B., Borda, M., Bot, A.: FPGA based hardware architectures for iterative algorithms implementations. In: International Conference on Telecommunications and Signal Processing, pp. 751–754 (2013)

17. Becher, A., Bauer, F., Ziener, D., Teich, J.: Energy-aware SQL query acceleration through FPGA-based dynamic partial reconfiguration. In: International Conference on Field Programmable Logic and Applications, pp. 1–8 (2014)
18. Dennl, C., Ziener, D., Teich, J.: On-the-fly composition of FPGA-based SQL query accelerators using a partially reconfigurable module library. IEEE Int. Symp. Field-Programma Custom Comput. Mach. **282**(1), 45–52 (2012)
19. Halstead, R.J., Sukhwani, B., Min, H., Thoennes, M., Dube, P., Asaad, S., Iyer, B.: Accelerating join operation for relational databases with FPGAs. In: Proceeding of the 2013 IEEE 21st Annual International Symposium on Field-Programmable Custom Computing Machines, pp. 17–20 (2013)

Skyline Join Query Processing
over Multiple Relations

Jinchao Zhang[1,2], Zheng Lin[1(✉)], Bo Li[1], Weiping Wang[1], and Dan Meng[1]

[1] Institute of Information Engineering, Chinese Academy of Sciences, Beijing, China
{zhangjinchao,linzheng,libo,wangweiping,mengdan}@iie.ac.cn
[2] University of Chinese Academy of Sciences, Beijing, China

Abstract. Skyline query on multiple relations, known as skyline join query processing, attracts much attention recently. However, most of the existing algorithms perform skyline join just on two relations. In this paper, we propose an efficient algorithm *Skyjog*, which is applicable for skyline join on two or even more relations. *Skyjog* divides each relation into two or three partitions. Based on the proposed group division approach, tuples generated by several join combinations of these partitions definitely are skyline points. *Skyjog* only has to examine tuples of other join combinations. Thus, *Skyjog* achieves performance efficiency by avoiding much skyline computation. Experiments demonstrate that *Skyjog* has an outstanding performance on all datasets, and outperforms the state-of-the-art skyline join algorithms on both two relations and more than two relations.

Keywords: Skyline join · Multiple relations · Group division

1 Introduction

Skyline query aims to find interesting points from the given dataset. Taking the classic seaside tourism problem [1] for example, a tourist traveling to the seaside is looking for a hotel that is not only cheap but also close to the beach. Those hotels with optimized dimensions of price and distance are the interesting ones, i.e., there does not exist a hotel that is both cheaper and closer to the beach than interesting hotels. The process of finding interesting hotels is skyline query, and the price and distance are coined as *skyline attributes*.

In some applications, *skyline attributes* belong to multiple relations, and the issue of computing skylines on multiple relations is termed as *skyline join* query processing. There exists many *skyline join* algorithms, while most of them are applicable for two relations. Two algorithms S^2J-M and S^3J-M [6] have been proposed for *skyline join* on more than two relations. These two algorithms are not efficient enough due to the complex computation procedure.

In this paper, we study the issue of *skyline join* query processing over multiple relations. Inspired by group division method in [3], we develop a novel approach that can directly identify skyline points on multiple relations. Based on this

© Springer International Publishing Switzerland 2016
H. Gao et al. (Eds.): DASFAA 2016 Workshops, LNCS 9645, pp. 353–361, 2016.
DOI: 10.1007/978-3-319-32055-7_29

approach, we propose an efficient algorithm *Skyjog*(Skyline join query process-ing by group division) for *skyline join* query on multiple relations. Each time, *Skyjog* selects two most left relations as the operands. Since tuples generated by some join combinations of partitions are guaranteed to be skyline points, *Skyjog* only determines tuples of other join combinations. Due to the skyline points identification without joining then calculating, *Skyjog* efficiently reduces the size of intermediate results, and avoids much skyline computation. Thus, *Skyjog* achieves an outstanding performance on all experiments.

2   Preliminaries

Let R denote a relation, and B denote the attribute set of R. Let τ and τ' denote tuples in R. We say that the tuple τ' is dominated by τ, denoted as $\tau \prec_B \tau'$, if following conditions are satisfied simultaneously, (1) $\tau.a_i \leq \tau'.a_i{}^1$, (2) $\exists a_j, \tau.a_j < \tau'.a_j$, where $a_i, a_j \in B$, and $\tau.a_i$ denotes the value of τ on the attribute a_i. The skyline query of R with respect to B is defined as $SKY_B(R)= \{\tau_t \mid \not\exists \tau_k \text{ s.t. } \tau_k \prec_B \tau_t, \tau_t, \tau_k \in R\}$. The attributes in B are termed as *skyline attributes*. For simplicity, we use $SKY(R)$ instead of $SKY_B(R)$ in the following.

In some cases, the *skyline attributes* are distributed over multiple relations. For example, for a skyline query on M relations $R_1, R_2, ..., R_M$, the set of *skyline attributes* consists of M disjoint sets, $B_1, B_2, ..., B_M$, where B_i is the *skyline attribute* set of R_i. Intuitively, skyline query on these relations can be calculated by joining all these M relations firstly, then performing the skyline query on the join result set with respect to the union set of $B_1, B_2, ..., B_M$. This query is formalized as $SKY_{B_1 \cup B_2 \cup ... \cup B_M}(R_1 \bowtie R_2 \bowtie ... \bowtie R_M)^2$, and we name this query as *skyline join* query since the skyline query is performed on a join result set. This approach for calculating *skyline join* is inefficient, since it performs skyline query on a large intermediate dataset that is generated by joining all M relations. Algorithms for *skyline join* should prune useless tuples before join operation, and reduce the skyline computation as much as possible.

As stated in [3], for *skyline join* on two relations, each relation is partitioned into groups in terms of the join attribute, i.e., tuple τ and tuple $\tau\prime$ are in the same group if the join attribute of them are the same. If τ is not dominated by other tuples of its group, then τ is a local skyline point. If τ is not dominated by other tuples of all groups, then τ is a global skyline point. Tuple τ belongs to one of three cases[3], (a) *LSS*, τ is both a local skyline point and a global skyline point. (b) *LSN*, τ is a local skyline point, but not a global skyline point. (c) *LNN*, τ is neither a local skyline point nor a global skyline point. Tuples belong to *LNN* are dominated by *LSN* tuples in the same group, and they cannot contribute to the finally skyline results. Thus, they should be pruned before the join operation.

---

[1] Small values are preferable in this paper.

[2] Join referred in this paper indicates equi-join operation.

[3] *LSS*, *LSN* and *LNN* are denoted as *LS(S)*, *LS(N)* and *LN(N)* in original paper.

3   Our Solution

3.1   The Amendment on Previous Work

Let $LSS(R_i)$ denote the LSS tuples of relation R_i, and $LSN(R_i)$ denote LSN tuples of relation R_i. The existing *skyline join* query algorithms in [3,11] are based on the property of $LSS(R_1) \bowtie LSS(R_2) \subseteq SKY(R_1 \bowtie R_2)$ (property 1) and $LSN(R_1) \bowtie LSS(R_2) \subseteq SKY(R_1 \bowtie R_2)$ (property 2), where R_1 and R_2 are joinable relations. However, we find that *property 2* is not always established. As shown in Fig. 1, the *HID* and *TID* are join attributes for R_1 and R_2 respectively, and the rest are *skyline attributes*. We note that the tuple *(A,5,3)* belongs to the $LSN(R_1)$, and the tuple *(A,5,10)* belongs to $LSS(R_2)$, but $(A,5,3) \cdot (A,5,10)$[4] is not a skyline tuple, since it is dominated by *(B,7,6)* · *(B,5,10)*.

R_1

HID	Price	Distance
A	4	2
A	5	3
B	7	6

R_2

TID	Rate	Quantity
A	5	10
B	5	10

$R_1 \bowtie R_2$

HID/TID	Price	Distance	Rate	Quantity
A	4	2	5	10
A	5	3	5	10
B	7	6	5	10

Fig. 1. An example for skyline join query

The *property 2* holds if we put a restriction on relations, *the vector consisting of all skyline attributes of the tuple is unique in inter-group*, i.e., for a tuple τ in group A, then there does not exist a tuples $\tau\prime$ in other groups that $\tau.a_i$ is equal to $\tau\prime.a_i$, for any $a_i \in B$. The following statement and datasets in experiments follow this restriction.

3.2   Foundation of Our Algorithm

As we know, join operation leads to an explosive increase in the cardinality of datasets. Besides that, *skyline join* query suffers the rapid growth of the number of dimensions, which is induced by the join operation as well. Intuitively, identifying skyline results before join operation is an effective approach for performance improvement. Based on the properties, calculating *skyline join* on two relations R_1 and R_2 is pretty straightforward. For each relation, it is partitioned into two groups, LSS and LSN (LNN is pruned, see Sect. 2). Tuples generated by three out of four join combinations, $LSS(R_1) \bowtie LSS(R_2)$, $LSS(R_1) \bowtie LSN(R_2)$, and $LSN(R_1) \bowtie LSS(R_2)$, are guaranteed to be skyline points. Only tuples in $LSN(R_1) \bowtie LSN(R_2)$ have to be checked further.

However, employing the group division approach for *skyline join* on more than two relations is a bit complex. For *skyline join* on M relations, which is $SKY(R_1 \bowtie R_2 \cdots \bowtie R_M)$, R_1 and R_M are head relation and rear relation

---

[4] The symbol '·' is the join operator for two tuples.

respectively, they only have the join relationships with one relation. The join attribute of R_1 is termed as *right join attribute(rj)*, since it will be used in the join operation with a relation on the right of R_1. Similarly, the join attribute of R_M is termed as *left join attribute(lj)*. Each of other relations has both *lj* attribute and *rj* attribute simultaneously. The computation for *skyline join* can be in the order of head to rear, or rear to head. We use the order of head to rear in this paper.

In order to utilize the group division approach, relations have to be partitioned in terms of a join attribute, this join attribute is termed as division attribute (DA). The join attribute used for current join operation is chosen as DA. The *skyline join* on M relations is shown in Fig. 2. In each step, the first two relations are selected as the left operand and right operand for join operation respectively. Then the intermediate result is served as left operand in next processing step. We notice that all relations excluding R_1 serves as the right operand in join operation. Thus, these relations use *lj* attribute as DA, only R_1 uses *rj* attribute for group division. Similar to *skyline join* on two relations, LNN of the head relation and rear relation are pruned, since tuples in LNN cannot contribute to the final results as well. However, tuples of LNN in other relations have a opportunity to generate a skyline point.

In each step, the left operand (denoted as L) contains two parts LSS and LSN, while the right operand (denoted as R) has one more part LNN, if this operand is not the rear relation. Tuples generated by $LSS(L) \bowtie LSS(R)$, $LSS(L) \bowtie LSN(R)$, and $LSN(L) \bowtie LSS(R)$ are still guaranteed to be skyline point in $L \bowtie R$, and belong to $LSN(L \bowtie R)$. Tuples generated by other join combinations, $LSS(L) \bowtie LNN(R)$, $LSN(L) \bowtie LSN(R)$ and $LSN(L) \bowtie LNN(R)$, have to be determine further, and we use U to denote the set of all these uncertained tuples. If tuple t in U belongs $LNN(L \bowtie R)$, then it will be discarded. Otherwise, t will be put in partition of $LSS(L \bowtie R)$ or $LSN(L \bowtie R)$, depending on if t is a global skyline point or not. Then $LSS(L \bowtie R) \cup LSN(L \bowtie R)$ will be the left operand in next step.

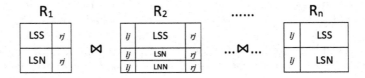

Fig. 2. Skyline join query on multiple relations

3.3   Algorithm Presentation

The pseudocode of proposed algorithm is illustrated in the Algorithm 1. The program starts with the preprocessing step (line 1–4). In this step, each relation is partitioned into three parts by DA, which is LSS, LSN, and LNN. We discard LNN of head relation and rear relation, and retain LNN of other relations. The first two relations are selected as the left operand and right operand respectively (line 6–7). Then the tuples generated by $LSS(lR) \bowtie LSS(rR)$, $LSS(lR) \bowtie LSN(rR)$, and $LSN(lR) \bowtie LSS(rR)$, are directly taken as skyline points (line 8). While tuples in $LSS(lR) \bowtie LNN(rR)$, $LSN(lR) \bowtie LSN(rR)$, and $LSN(lR) \bowtie LNN(rR)$, are identified which partition they belong to (line 10). Then, the intermediate relation is obtained by merging results of the last two steps(line 11–12), and this relation is inserted as the head relation (line 13).

Algorithm 1. Skyjog
Input: n relations in chain queries, $R_1, \cdots, R_n$
Output: skyline result of $R_1 \bowtie R_2 \bowtie \cdots \bowtie R_n$

---

1. **for** $i = 1$ to n
2. PR_i=preprocess(R_i)
3. $list$.push(PR_i)
4. **end for**
5. **while** $list$.size > 1
6. $lR=list$.removeHead
7. $rR=list$.removeHead
8. $tR.LSS=(lR.LSS \bowtie rR.LSS) \cup (lR.LSS \bowtie rR.LSN) \cup (lR.LSN \bowtie rR.LSS)$
9. $tP=(lR.LSS \bowtie rR.LNN) \cup (lR.LSN \bowtie rR.LSN) \cup (lR.LSN \bowtie rR.LNN)$
10. tT=groupDivision($tR.LSS,tP$)
11. $tR.LSS=tR.LSS \cup tT.LSS$
12. $tR.LSN=tT.LSN$
13. $list$.insert($0,tR$)
14. **end while**
15. output($list$.head)

---

4   Performance Evaluation

Experiments in this section are conducted on a server with a 2.2 GHz CPU, 32 GB RAM. All algorithms are implemented in Java. We modify the tool written by Börzsöny [1] to generate datasets based on independent, correlated and anti-correlated distributions. The number of relations involved in each query varies from 2 to 4. The number of dimensions ranges from 2 to 4. The join rate[5] ranges from 0.1 to 0.5. The cardinality of each dataset is fixed to 100 k, since similar results are obtained on datasets with different cardinality.

Algorithms Compared. $S^2 J$ and $S^3 J$ are two state-of-the-art algorithms for *skyline join* on two relations, $S^2 J$-M and $S^3 J$-M are two state-of-the-art algorithms for *skyline join* on more than two relations. We compare *Skyjog* with them in the experiments.

Evaluation Measures. Algorithms are evaluated by metrics of execution time, data materialized and dominance checks. Data materialized indicates the number of intermediate results generated by join operation. A dominance check determines the dominance relationship between two data points.

---

[5] Join rate indicates the proportion of the dataset that will be involved in join result.

4.1   Evaluation of Queries on Two Relations

Experiments in this subsection are carried out on two 3-dimensional relations, two relations involved in a query follow the same distribution. The execution time of the algorithms is shown in Fig. 3.

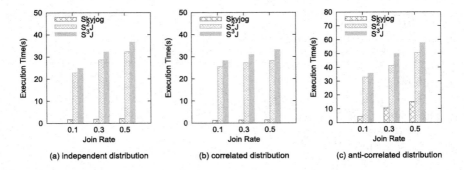

(a) independent distribution   (b) correlated distribution   (c) anti-correlated distribution

Fig. 3. Skyline join query on 2 relations

Table 1. Data materialized and dominance checks on 2 independent distributions

(a) Data Materialized

Join Rate	0.1	0.3	0.5
Skyjog	15%	15%	16%
S^2J	100%	100%	100%
S^3J	59%	59%	83%

(b) Dominance Checks

Join Rate	0.1	0.3	0.5
Skyjog	69%	70%	49%
S^2J	100%	100%	100%
S^3J	84%	78%	59%

As expected, the *Skyjog* algorithm outperforms other alternatives on all datasets. It is about one order of magnitude faster than S^2J and S^3J on independent and correlated distributions, and is about 3 times faster than S^2J and S^3J on anti-correlated distributions. With the growth of join rate, the execution time for all algorithms increases, and obviously increases on anti-correlated distributions. Since the cardinality of *skyline join* result on anti-correlated distributions severely increases with the growth of join rate, and the algorithms have to spend more time for skyline calculation.

Benefiting from group division, *Skyjog* can directly identify lots of skyline results, and avoids considerable computation. We take the query on independent distributions as an example, and investigate the metrics of data materialized and dominance checks. As shown in Table 1, we use S^2J as the baseline (100 %), and compare other algorithms with it. We see that data materialized of *Skyjog* is far less than that of S^2J and S^3J. *Skyjog* also has an advantage in terms of metric of dominance checks. Thus, we understand why *Skyjog* has an outstanding query performance.

4.2   Evaluation of Queries on More Than Two Relations

The algorithms are evaluated on 3 and 4 relations in this subsection. We aim to study the effect of dimensionality, thus set join rate as 0.1, and only vary the number of dimensions of each relation between 2 to 4. The execution time of the algorithms is shown in Figs. 4 and 5.

We see that *Skyjog* outperforms S^2J-M and S^3J-M on all datasets. It is at least one order of magnitude faster than S^2J-M and S^3J-M on independent and correlated distributions. For anti-correlated distributions, *Skyjog* is 2 to 5 times faster than S^2J-M and S^3J-M. On correlated distributions, the execution time of *Skyjog* barely increases with the growth of dimensionality. However, with the

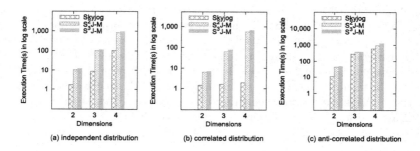

Fig. 4. Skyline join query on 3 relations

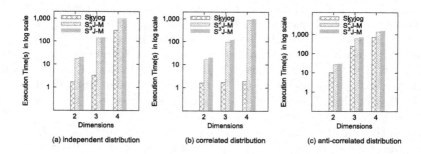

Fig. 5. Skyline join query on 4 relations

Table 2. Performance analysis on 4 independent distributions

(a) Data Materialized

Dimension	2	3	4
Skyjog	1%	4%	14%
$S^2J - M$	100%	100%	100%
$S^3J - M$	57%	61%	86%

(b) Dominance Checks

Dimension	2	3	4
Skyjog	68%	105%	127%
$S^2J - M$	100%	100%	100%
$S^3J - M$	90%	91%	91%

rise of dimensionality, the execution time of other algorithms on correlated distributions rapidly increases. The largest performance gap between *Skyjog* and these two algorithms occurs on datasets with 4 dimensions. As shown in Figs. 4(b) and 5(b), *Skyjog* is about two orders of magnitude faster than S^2J-M and S^3J-M. This can be explained by that group division prunes most of the points before join operation, while S^2J-M has to perform the join operation on all outer relations(see [6] for details) without any point pruning.

On independent and anti-correlated distributions, the execution time of all algorithms increases with the growth of dimensionality. Since the cardinality of skyline results rapidly increases with the growth of dimesionality, and the group division starts being less effective, as a result, *Skyjog* has to generate more results, and takes more time on dominance checks.

Table 2 summarizes the gains in terms of percentages on 4 independent relations. We use $S^2J - M$ as the baseline algorithm, and compare other algorithms with it. We see that, with the growth of dimensions, the number of dominance checks for *Skyjog* increases, and becomes larger than that of S^2J-M on datasets with dimensions of 3 and 4 (Table 2(b)). However, *Skyjog* generates very few intermediate results (Table 2(a)) compared to S^2J-M. Due to the stop condition, the intermediate results generated by S^3J-M is less than S^2J-M, and S^3J-M also performs less dominance checks than S^2J-M.

5   Related Work

The *skyline join* query processing over multiple relations was firstly studied by Jin et al. [3] in 2007. Inspired by this work, Sun et al. [9] proposed two algorithms for *skyline join* queries. Jin et al. [4] developed nonblocking algorithms for skyline queries on equi-joins. Raghavan and Rundensteiner [7] proposed a progressive algorithm which computes the *skyline join* queries at multiple levels of abstraction. Based on this work, a framework termed as *SKIN* [8] was proposed in a row in the literature of [7]. Bhattacharya and Teja [10] developed several efficient algorithms for aggregate *skyline join* queries. Vlachou et al. [11] proposed an algorithm *sort-first-skyline-join (SFSJ)* which is inspired by *SFS* algorithm [2]. Nagendra and Candan [5] proposed two algorithms, S^2J and S^3J, using layer/region pruning (*LR-pruning*) approach. In the following work [6], they proposed another two algorithms, S^2J-M and S^3J-M, for *skyline join* queries over more than two relations.

6   Conclusion

This paper studies the issue of *skyline join* query processing. Firstly, we fix up the property $LSN(R_1) \bowtie LSS(R_2) \subseteq SKY(R_1 \bowtie R_2)$ proposed in [3] by putting a restriction on datasets. Inspired by the properties, we devise a group division approach which is applicable for *skyline join* on two and more than two relations. Based on this, we propose an efficient *skyline join* algorithm *Skyjog*. Due to the skyline points identification without joining then calculating, *Skyjog* efficiently reduces the size of intermediate results, and avoids much skyline computation. The extensive experiments demonstrate that *Skyjog* performs well on various datasets, and outperforms the state-of-the-art algorithms on all experimental datasets.

Acknowledgments. This work is supported by the National KeJiZhiCheng Project (2012BAH46B03), the National HeGaoJi Key Project (2013ZX01039-002-001-001), the National Natural Science Foundation of China (61502478), and "Strategic Priority Research Program" of the Chinese Academy of Sciences (XDA06030200).

References

1. Borzsony, S., Kossmann, D., Stocker, K.: The skyline operator. In: Proceedings of International Conference on Data Engineering, pp. 421–430. IEEE (2001)
2. Chomicki, J., Godfrey, P., Gryz, J., Liang, D.: Skyline with presorting. In: Proceedings of International Conference on Data Engineering, pp. 717–719. IEEE (2003)
3. Jin, W., Ester, M., Hu, Z., Han, J.: The multi-relational skyline operator. In: Proceedings of International Conference on Data Engineering, pp. 1276–1280. IEEE (2007)
4. Jin, W., Morse, M.D., Patel, J.M., Ester, M., Hu, Z.: Evaluating skylines in the presence of equijoins. In: Proceedings of International Conference on Data Engineering, pp. 249–260. IEEE (2010)
5. Nagendra, M., Candan, K.S.: Skyline-sensitive joins with LR-pruning. In: Proceedings of the International Conference on Extending Database Technology, pp. 252–263. ACM (2012)
6. Nagendra, M., Candan, K.S.: Efficient processing of skyline-join queries over multiple data sources. ACM Trans. Database Syst. **40**(2), 10 (2015)
7. Raghavan, V., Rundensteiner, E., et al.: Progressive result generation for multi-criteria decision support queries. In: Proceedings of International Conference on Data Engineering, pp. 733–744. IEEE (2010)
8. Raghavan, V., Rundensteiner, E.A., Srivastava, S.: Skyline and mapping aware join query evaluation. Inf. Syst. **36**(6), 917–936 (2011)
9. Sun, D., Wu, S., Li, J., Tung, A.K.: Skyline-join in distributed databases. In: Proceedings of International Conference on Data Engineering Workshop, pp. 176–181. IEEE (2008)
10. Teja, A.: Aggregate skyline join queries: skylines with aggregate operations over multiple relations. In: Management of Data, p. 15 (2010)
11. Vlachou, A., Doulkeridis, C., Polyzotis, N.: Skyline query processing over joins. In: Proceedings of the 2011 ACM SIGMOD, pp. 73–84. ACM (2011)

Detect Redundant RDF Data by Rules

Tao Guang[1,2(✉)], Jinguang Gu[1,2], and Li Huang[1,2]

[1] College of Computer Science and Technology,
Wuhan University of Science and Technology, Wuhan , 430065, China
522307672@qq.com
[2] Hubei Province Key Laboratory of Intelligent Information Processing
and Real-Time Industrial System, Wuhan , 430065, China

Abstract. The development and standardization of semantic web technologies have resulted in an unprecedented volume of RDF datasets being published on the Web. However, data quality exists in most of the information systems, and the RDF data is no exception. The quality of RDF data has become a hot spot of Web research and many data quality dimensions and metrics have been proposed. In this paper, we focus on the redundant problem in RDF data, and propose a rule based method to find and delete the semantic redundant triples. By evaluating the existing datasets, we prove that our method can remove the redundant triples to help data publisher provide more concise RDF data.

Keywords: RDF · Data quality · Semantic redundancy · Rule

1 Introduction

The Semantic Web is a proposal to make the web more accessible to computers. In 1999 the W3C issued a recommendation of a mete data model and language to serve as the basis for such infrastructure, the Resource Description Framework (RDF) [1]. As time passed, RDF evolved and increasingly gained attraction from both researchers and practitioners as a data model apt to represent the first layer of semantics on the Web [2]. From the Linking Open Data Cloud state [3], we can see that there are about 1,014 datasets which contain more than 30 billion triples. Although gathering and publishing such data is certainly a step in the right direction, data is only as useful as its quality. Existing research [4–6] shows that many RDF datasets contains quality problems. However, they pay litter attention to the redundant problem of RDF data. Data redundancy means the wasted space used to represent certain meaning in either stand-alone data space or the Web of Data environment. Redundancy in RDF data has effects on a wide range of applications including query answering (in SPARQL endpoints), OBDA, data publishing, and ontology reasoning. We need more disk space and longer time to download a dump file from the web. This can hamper the realization of scalable RDF-based knowledge bases. Data publishers should pay attention to this problem.

We can categorize the RDF data redundancy into syntactic redundancy and semantic redundancy. Syntactic redundancy relates to the serialization of RDF data. To deal with it, RDF compression techniques have been proposed, such as RDF serializations techniques, i.e. HDT serialization [7] and K2-triples [8]. Semantic redundancy relates to the

© Springer International Publishing Switzerland 2016
H. Gao et al. (Eds.): DASFAA 2016 Workshops, LNCS 9645, pp. 362–368, 2016.
DOI: 10.1007/978-3-319-32055-7_30

semantic of RDF data. For a RDF data, if some triples can be removed without causing any change in its meaning then these triples are semantic redundancy. For example, given the following two RDF datasets:

D_1: *(?U, foaf:lastName, ?y), (?U, rdf:type, foaf:person)*
D_2: *(?U, foaf:lastName, ?y)*

In OWL2RL [9] there is a rule: if *(?p, rdfs:domain, ?c) (?x, ?p, ?y)*, then *(?x, rdf:type, ?c)*. And in the FOAF ontology[1], there is: *(foaf:lastName, rdfs:domain, foaf:person)*. Now consider the D_1 and D_2, we can see that D_1 is larger than D_2 but is not semantically richer. This means that D_1 contains redundant triples, D_1 is semantic redundancy. Pichler et al. [10] study the problem of redundancy elimination on RDF graph in the presence of rules and constrains, but their mainly work is a fine-grained complexity analysis which has no experiment. In this paper, we focus on semantic redundancy in RDF data, providing a complete experiment which uses rules and inference to remove the redundant triples in RDF data.

2   Redundancy Detection System Implementation

2.1   Entity Description Pattern

Our main purpose is to detect the redundant triples in a RDF dataset, if we know the relation between some triples then the compute time will be less. In our system, we use an entity description pattern to group the triples of RDF dataset.

In a RDF dataset, a non-literal node corresponds to an URI resource is an entity. For each entity in the dataset, it may be the subject of a triple, or it may be the object of a triple. We can extract and group triples in dataset each of which has entity e as its subject or object, these triples are the entity description block (EDB) of entity e. For an EDB, it can be summarized by a notion of entity description pattern which short name is EDP. For an entity e, its EDP is a tuple $p_e = (C_e, A_e, R_e, V_e)$, where

- $C_e = \{c_i| < e, rdf:type, c_i >\in G\}$ is called as the class component;
- $A_e = \{p_i| < e, p_i, l_i >\in G, l_i \ is \ a \ literal\}$ is called as the attribute component;
- $R_e = \{r_i| < e, r_i, o_i >\in G, o_i \ is \ a \ URI \ resource \ or \ blank \ node\}$ is called as the relation component;
- $V_e = \{v_i| < s_i, v_i, e >\in G\}$ is called as the reverse relation component.

The number of p_e in a RDF dataset is often much less than the number of entities, cause many entities have the same p_e. By grouping the entities which have the same p_e, we can make a RDF dataset to be a set of EDPs. However, p_e only contains the type information and predicate information of an entity, although the p_e of entities is the same, their object information (l_i, o_i, s_i) is always different and large, so we store them in a file which filename is the hash of p_e. By generating the EDPs of a RDF dataset, we can group the relational triples together to help detect the redundancy more quickly.

---

[1] http://xmlns.com/foaf/spec/.

2.2   EDPRule

Inference is an important characteristic of RDF data. Rule-based inference is becoming quite popular by many RDF stores and query engines. In this paper, we choose OWL2RL [9] as the base rule of the system. OWL2RL is the W3C standard and is aimed at applications that require scalable reasoning without sacrificing too much expressive power. We can get OWL2RL rules from the ontologies of the RDF data easily. Since we group the RDF data into several EDPs, we observe the OWL2RL and find that its many axioms can be transformed into EDP related rules. We call these rules EDPRule. Let us take *prp-dom* rule and *prp-spo₁* rule as an example to generate EDPRule:

prp-dom rule: *if (?p, rdfs:domain, ?c) (?x, ?p, ?y), then (?x, rdf:type, ?c),*
prp-spo₁ rule: *if(?p₁, rdfs:subPropertyOf, ?p) (?x, ?p₁, ?y), then (?x, ?p, ?y).*

From *prp-dom* rule, we can get an EDPRule easily:

EDPRule: if (?p, rdfs:domain, ?c) and R_e or A_e has ?p, then ?c in C_e is redundant

Then combine with *prp-spo₁* rule, we can extend this EDPRule to:

EDPRule: if (?p, rdfs:domain, ?c) and R_e or A_e has ?p or its sub property, then ?c in C_e is redundant

This EDPRule can also combine with other rules in OWL2RL to become a more complex rule. The central idea of transforming OWL2RL into EDPRule is to merge as many rules in OWL2RL as possible to help detect more redundant triples.

Let *ceq* be the equivalent classes of class *c*, *csup* be the super classes of *c*, *ceqsub* be the union of *c*, *ceq* and *csup*, *peq* be the equivalent properties of property *p*, *psub* be the sub properties of *p*, *psup* be the super properties of *p*, *peqsub* be the union of *p*, *peq* and *psub*, *peqsup* be the union of *p*, *peq* and *psup*, *invp* be the inverse properties of *p*, *invpeq* be the equivalent properties of *invp*, *invpsub* be the sub properties of *invp*, *invpeqsub* be the union of *invp*, *invpeq* and *invpsub*, *invpsup* be the super properties of *invp*, *invpeqsup* be the union of *invp*, *invpeq* and *invpsup*. We get the EDPRules from OWL2RL as follows:

1. *if C_e has c, then ceq and csup in C_e are redundant*
2. *if R_e (A_e) has p, R_e (A_e) has peq or psup, and their object information are the same, then peq and psup in R_e (A_e) are redundant*
3. *if R_e has peqsub, V_e has invpeqsup, and their object information are the same, then invpeqsup in V_e are redundant*
4. *if (p, rdfs:domain, c) and R_e (A_e) has peqsub or V_e has invpeqsub, then ceqsub in C_e are redundant*
5. *if (p, rdfs:range, c) and R_e (A_e) has invpeqsub or V_e has peqsub, then ceqsub in C_e are redundant*
6. *if (p, rdf:type, owl:SymmetricProperty), R_e has peqsub, V_e has peqsup, and their object information are the same, then peqsup in V_e are redundant*

If we want to get the EDPRules of an ontology file, we just need to parse it, get its class hierarchy, property hierarchy and the attributes of each property. Then follow the above six descriptions to generate its EDPRules.

2.3   System Framework

The framework (cf. Fig. 1) consisted of three components, including getting the EDP representation of RDF data, generating EDPRule from the ontologies of RDF data, using EDPRule and EDP to detect redundant triples. We describe the main components of the framework as follows:

1. **EDP Generate** component parse the RDF data, generate the EDP of each entity, then merge the entities which have the same EDP. By doing this, we can group the triples in several EDPs.
2. **EDPRule Generate** component iterate the ontology files of the RDF data, analysis them and get their EDPRules. Finally set all the EDPRules to form an EDPRule array which is called the EDPRules of the RDF data.
3. **Redundancy Detect** component just need to make EDP of step 2 matches with each of EDPRule of step 3, if the EDP satisfies the EDPRule, then delete the redundant triples and export the other triples to the concise RDF data.

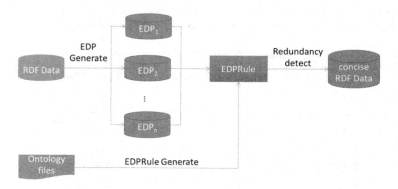

Fig. 1. Redundancy detection framework

3   Redundancy Analysis Result on Linked Open Data

3.1   Datasets

We select the datasets for analysis from Linked Open Data cloud diagram[2]. There are nine colors in the diagram which represents the datasets of different domains. We randomly select 6 datasets from these domains. The main purpose is to make our sample as representative as possible. The general information about datasets is listed in Table 1.

---

[2] http://lodcloud.net/versions/2014-08-30/lodcloud_colored.svg.

Table 1. General information about datasets

Dataset	Domain	#Triples	#EDP
Web DogFood	Publications	360,012	3253
Geospecies	Life sciences	2,201,532	1159
SISVU	Cross-domain	1,175,150	375
DBtune	Media	1,047,950	37
EARTh	Geographic	251,396	47
BFS	Government	1,705,130	391

In the table, *#Triples* is the total triples number and *#EDP* is the EDP number of the dataset. From the table, we can find that the EDP number is much less than the number of triples.

3.2 Result

Table 2 gives the result of our experiments. In the table, *#RTriples* is semantic redundant triples; *#RRatio* is the redundancy ratio, i.e. $\frac{\#RTriples}{\#Triples}$; and *#EDPRules* is the number of EDPRule of dataset. Average 20 % of triples are redundant in all analyzed datasets. The most one is BFS, which has 32.77 % redundant triples. Since #EDP in Table 1 is also small, as shown in Table 3 the redundancy detection is fast and easy.

Table 2. Semantic redundancy result

Dataset	#RTriples	#RRatio	#EDPRules
Web DogFood	37,761	10.49 %	2540
Geospecies	100,429	4.56 %	2779
SISVU	211,829	18.03 %	120
DBtune	205,676	19.63 %	411
EARTh	70,207	27.93 %	2083
BFS	558,754	32.77 %	129

3.3 Time Evaluation

Table 3 is the running time of our framework.

T_{load} is the time to load RDF files from the hard disk; T_{EDP} is the time to generate EDP; $T_{EDPRule}$ is the time to generate EDPRule and T_R is the time to detect and remove redundancy. We can see that the T_{load} is the longest. Because the size of the RDF file is large, i.e. Geospecies (178M), SISVU (151M), BFS(152M). If we directly load them, it will lead to memory overflow. So we first split the RDF file into several files, the size of each is 20M, then loading them one by one and the time of using Jena to load a RDF file of 20M is about 7 s. This is the reason why T_{load} is long. Since generating EDP is necessary, In the future we intend to use distribute solution to generate the EDP of a

Table 3. Running time

Dateset	$T_{load}(s)$	$T_{EDP}(s)$	$T_{EDPRule}(s)$	$T_R(s)$
Web DogFood	12.6	2.7	11.3	1.4
Geospecies	50.4	6.4	16.3	1.3
SISVU	52.8	10.4	0.3	1.0
DBtune	33.8	13.6	4.2	1.1
EARTh	8.1	4.0	2	0.8
BFS	41.2	8.7	8.5	2.7

RDF file. From T_{EDP} we can see that if we use distribute solution the time to load a RDF File and generate its EDP will be less than 10 s. From the table we can also find that there are significant differences between the $T_{EDPRule}$ of the six RDF datasets, the longest is 11.3 s and the shortest is 0.3 s. $T_{EDPRule}$ contains the time to load ontology files and the time to analysis them. The more ontology files a RDF dataset uses, the longer the $T_{EDPRule}$ is. We are going to establish a database to store the EDPRule of ontology which we have analysed. When we analyze a ontology which is in the database, we can get its EDPRule from the database. In the future, the time of $T_{EDPRule}$ can be ignored. What we are most concerned is T_R, from the table we can find that T_R is not more than 1.5 s, so our method can detect and remove redundant triples fast.

4 Conclusion and Future Work

In this paper, we proposed a rule based approach to deal with the semantic redundancy of RDF dataset. First, we introduce the necessity of solving the RDF data redundancy, and then we introduced the framework to detect redundant triples. Finally, evaluation result showed that our approach can detect and remove the redundant triples of RDF dataset.

In the future work, since the time to generating the EDPs for a RDF dataset is a little long, we will optimize our program with a distribute solution. And we will study other quality problems in RDF data.

Acknowledgement. This work was partially supported by a grant from the NSF (Natural Science Foundation) of China under grant number 60803160 and 61272110, the Key Projects of National Social Science Foundation of China under grant number 11&ZD189, and it was partially supported by a grant from NSF of Hubei Prov. of China under grant number 2013CFB334. It was partially supported by NSF of educational agency of Hubei Prov. under grant number Q20101110, and the State Key Lab of Software Engineering Open Foundation of Wuhan University under grant number SKLSE2012-09-07.

References

1. Hayes, P.: RDF semantics. Technical report, W3C. W3C recommendation, February 2014. http://www.w3.org/TR/2014/REC-rdf11-mt-20140225/
2. W3C Data Activity. http://www.w3.org/2013/data/
3. Bizer, C., Paulheim, H.: State of the LOD Cloud 2014 (2014)
4. Zaveri, A., Rula, A., Maurino, A., Pietrobon, R., Lehmann, J., Auer, S.: Quality assessment methodologies for linked open data. Semantic Web (2013)
5. Acosta, M., Zaveri, A., Simperl, E., Kontokostas, D., Auer, S., Lehmann, J.: Crowdsourcing linked data quality assessment. In: Alani, H., et al. (eds.) ISWC 2013, Part II. LNCS, vol. 8219, pp. 260–276. Springer, Heidelberg (2013)
6. Mendes, P.N., Bizer, C., Young J.H., Miklos, Z., Calbimonte J.P., Moraru, A.: Conceptual model and best practices for high-quality metadata. Delivery 2.1 of PlanetData, FP7 project 257641 (2012)
7. Fernández, J.D., Martínez-Prieto, M.A., Gutiérrez, C., Polleres, A., Arias, M.: Binary RDF representation for publication and exchange (HDT). Web Semant. Sci. Serv. Agents World Wide Web **19**, 22–41 (2013)
8. Lvarez-García, S., Brisaboa, N.R., Fernández, J.D., Martínez-Prieto, M.A.: Compressed k2-triples for full-in-memory RDF engines. ArXiv preprint (2011)
9. Motik, B., Grau, B.C., Horrocks, I., Wu, Z., Fokoue, A., Lutz, C.: OWL 2 Web ontology language profiles, 2nd edn. W3C Recommendation (December 2012)
10. Pichler, R., Polleres, A., Skritek, S., Woltran, S.: Redundancy elimination on RDF graphs in the presence of rules, constraints, and queries. In: Hitzler, P., Lukasiewicz, T. (eds.) RR 2010. LNCS, vol. 6333, pp. 133–148. Springer, Heidelberg (2010)

MoI 2016

Behavior-Based Twitter Overlapping Community Detection

Lixiang Guo[✉], Zhaoyun Ding, and Hui Wang

College of Information Systems and Management,
National University of Defense Technology, Changsha, 410073, China
{guolixiang10,zyding,huiwang}@nudt.edu.cn

Abstract. In this paper, we try to cluster twitter users into different communities. These communities can be overlapping based on their interests. The paper proposed a RWC (relation-weight-clustering) model to construct twitter users' network. This model takes twitter users' "@" and "RT@" behaviors into account. By counting their "@" and "RT@" frequency, the relation strength can be then descripted. Using SVM, we can get the users interest vector by analyzing their tweets. And the common interest vector between two users is calculated according to their common interests. Using community detection algorithm to resolve the relation-nodes-based network, the overlapping communities are formed with modularity of 0.682.

Keywords: Overlapping community detection · Interest space · RWC model

1 Introduction

Many more social web services like Twitter, Facebook, LinkedIn, etc. which are based on social network have emerged in the latest decade. It has attracted many researchers to investigate the mass data generated by them every day.

Community detection has been a hotspot in the field of social network research. In other words, the community detection technology enables us to find community structure in the social network and to get a deep insight of relations or interests among nodes.

After Michelle Girvan and Mark Newman proposed the concept of modularity [1] in 2002, community detection really took off. Many algorithms have been put forward aimed to optimizing the modularity function. A typical one of them is FN algorithm, a greedy optimization method. It views every nodes as small independent communities at the beginning. And then combine two communities as a new one, where these two communities are really linked in the network and the value of the modularity increases most or deceases least. After the end of iteration, all the nodes become a community. The modularity is calculated in each iteration. Then the community partition which is corresponding to the largest modularity value is the approximately optimal community structure. Besides, the modularity is often used to evaluate the quality of community structure generated from other community detection algorithms.

© Springer International Publishing Switzerland 2016
H. Gao et al. (Eds.): DASFAA 2016 Workshops, LNCS 9645, pp. 371–376, 2016.
DOI: 10.1007/978-3-319-32055-7_31

In fact, users in the Twitter-liked social network may have kinds of interests. This means a user can belong to more than one interest-based community. Hence the community structure should be overlapping. To detect overlapping community, Ahn et al. [2] proposed LCA algorithm which reinvented communities as groups of links rather than nodes. The groups of links were mapped to nodes at last. Then the overlapping community was gotten. Zhou et al. built an R-C model [3] taking the link similarity into consideration to improving LCA.

The work of Zhou et al. does not consider the relation strength between users. This might lead to unreasonable community structure. On the foundation of their work, we take the relation strength into account. And the community detected is more reasonable.

In this paper, we collect tweets from 47360 Twitter users. User interest space is built through analyzing their tweets' contents. By counting the frequency of their "@" and "RT@" behaviors, the relation strength can be easily got. Finally, the modularity of the overlapping community is 0.682.

2   Relation-Weight-Clustering Model

The relation link between two users is represented as a common interest vector. Then the weight is added to the common interest vector. We view the weighted common interest vector as clustering object. Using fast optimizing algorithm [4], the relation links are clustered into several groups. Finally, the relation links are mapped to user nodes, which is corresponding to user communities.

2.1   User Interest Vector Construction

A user may usually have different interests. It means that a user may belong to different interest-oriented communities. We use support vector machine (SVM) to gain the users' interests. A user's interest vector I is an n-dimension vector, where $I = (w_1, w_2, \ldots, w_n)$ and n is the number of interests. Each dimension of I is a specific interest and its value is the possibility of the user's tweets on this interest.

2.2   Relation Link Interest Vector Construction

We assume that two users become friends for sharing common interests. Based on this assumption, we use a vector C to represent the relation link between two users. And C is defined as

$$C = I_i \cap I_j, \tag{1}$$

where

$$I_i \bigcap I_j = (\min\{w_{i1}, w_{j1}\}, \min\{w_{i2}, w_{j2}\}, \ldots, \min\{w_{in}, w_{jn}\}). \tag{2}$$

Considering that the interaction frequency reflects the relation strength intuitively, we think about adding an interaction-related factor ω to C. The definition of ω is

$$\omega = \max\{\alpha_1 \frac{\#U_1@U_2}{\#U_1@} + \alpha_2 \frac{\#U_1RT@U_2}{\#U_1RT@}, \quad \alpha_1 \frac{\#U_2@U_1}{\#U_2@} + \alpha_2 \frac{\#U_2RT@U_1}{\#U_2RT@}\}, \tag{3}$$

where $\#U_1@U_2$ is the times of user U_1 "@" user U_2 in all tweets of U_1 and vice versa. $\#U_1@$ is the times of all the "@" behavior of U_1's tweets and vice versa. And "RT@" denotes retweet behavior. α_1 and α_2 are the weight which are satisfied

$$\alpha_1, \alpha_2 > 0, \alpha_1 + \alpha_2 = 1. \tag{4}$$

Therefore the weighted relation link interest vector can be

$$C_w = \mu \cdot \omega \cdot C, \tag{5}$$

where μ is an alterable factor for adjusting the scale of C_w to an appropriate level.

2.3   The Relation-Based Network Construction

We use R-C network model [3] as the basic model. In this network, the nodes are relation links as above. There is an edge between two relation links if and only if they share common user. The weight of edge can be regarded as the similarity of those two relation links. Hence it can be defined as

$$W(C_{w1}, C_{w2}) = \frac{C_{w1} \cdot C_{w2}}{|C_{w1}|^2 + |C_{w2}|^2 - C_{w1} \cdot C_{w2}}. \tag{6}$$

The equation above is *Tanimoto coefficient* (also called *Extended Jaccard coefficient*). It is easy to find that $W(C_{w1}, C_{w2})$ is between zero and one. A larger $W(C_{w1}, C_{w2})$ means that the two relation links are more similar.

3   Experiment and Analysis

3.1   Data Collection

We have collect 47360 Chinese twitter users' profiles and their tweets in September, 2015 by using twitter API. And the data are stored in the MySQL database.

3.2   Data Processing

Twitter User Interest Vector. For a twitter user, we analyze every tweets of him. Those tweets with too many non-Chinese characters are filtered out.

At first, we decide to classify these tweets into six categories (*A, B, C, D, E* and *Other*). Actually, tweets which are in *Other* group are also neglected. Using SVM, the rest of tweets are attached with a unique label.

Then, a user's interest vector can be represent as

$$I = (\frac{\#A}{total}, \frac{\#B}{total}, \frac{\#C}{total}, \frac{\#D}{total}, \frac{\#E}{total}),$$ (7)

where $total = \#A + \#B + \#C + \#D + \#E$.

Twitter Users Relation Link Interest Vector Construction. We extract "@" and "RT@" relations from tweets. If two users have "@" or "RT@" behavior each other, there will be a link between them. And the relation link interest vector can be calculated by using (5).

Table 1 shows a pair of users with its weighted interest vector. The adjusting factor is set to 1000.

Table 1. Some examples for relation link interest vector

u1(id)	u2(id)	Weight	Relation interest vector	Weighted vector
127262132	2236766378	0.136	0.17, 0.08, 0.05, 0.2, 0.22	23.12, 10.88, 6.8, 27.2, 29.92
1265070655	1862357449	0.196	0.14, 0.05, 0.13, 0.18, 0.18	27.44, 9.8, 25.48, 35.280003, 35.280003
1265070655	2833539408	0.043	0.14, 0.02, 0.44, 0.08, 0.15	6.02, 0.85999995, 18.92, 3.4399998, 6.4500003
870309318	1618790083	0.034	0.15, 0.04, 0.4, 0.12, 0.14	5.1000004, 1.36, 13.6, 4.08, 4.76
870309318	16865364	0.027	0.15, 0.04, 0.19, 0.12, 0.16	4.05, 1.0799999, 5.13, 3.24, 4.3199997
870309318	145440266	0.048	0.15, 0.03, 0.53, 0.11, 0.11	7.2000003, 1.4399999, 25.439999, 5.2799997, 5.2799997
870309318	633328589	0.014	0.15, 0.04, 0.41, 0.12, 0.16	2.1000001, 0.56, 5.74, 1.68, 2.24
18190842	2197807908	0.014	0.24, 0.06, 0.11, 0.2, 0.19	3.36, 0.84, 1.54, 2.8, 2.6599998

3.3 Twitter User Relation-Link Network Construction

There is an edge between two relation links if and only if they share common user. Then the weighted undirected network is constructed as in Table 2 below.

Table 2. Some examples for twitter user relation-link network structure

LinkId_1	LinkId_2	tanimotoSim[a]
3	4	0.293
4	14006	0.996
8	15856	0.3
8	15865	0.296
9	54000	0.486

[a]"tanimotoSim" is the tanimoto coefficient.

3.4 Community Detection

Clustering the Relation-Links into Groups. Using maximizing modularity method in [4], the twitter user relation-links are partitioned into 471 communities with modularity of 0.682.

Mapping the Relation-Links to Twitter Users. *Rule 1.* If some relation-links are in the same group, users attached to these relation-links belong to a group.

Based on rule 1, the final twitter users overlapping communities are detected as in Table 3. The numbers in the community column are the IDs of communities. That a user corresponds to several IDs means the user belongs to these communities at the same time.

Table 3. Some samples for overlapping communities of twitter users

userId	Community
1001077530	153
100122533	454, 115, 232, 169, 104, 25, 236, 88, 360
1001268486	152, 115, 360
100172757	171
100175420	117, 48, 128, 115, 55, 169, 33, 168, 277, 32, 40, 358, 104, 88, 141
100176531	104
100198190	171, 470, 148, 241, 136, 55, 195, 400, 169, 168, 360, 141
100233785	262, 300, 115
100253361	115, 188, 88
100506067	354, 183, 40

4 Conclusion

This paper does not take friendship relation among twitter users as the source of basic network. We take the "@" and "RT@" behaviors among twitter users as the basic composition of network instead. It is reasonable to do that because the online interactive behavior ("@" and "RT@") can reveal the common interests even they are strange each other in the real world. Using RWC model, we find the overlapping communities based on interest.

This method can be applied to different scene to get high quality and reasonable communities in the social network.

References

1. Newman, M.E., Girvan, M.: Finding and evaluating community structure in networks. Phys. Rev. E: Stat. Nonlin. Soft Matter Phys. **69**(2Pt2), 026113–026113 (2004)
2. Ahn, Y.Y., Bagrow, J.P., Lehmann, S.: Link communities reveal multiscale complexity in networks. Nature **466**(7307), 761–764 (2010). doi:10.1038/nature09182
3. Zhou, X.P., Liang, X.: User community detection on micro-blog using R-C model. J. Softw. **25**(12), 2808–2823 (2014)
4. Blondel, V.D., Guillaume, J.L., Lambiotte, R., Lefebvre, E.: Fast unfolding of communities in large networks. J. Stat. Mech. Theory Exp. **30**(2), 155–168 (2008)

Versatile Safe-Region Generation Method for Continuous Monitoring of Moving Objects in the Road Network Distance

Yutaka Ohsawa[(✉)] and Htoo Htoo

Graduate School of Science and Engineering, Saitama University, Saitama, Japan
`ohsawa@mail.saitama-u.ac.jp`

Abstract. This paper proposes a fast safe-region generation method for several kinds of vicinity queries including distance range queries, set k nearest neighbor (NN) queries, and ordered kNN queries. When a user is driving a car on a road network, he/she wants to know objects located in a vicinity of the car. However, the result is changing according to the movement of the car, and therefore, the up-to-date result is always expected, and requested to the server. On the other hand, frequent requests for updating results to the server cause heavy loading. To cope with this problem efficiently, the idea of safe-region has been proposed. This paper proposes a fast generation method of the safe-region applicable to several types of vicinity queries. Through experimental evaluations, the proposed algorithm achieves a great performance in terms of processing times, and is one or two orders of magnitude faster than existing algorithms.

1 Introduction

With the availability of wireless communication and enormous utility of mobile devices, continuous queries over moving objects have been an explosive increase in the demand for various location based services (LBS) applications. A continuous query continuously monitors over data objects which meets a specific query condition to a query point. Contrary, when a query is answered only one time over static objects, this type of query is known as a snapshot query.

Various algorithms for continuous queries in Euclidean distance have been actively researched in the literature. In practice, moving objects including cars and pedestrian are moving on a road network in real world scenarios, and henceforward, queries based on distance or travel time on road network is more preferable than in Euclidean distance.

Some literature proposed algorithms adaptable to the road network distance, however, they are targeting to an individual type of query, for example, k nearest neighbor (kNN) query, reverse k nearest neighbor (RkNN) query, and distance range query. The aim of this study is to develop a framework adaptable to versatile vicinity queries, including set-kNN, ordered-kNN, RkNN, and distance range query.

© Springer International Publishing Switzerland 2016
H. Gao et al. (Eds.): DASFAA 2016 Workshops, LNCS 9645, pp. 377–392, 2016.
DOI: 10.1007/978-3-319-32055-7_32

In applications for continuous queries, the client-server model is configured in general. In this architecture, when a moving object sends a query to the server, a thread to manage each moving object is initiated at the server side. Then the server sends back its query result to the moving object. The moving object changes the location continuously, and thus, the query result becomes useless according to the location changes. To cope with this problem, three types of update methods have been proposed; (1) update periodically, (2) update by the fixed distance move, (3) set safe-region.

The periodical update and the update by the fixed distance move are apt to overlook the changes of the result or repeat futile queries to get the same result. Hence, safe-region methods have been proposed to solve these problems. The safe-region is an area on the road network in which the query result remains unchanged. As long as a moving object is inside the safe-region, a new query request to the server is not necessary. On the other hand, when the moving object leaves the region, the query result becomes different at the location, and therefore the moving object requests the server to send a new query result and the safe-region.

Figure 1 shows an example of a safe-region of ordered-3NN query at q by thick line. The result of the ordered-3NN of q are data points a, b, and c. In this case, when a moving object leaves the safe-region, it issues a new query to the server, and gets a new ordered-3NN data objects and the safe-region.

The safe-region approaches have been actively studied as described in the next section, however, these algorithms have been developed for each individual query type. This paper proposes a fast algorithm applicable to versatile vicinity queries in the road network distance. The contributions of the paper are summarized as follow:

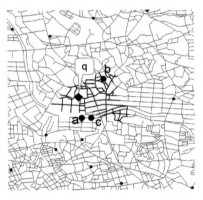

Fig. 1. Example of a safe-region

- we propose a generation method of versatile safe-region on demand in the road network distance which is applicable to vicinity queries including set-kNN, ordered-kNN, reverse-kNN, and distance range query.
- we evaluate our proposed method comparing to existing works, and show that the proposed method has a great performance in terms of processing time, and is one or two orders of magnitude faster than existing approaches.

The rest of the paper is organized as follow. In Sect. 2, related works of continuous queries are discussed. Section 3 presents the basic principles in safe-region generation for vicinity queries. In Sect. 4 discusses how to improve the efficiency for continuous queries, the generalization of the algorithm for various type of vicinity queries, and the determination of the border points on the road network edge. Section 5 describes evaluations of the performance of the proposed method. Section 6 concludes this paper and describes future works.

2   Related Work

Continuous queries for the moving objects have been actively researched since 2000s. They can be classified into three main categories based on (1) query types, (2) Euclidean distance or road network distance, and (3) mobility nature of queries and data objects.

In the literature, variety of continuous query types have been researched, consisting of range query [1,2], kNN query [3], RNN query [4,5], spatial semi-join query [6], path NN query [7], skyline query [8].

In continuous queries for moving objects, researches have been mainly focused on Euclidean distance in the pioneer studies. However, the movement of cars and humans are constrained on a road network in practice. To the best of our knowledge, Mouratidis et al. [3] first proposed a continuous query method in the road network distance. In their approach, k nearest neighbors are continuously monitored on road network, where the distance between a query and a data object is determined by the length of the shortest path connecting them.

In real world scenarios, continuous queries can be categorized into three groups depend on the mobility of a query and/or data objects [9].

(1) moving query objects querying static data objects.
(2) static query objects querying moving data objects.
(3) moving query objects querying moving data objects.

For an instance in case (1), a user queries a convenience store while driving a car. In case (2), a person queries to get the closest available taxi among taxis running around. Cheema et al. [10] proposed an algorithm for the reverse k nearest neighbor query (RkNN) based on (2). Researches on mobility of both queries and data objects (case (3)) have been introduced by Stojanovic et al. [9], Cheema et al. [11] and Liu et al. [12]. These types of queries are necessary, for example, a car queries to nearest rival cars within a certain region while driving in a car race.

Continuous queries are generally based on the client-server model, and the task of a server is to continuously compute and update the result of each query according to the location changes of the moving objects. Consequently, queries are repeated periodically or by moving a certain distance. However, if the frequency of updates becomes high, the performance in monitoring declines.

To overcome overloads at the server side, Prabhakar et al. [2] proposed a safe-region method. When a moving object queries for kNN or range query, the server generates a safe-region in which the query result remains unchanged. By the time the moving object leaves the safe-region, a new query result and the safe-region are requested to the server. Thereafter, various continuous query methods based on safe-regions have been proposed focusing on Euclidean distance. Yiu et al. [13] proposed efficient algorithms for RkNN queries in road network distance. Among them, the Eager algorithm can be applied to safe-region generations for static RkNN queries.

Alternatively, Cheema et al. [14] proposed an efficient and effective monitoring technique based on the safe-region for range queries: they used the term of safe-zone instead of safe-region. They also proposed safe-region generation method for continuous RkNN queries. Although safe-region generation methods have been actively researched, these algorithms were proposed for an individual query type. Moreover, these algorithms are based on a similar methodology in region expansions. In region expansions, the region is gradually expanded verifying the query condition from the query point. The most time consuming step is the verification for the query condition at each network node.

This paper first proposes a safe-region generation method applicable to versatile vicinity queries; including set-kNN, ordered-kNN, reverse-kNN, and distance range queries. Secondly, the performance efficiency is improved in the verification process, which checks whether the query condition is satisfied or not, by applying incremental Euclidean restriction (IER) [15] and the idea of the SSMTA* algorithm [16]. As the result, the method presented in this paper improves the processing time to generate safe-region for versatile vicinity queries by one or two orders of magnitude.

3   Basic Principles in Safe-Region Generation

3.1   Safe-Region for Vicinity Queries

First, we give definitions of the safe-regions for three types of vicinity queries related to nearest neighbor queries.

Definition 1 (Safe-region for Set-kNN Query). *Safe-region is a region where kNN queries invoked anywhere in the region give the same set of data points, ignoring the order in distances.*

Definition 2 (Safe-region for Ordered-kNN Query). *Similar to set-kNN query, however, the order of the distances in the kNN result is considered.*

Definition 3 (Safe-region for RkNN Query). *Safe-region is a region where kNN query result invoked anywhere in the region always contains a specified given data point.*

For an example, in Fig. 2(a), $p1 \sim p4$ belong to a data objects set P, and $a \sim d$ are query points. Here, we deal with 2NN of each point. For a, the 2NN is $p1$ and $p2$, and for b, they are $p2$ and $p1$. Therefore, when the order of 2NN is considered for the ordered-kNN query, the result of their 2NN are different. However, the result set of the 2NN queries are the same with $\{p1, p2\}$ in any order for the set-kNN query. This is the difference between the set-kNN and the ordered-kNN queries. Furthermore, the 2NN of c is $p4$ and $p1$, as the result, all 2NN of a, b and c contain $p1$. As a consequence, $p1$ is included in 2NN results invoked from a, b and c. Therefore, when $p1$ is specified as a query point, 2NN queries invoked at a, b, and c include $p1$ in their results. Therefore, the safe-region of R2NN of $p1$ includes these three data points. In set-kNN and ordered-kNN queries, the space is partitioned into non-overlapping regions. Contrary, the space is divided into mutually overlapping regions in RkNN query (except when k is one), and

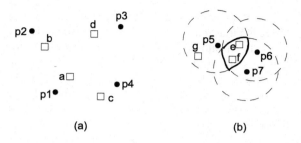

Fig. 2. Examples of vicinity queries

therefore, generated safe region for RkNN is larger than in the ordered and set-kNN queries. Hence, to specify a data object as the query point is necessary for RkNN query.

Definition 4 (Safe-region for Distance Range Query). *Safe-region is a region where distance range query, with a given distance D and a query point, invoked anywhere in the region contain the same set of data points.*

Figure 2(b) shows an example of the distance range query. In the figure, e, f and g are query points. $p5$, $p6$ and $p7$ are data objects and circles in dotted line show the areas centered at each data object and the radiuses are D. The area shown by bold lines shows the region in which the distance range query result is $p5$, $p6$ and $p7$. The objects e and f are included in this area, therefore, the distance range query results invoked from these two points give the same result $\{p5, p6, p7\}$, and the region in which these two points lie is the safe-region of the distance range query.

Definition 5 (Safe-region for Vicinity Queries). *A query on a road network to find the objects in P whose result is the same. This type of queries includes set-kNN, ordered-kNN, RkNN and distance range query. Especially, a safe region for set-kNN is denoted by SR_s, for ordered-kNN by SR_o, for RkNN by SR_r, and for distance range query by SR_d.*

For the distance range query SR_d, its query condition is considered on a given distance, and different from other nearest neighbor queries. Hence, the relationship among vicinity queries based on kNN queries satisfies the following relationship:

$$SR_o \subseteq SR_s \subseteq SR_r$$

3.2 Basic Principle of the Proposed Method

In this paper, a large road network is considered and modeled as a directed graph $G(V, E, W)$, where V is a set of nodes (intersections), E is a set of edges (road segments), and W is a set of edge weights and $w(\in W) \geq 0$ stands. w is assumed as the length of edge in the rest of the paper.

We define a safe-region (SR) as a region that gives the same query result. The generation of SR is emerged based on Voronoi region. Hence, SR for 1NN, in other words, order-1 Voronoi region (VR) is discussed in this section. The order-1 Voronoi region is the simplest SR. If Voronoi region in the road network distance is to be considered, VR(q) is the set of road segment in which $q(\in P)$ lies as the nearest object. An edge on G may belong to plural VRs, and the edge is exactly partitioned into plural regions to be belonged to respective VRs.

Figure 3 illustrates a road network. In this figure, white circles are road network nodes, black circles are data objects in P and the numbers attached along road segments are distances. The area bounded by the symbol × shows the order-1 Voronoi region in which q is included as the nearest object.

The algorithm described in the next subsection can be applied to several types of vicinity queries. The basic principle how to generate SR for these queries is followed by two steps below:

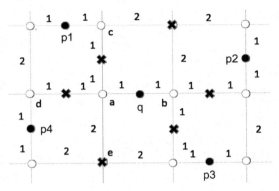

Fig. 3. The order-1 VR

(1) Gradually expanding the search region from a query point q to adjacent nodes as the similar way in Dijkstra's algorithm.
(2) At every visited node in step (1), verifying whether expanded nodes meet the query condition or not. If the result of verification is true, the node is expanded. Contrary, if the result is false, the node is not expanded anymore.

In the above step (1), a best-first search is applied from a query point to a current noticed node referring to a priority queue (PQ). In the step (2) after verifying whether the current node meets the query condition, all adjacent nodes to the current node are inserted into PQ. Then, the further search is proceeded from these nodes ahead. Contrary, if the query condition is not satisfied at the current node, node expansions from this node ahead are terminated. The query condition differs from each query type. The detail is described in Sect. 4.2.

3.3   Basic Method

In this subsection, an algorithm how to generate the safe region is described. The process in this algorithm is to generate a region by expanding the search area gradually while the query condition is satisfied. The search starts at q, and it is controlled by a best-first search using a priority queue (PQ). The record format in PQ is as follow:

$$< cost, n, p, \ell(p, n) >$$

Here, n represents a current node, $cost$ is the cost (the road network distance from q to n), p is the previously visited node to n, and $\ell(p, n)$ is the road segment between p–n. The cost is assumed in the road network distance, and the length of $\ell(p, n)$ is expressed as $d_N(p, n)$ for the road network distance.

At first, for two end nodes of the road segment on which q exists, the following records are inserted into PQ.

$$< 1, a, q, \ell(a, q) >, < 1, b, q, \ell(b, q) >$$

Then, the record with the minimum cost is dequeued from PQ. Thus, $< 1, a, q, \ell(a, q) >$ is dequeued from PQ. Moreover, the nearest neighbor (NN) of a is searched and checked whether the NN is q or not. If q is the NN of a, a also lies in SR(q) where q is a generator of SR.

In the NN search, the incremental Euclidean restriction (IER) [15] framework is used to find the NN faster. In IER framework, the NN candidates are incrementally searched in Euclidean distance, and verified these candidates in road network distances by A* algorithm. The verification process invokes next NN search while the Euclidean distance of the next NN candidate is smaller than or equal to the road network distance to the current NN candidate.

To expand the search range, records for adjacent nodes to a, those are c, d and e in Fig. 3, are enqueued into PQ.

$$< 3, c, a, \ell(c, a) >, < 3, d, a, \ell(d, a) >, < 3, e, a, \ell(e, a) >$$

In the next time, $< 1, b, q, \ell(b, q) >$ is dequeued from PQ and the similar process is performed at the node b.

Then, let $< 3, c, a, \ell(c, a) >$ be the next dequeued record from PQ. Similarly, NN of c is checked whether q is NN of c. In this case, the result is false, therefore c does not lie in $SR(q)$. At this point, a border point is determined on the road segment $\ell(c, a)$. The same process is performed to the next dequeued record $< 3, d, a, \ell(d, a) >$. When the record $< 3, e, a, \ell(e, a) >$ is dequeued, e can be considered as a border point because e is equidistant from q, p_3 and p_4. By repeating the process, SR(q) where q is the generator can be generated.

Algorithm 1 shows the above process in pseudocode. The PQ in lines 2 and 3 is the priority queue (heap) to control the region expansion, and in these lines, initial records at q are inserted into PQ. In line 4, a closed set (CS) is prepared for once checked road segments to avoid duplicated checks. R in line 5 is the result set of road segments included in the safe-region. In line 6, INITIALSET function is called to search NN of q and the result is assigned into T. The function INITIALSET is differently implemented depending on the query type. In the case of 1NN query, the 1NN of q is set to T (in this case, only one data object). The detail of this function for general query types is explained in Sect. 4.2.

Line 7 to 22 perform the following process. Initially, a record with the minimum cost is dequeued from PQ, and the road segment of the record $r.\ell$ is checked whether it is already registered in the CS. If $r.\ell$ is in the CS, it means that the segment has already been checked, and the rest steps are skipped. In line 12, $r.\ell$ is added into the closed set. In line 13, the current node $r.n$ is checked whether the node meets the query condition. As in the example of SR(q), the query condition is that NNs of current node $r.n$ are same as objects in T. If the result of VERIFY is true, the node ($r.n$) is expanded. Hence, adjacent road segments of $r.n$ are searched by referring to the adjacency list, and records for all adjacent road segments are created. Then, these records are enqueued into PQ. Moreover, the whole edge ($r.\ell$) is added into the result set R as shown in line 18. However, if the VERIFY result is false, the function ADDWITHCHECK calculates the part of the edge where the verify condition is still satisfied, and the result part is inserted into R. The detail, how to determine the part of the edge, is discussed in Sect. 4.3.

Algorithm 1. Safe-region generation: SRG

1: **function** SRG(q)
2:     $PQ.enqueue(< d_N(n_1,q), n_1, q, \ell(n_1,q) >)$
3:     $PQ.enqueue(< d_N(n_2,q), n_2, q, \ell(n_2,q) >)$
4:     $CS \leftarrow \emptyset$
5:     $R \leftarrow \emptyset$
6:     $T \leftarrow$ INITIALSET(q)
7:     **while** $PQ.size() > 0$ **do**
8:         $r \leftarrow PQ.deleteMin()$
9:         **if** CS contains r **then**
10:             **continue;**
11:         **end if**
12:         $CS \leftarrow CS \cup r.\ell$
13:         **if** VERIFY($r.n$,T) **then**
14:             $ns \leftarrow AdjacentNode(r.n)$
15:             **for all** $n \in ns$ **do**
16:                 $PQ \leftarrow PQ \cup < r.d + d_N(r.n,n), n, r.n, r.\ell >$
17:             **end for**
18:             $R \leftarrow R \cup$ ADD($r.\ell$)
19:         **else**
20:             $R \leftarrow R \cup$ ADDWITHCHECK($r.\ell, q, T$)
21:         **end if**
22:     **end while**
23:     **return R**
24: **end function**

---

4 Improving Efficiency and Generalization

4.1 Improvement in the Processing Time

The most time consuming step in Algorithm 1 is at the VERIFY procedure called for every node in the region expansion. For a visiting node n, kNN candidates of n are searched in Euclidean distance, and these candidates are verified in the road network distance. To verify in the road network distance, the A* algorithm can be applied. However, in general SR, at least k number of objects are searched as targets at every node n. In such condition, even if A* algorithm is fast for a single query, the total processing time becomes long due to repeated searching in adjacent regions.

To improve in the efficiency in terms of processing time for the distance calculation, the idea of the single source multi- targets A* (SSMTA*) [16] algorithm has been applied to the proposed method. The original SSMTA* algorithm concurrently finds the shortest paths from a source node to multiple target nodes efficiently. Contrary, when it is applied to SR generation, the target points are changing sequentially.

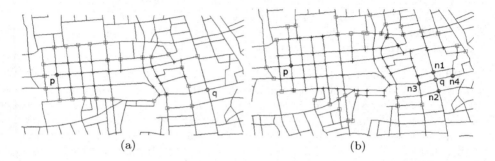

(a)   (b)

Fig. 4. The distance calculation by SSMTA* algorithm

In Fig. 4(a), q is a query point in SR, and p is a kNN candidate searched in Euclidean distance. The A* algorithm is applied to find the road network distance from p to q. In the A* algorithm, a priority queue PQA and a closed set CSA are used. The record format in PQA and CSA is the following (these are different from CS and PQ introduced in Sect. 3).

$$< c, v, d >   \quad (1)$$

Here, v is a current node in A* search, d shows the distance on a road network between p and v. The first item c is the lower bound distance between p and q, that is $c = d_N(p, v) + d_E(v, q)$ where $d_N(p, v)$ is the road network distance between p and v, and $d_E(v, q)$ is the Euclidean distance between v and q.

The priority queue PQA stores these records and returns a record in ascending order of c value. The current nodes (v) in PQA are called wave-front, and their distances from p have not been determined yet. Once a record is dequeued from PQA, it is registered into CSA. Since the distance d in a record in CSA has already been determined, it shows the shortest path length between p and v. The symbols □ in Fig.4 show the nodes in PQA, and the symbols + show the nodes in CSA.

Figure 4(b) shows the region expansion from q to neighboring nodes. In this figure, the search region in Algorithm 1 is enlarged from q to $n1 \sim n4$ gradually. Every time a new neighboring node is investigated, the function VERIFY needs to calculate the road network distance from p to the node. If A* algorithm is used in this check, almost the same road network nodes are repeatedly processed. To improve the efficiency of the verification process, we reuse the contents of PQA and CSA in the verification of the neighboring nodes.

When a new data object p' becomes a candidate of kNN of a node (in the first step, it is q), the distance between p' and q is obtained by applying A* algorithm. And then, the contents in PQA and CSA are kept for the next search. When a neighbor node (n) becomes a target, the distance between q and n is obtained by one of the following two cases.

case 1 If n has already been in CSA, the distance between q and n can be obtained by referring to d value in the record, that is $n.d$.

case 2 Otherwise, recalculate c value of all records in PQA for a new target point n, and then resume the search by A* algorithm.

When n has been included in CSA, we can obtain $d_N(p, n)$ by case 1. In this case, the search area is not expanded at all. Otherwise, the c values in PQA are recalculated by the equation $c = d_N(p, v) + d_E(v, n)$. This process is necessary because the target node (n) is changed. By referring to updated PQA, the distance search targeting to n is started again. The basic A* algorithm (called pair-wise A* algorithm) finds the shortest path from p to n repeatedly every time the target point is changed. Comparing to it, it is realized that the processing time can be considerably reduced in the improved method. Moreover, the update process for all records in PQA is taken place in the memory and it does not take long processing time.

Figure 4(b) shows the wave-front of PQA and contents of CSA for searching the shortest paths targeted to 5 adjacent nodes (q and $n1 \sim n4$). Comparing to Fig. 4(a), the region of expanded nodes in PQA and CSA is larger, however, the total number of nodes are smaller than repeatedly invoking pair-wise A* algorithm. By pair-wise A* algorithm, the similar node expansion to Fig. 4(a) must be repeated five times, and then, the total number of nodes becomes about five times for Fig. 4(a).

4.2   Generalized SR

The algorithm for SRG shown in Algorithm 1 is applicable to a variety of kNN queries including set-kNN, ordered-kNN and RkNN. To adapt these queries, three functions, INITIALSET(q), VERIFY and ADDWITHCHECK are needed to prepare for an individual query. Among them, ADDWITHCHECK is described in Sect. 4.3.

In set-kNN query and ordered-kNN query, INITIALSET(q) returns kNN data objects to q as the result. In set-kNN query, VERIFY(n,T) returns true if kNN results at node n are exactly the same (order omitted) as the objects in the set T. In ordered-kNN query, VERIFY returns ture when kNN results at node n is exactly the same as the objects in T in a sorted order.

In RkNN query, 1NN to q in data objects is searched by the result of INITIALSET(q), and returns the data object (let s_q be a specified query point). This means that T holds only one data point s_q. The VERIFY(n,T) returns true if s_q is included in kNN results of n.

The distance range query is also included as a variation of the vicinity query. In this query, INITIALSET(q) searches the data objects located in the range whose distance from q is less than or equal to D (the radius of the range), and they are set to T. In VERIFY, the distances from the current node n to each object in T are investigated. If all distances do not exceed D and do not include any other data object except objects in T, it returns true. Otherwise, it returns false.

4.3   The Borders Determination on Edges

The safe-region is a collection of road network edge segment on which the query condition is satisfied at any part. In Sect. 4.1, we described that the verification of the query condition was checked at road network nodes. If the query condition is satisfied at a node, it is also satisfied on the whole edge ended at the node. Therefore, the whole edge is added into the safe-region by ADD in Algorithm 1. On the contrary, when the query condition is not satisfied at a node, the border of the safe-region exists on the edge. How to determine the border points on the road network have been studied by Cho et al. [17,18].

Figure 5(a) shows a part of a road network, and in this example VERIFY returns 'true' at node A, but it returns 'false' at node B. Then, a border of the safe-region exists on the edge labeled LinkL (shown by bold line). In this figure, 'd', 'e', and 'g' are data objects.

Figure 5(b) shows the distance between each object and node A (left vertical axis), and between each object and node B (right vertical axis). The horizontal axis shows the normalized position on the edge. The lines in this figure show the distance change from each data object. The absolute value of all gradient of lines are the same. If there are plural objects existed on the left side of A, plural lines with the same gradient but different intercept appear. The similar situation appears for the right side of B. If there are plural objects on the edge having 'e', lines similar with 'e' but having different peak positions and intercepts appear. The border point on LinkL can be determined easily to find the nearest position from A and beginning at the position where the query condition is not satisfied.

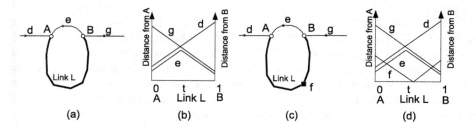

Fig. 5. The borders determination on edges

Figure 5(c) shows another case, a data object 'f' is on LinkL. In this case, a new line type appears in Fig. 5(d), the distance between A and 'f' decreases according to the move toward B, and it becomes zero at 'f', and then it increases according to the move toward B. The determination of the border position becomes complex a little, however, the method how to determine is the same with the case Fig. 5(a) and (b).

5   Experimental Results

This section presents evaluations of the proposed method comparing to existing works. Algorithms were implemented by Java language. The computer used for this evaluation was Intel Core i7 4770 CPU (3.4 GHz). Table 1 shows the road networks used in this experiment. MapA is a road network of the center part of a city, and MapB includes the center of a city and rural area. Data points to be searched were generated by pseudo-random sequence on the road network edges with various densities. For example, the density of 0.001 means a data point exists once 1,000 road edges. For the moving paths of an object, we used both real paths and randomly generated paths for moving objects in experiments. To generate a path on a road network, we randomly set a start point s and a destination point e, and a moving object was started a move from s to e via the shortest route. When the moving object arrived at e, a new destination point e was set and the moving object continuously moved to e from the current location. By repeating this process, paths for moving objects were generated. Besides, we prepared 100 real paths. It took about 30 min for the moving object to move on each path.

Table 1. Road maps

Map-name	No. of node	No. of link	Area-size
MapA	16,284	24.914	168 km^2
MapB	109,373	81,233	284 km^2

Figure 6(a) shows the processing time to generate a safe-region for set-kNN queries. In this figure, 'Basic' shows the processing time of the basic algorithm described in Sect. 4.1 (it is an existing algorithm) and 'Prop' shows the processing time of the proposed method. 'A' and 'B' in the parentheses correspond to 'MapA' and 'MapB' respectively. The horizontal axis shows k, the number of nearest neighbors to be searched. The density of the data points were set to 0.005. As shown in this figure, the proposed method requires less than one-tenth to the basic method. Figure 6(b) shows the processing time when the density of the data points was varied. The value of k was fixed to 5 in this experiment.

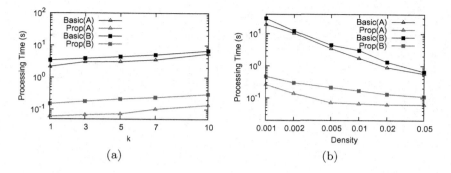

Fig. 6. The processing time of a safe-region for set-kNN

Figure 7 shows the processing time to generate a safe-region for ordered-kNN queries. The size of the safe-region in ordered-kNN queries becomes smaller than in set-kNN queries, because the order of the distances is also considered in ordered-kNN queries. Accordingly, the processing time in both proposed method and the basic method becomes faster in this case. However, the proposed algorithm still outperforms the basic algorithm.

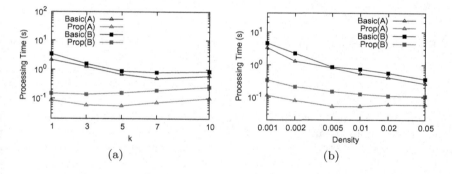

Fig. 7. The processing time of a safe-region for ordered-kNN

Figure 8 compares the processing time for reverse-kNN queries. In this experiment, the nearest-neighbor (p) to the current position of the moving object is first searched, and then the region where p is included in the kNN set is retrieved. This means that p is always included in the kNN of the moving object which moves inside the safe-region. The size of the safe-region of reverse-kNN becomes the biggest among three types of nearest neighbor queries. Therefore, the processing time is also longer than set-kNN and ordered-kNN queries. However, the processing time of the proposed algorithm is one to two orders of magnitude shorter than the basic algorithm.

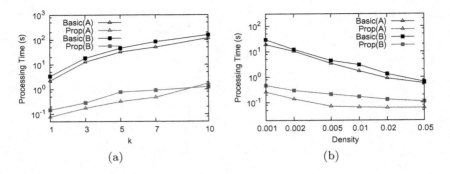

Fig. 8. The processing time of a safe-region for reverse-kNN

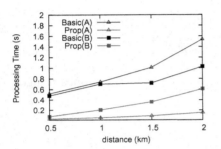

Fig. 9. Distance range query

Fig. 10. Average distance between queries

Figure 9 shows the processing time in generating the safe-region for distance range query when distances D vary. In this experiment, the density of data points was fixed to 0.005. When D is larger, the processing time becomes longer. However, the proposed algorithm is also faster than the basic algorithm.

In Fig. 10, we measure the communication cost to the server in terms of traveling distance by a moving object. When a moving object is moving along the route and it reaches the end of the safe-region, a new query result and its safe-region are provided by the server. The frequency of query requests depends on the traveling distance across the safe-region. In this figure, the vertical axis shows that the average traveling distance of a moving object within a safe-region, and the horizontal axis shows the density of data points for three types of queries. In this figure, the value of k is set to 5 for SkNN (set-kNN) and OkNN (ordered-kNN) queries, and D is 1.5 km for distance range query. As shown in this figure, the average traveling distance decreases according to the density increase. When the data density increases, the distance within the safe-region becomes shorter and frequent generation of new safe regions are requested to the server. Consequently, the communication cost becomes higher in such situation.

6 Conclusion

In this paper, a versatile safe-region generation method for continuous queries over moving objects in road network distance is proposed. In the safe-region generation process, repeated expansions over same road network nodes are avoided in the proposed method, and it improves the efficiency. Moreover, in evaluations, the proposed method is applied to continuous vicinity queries including set-kNN query, ordered-kNN query, reverse-kNN query, and distance range query. Comparing to existing works, our proposed method has a great efficiency in terms of processing time, and shows that two orders of magnitude faster than existing approaches especially when data objects are sparsely distributed. To apply the proposed method to more complicated spatial queries is future works.

Acknowledgments. The present study was partially supported by the Japanese Ministry of Education, Science, Sports and Culture (Grant-in-Aid Scientific Research (C) 15K00147).

References

1. Gedik, B., Liu, L.: MobiEyes: distributed processing of continuously moving queries on moving objects in a mobile system. In: Bertino, E., Christodoulakis, S., Plexousakis, D., Christophides, V., Koubarakis, M., Böhm, K. (eds.) EDBT 2004. LNCS, vol. 2992, pp. 67–87. Springer, Heidelberg (2004)
2. Prabhakar, S., Xia, Y., Kalashnikov, D., Aref, W., Hambrush, S.: Query indexing and velocity constrained indexing: scalable techniques for continuous queries on moving objects. IEEE Trans. Comput. **51**(10), 1124–1140 (2002)
3. Mouratidis, K., Yiu, M.L., Papadias, D., Mamoulis, N.: Continuous nearest neighbor monitoring in road networks. In: Proceedings of the 32nd VLDB, pp. 43–54 (2006)
4. Bentis, R., Jensen, C.S., Karčlauskas, G., Šaltenis, S.: Nearest and reverse nearest neighbor queries for moving objects. VLDB J. **15**(3), 229–250 (2006)
5. Xia, T., Zhang, D.: Continuous reverse nearest neighbor monitoring. In: Proceeding of the 22nd International Conference on Data Engineering, p. 77 (2006)
6. Iwerks, G.S., Samet, H., Smith, K.P.: Maintenance of spatial semijoin queries on moving points. In: Proceedings of VLDB (2004)
7. Chen, Z., Shen, H.T., Zhou, X., Yu, J.X.: Monitoring path nearest neighbor in road networks. In: SIGMOD 2009, pp. 591–602 (2009)
8. Huang, Y.K., Chang, C.H., Lee, C.: Continuous distance-based skyline queries in road networks. Inf. Syst. **37**, 611–633 (2012)
9. Stojanovic, D., Papadopoulos, A.N., Predic, B., Djordjevic-Kajan, S., Nanopoulos, A.: Continuous range monitoring of mobile objects in road network. Data Knowl. Eng. **64**, 77–100 (2007)
10. Cheema, M.A., Lin, X., Zhang, W., Mhang, Y.: Influence zone: efficiently processing reverse k nearest neighbors queries. In: Proceeding ICDE, pp. 577–588 (2011)
11. Cheema, M.A., Zhang, W., Lin, X., Zhang, Y., Li, X.: Continuous reverse k nearest neighbors queries in Euclidean space and in spatial networks. VLDB J. **21**, 69–95 (2012)

12. Liu, F., Do, T.T., Hua, K.A.: Dynamic range query in spatial network environments. In: Bressan, S., Küng, J., Wagner, R. (eds.) DEXA 2006. LNCS, vol. 4080, pp. 254–265. Springer, Heidelberg (2006)
13. Yiu, M.L., Papadias, D., Mamoulis, N., Tao, Y.: Reverse nearest neighbor in large graphs. IEEE Trans. Knowl. Data Eng. 18(4), 1–14 (2006)
14. Cheema, M.A., Brankovic, L., Lin, X., Zhang, W., Wang, W.: Continuous monitoring of distance based range queries. IEEE Trans. Knowl. Data Eng. 23, 1182–1199 (2011)
15. Papadias, D., Zhang, J., Mamoulis, N., Tao, Y.: Query processing in spatial network databases. In: Proceedings of 29th VLDB, pp. 790–801 (2003)
16. Htoo, H., Ohsawa, Y., Sonehara, N., Sakauchi, M.: Incremental single-source multi target A* algorithm for LBS based on road network distance. IEICE Trans. Inf. Syst. **E96–D**(5), 1043–1052 (2013)
17. Cho, H.J., Kwon, S.J., Chung, T.S.: A safe exit algorithm for continuous nearest neighbor monitoring in road networks. Mobile Inf. Syst. 9, 37–53 (2013)
18. Cho, H.J., Chung, C.W.: An efficient and scalable approach to CNN queries in a road network. In: Proceedings of the 31st International Conference on Very Large Data Bases, pp. 805–876 (2005)

Author Index

Balakrishna, Mithun 89
Borovina Josko, João Marcelo 307

Cai, Yi 3, 98, 112, 126
Cai, Zhipeng 179
Chen, Wei 15, 57
Chen, Wen Hao 3
Chen, Xiaoli 194
Chen, Xin 134
Chen, Yujun 223
Chen, Zhenhong 98, 112
Cheng, Ruhong 238
Cheng, Xiang 27, 142
Cui, Wenjuan 134

Deng, Kejun 285
Ding, Zhaoyun 371
Du, Bowen 223
Du, Yi 134

Erekhinskaya, Tatiana 89

Fan, Hongjie 285
Feng, Shi 65
Ferreira, João Eduardo 307

Gao, Hong 179, 294, 321, 329, 343
Gao, Yitong 294
Gu, Jinguang 362
Guang, Tao 362
Guo, Danhuai 134
Guo, Lixiang 371
Guo, Yike 126

Htoo, Htoo 377
Hu, Yawei 238, 254
Huang, DongPing 98, 112
Huang, Guoyan 81
Huang, Li 362
Huang, Songping 126

Jiang, Le 27

Lai, Kin Keung 3
Leung, Ho-fung 98, 112
Li, Bo 353
Li, Jianbo 254
Li, Jianhui 134
Li, Jianzhong 179, 294, 321
Li, Qing 160
Li, Shuchen 27, 142
Li, Xiangsheng 43, 168
Li, Xiao 142
Lin, Zheng 353
Liu, An 238, 254
Liu, Junfei 285
Liu, Weilin 223
Liu, Wenyin 160
Liu, Zhongqiang 15
Luo, Wenfeng 285

Ma, Yun 160
Mao, Hualin 238, 254
Mao, Yingchi 194
Meng, Dan 353
Mo, Biyun 81, 168
Moldovan, Dan 89

Nguyen, Minh Hieu 268
Nguyen, Thanh Trung 268

Ohsawa, Yutaka 377
Oikawa, Marcio Katsumi 307

Pang, Jianhui 43, 168
Pu, Juhua 223

Rao, Yanghui 43, 168

Shi, Shengfei 329, 343
Su, Sen 27, 142

Tang, Xiaolan 223
Tang, Yan 209
Tatu, Marta 89

Wang, Chu 65
Wang, Daling 65

Wang, Fu Lee   168
Wang, Hongzhi   179, 294, 321, 329, 343
Wang, Hui   371
Wang, Pengfei   134
Wang, Weiping   353
Wang, Yanzheng   343
Wu, Chao   126
Wu, Qingyuan   81

Xiao, Mingjun   238, 254
Xie, Haoran   43
Xu, Zhuoming   194, 209

Yang, Donghua   179
Yang, Feng   15

Yang, Jianyu   142
Yang, Kai   98, 112
Yang, Zhenguo   160
Yue, Mingliang   329

Zhang, Jinchao   353
Zhang, Kaiqi   179, 321
Zhang, Yan   294
Zhang, Yi   223
Zhang, Yifei   65
Zhao, Mei   126
Zhao, Xiangyu   15, 57
Zhou, Yuanchun   134
Zhu, Endong   81
Zhuang, Yuanhang   209

Printed in the United States
By Bookmasters